ENVIRONMENTAL PROTECTION

ENVIRONMENTAL PROTECTION

Arvind Kumar
Shamshun Nehar

2024

Daya Publishing House®
A Division of
Astral International Pvt. Ltd.
New Delhi – 110 002

First Publishes, 2007

Reprinted, 2024

ISBN: 978-93-5919-124-9 (HB)

Disclaimer:

Every possible effort has been made to ensure that the information contained in this book is accurate at the time of going to press, and the publisher and author cannot accept responsibility for any errors or omissions, however caused. No responsibility for loss or damage occasioned to any person acting, or refraining from action, as a result of the material in this publication can be accepted by the editor, the publisher or the author. The Publisher is not associated with any product or vendor mentioned in the book. The contents of this work are intended to further general scientific research, understanding and discussion only. Readers should consult with a specialist where appropriate.

Every effort has been made to trace the owners of copyright material used in this book, if any. The author and the publisher will be grateful for any omission brought to their notice for acknowledgment in the future editions of the book.

Published by : **Daya Publishing House®**
A Division of
Astral International Pvt. Ltd.
– ISO 9001:2015 Certified Company –
4736/23, Ansari Road, Darya Ganj,
New Delhi-110 002
Ph. 011-43549197, 23278134
E-mail: info@astralint.com
Website: www.astralint.com

Printed at : **Replika Press Pvt. Ltd.**

Preface

The sprawling urbanization and growing pressure on the natural resources have been questioning the quality of environment. The growing population and corresponding increase of production sector has not been giving any scope of resilience even for renewable 'commons'. The prime components of environment the water, the air, the soil and biota have also been put to diversified uses making their quality questionable. This problem is much severe in the developing countries like India, where the population is growing unabatedly multiplying the pressure on the elements of environment.

In this context, it is strongly felt that there is a need to deal the problems on region basis, because, the nature and magnitude of the problem changes from one corner to the other. In order to uphold the comprehensive quality of the environmental of a country or continent, specific problems in the region must be dealt case-wise and region-wise. In India, the pace of the environmental problems has been increasing unchecked in all sectors. But the intensity of the problems faced by the country is different in different regions. The solutions in order to protect the quality of environment come out only when we adopt strategies like "Education, Research, Extension, Consultancy and Training". With these few words of introduction, we invited the scientists, researchers and industrialists from all corners of the country to identify and delineate the regional problems and evolve strategies for environmental protection as well as for overcoming the problem of environmental degradation which will help the nation to march towards a sustainable development.

This book is the unique compilation of esteemed articles of internationally acknowledged experts in the field of Environmental sciences with the intention of providing a sufficient depth of the subject to satisfy the needs at a level which will be comprehensive and interesting. This book will be useful to the students, research scholars, scientists in the field of Environmental management and ecoplanners, politicians and other people with similar interest.

My special thanks and appreciation go to the scientists whose contributions have enriched this volume. I wish to express my sincere gratitude to Dr. P.C. Hembram, Hon'ble Vice Chancellor, S.K.M.

University, Dumka, Professor M.C. Dash, Hon'ble former Vice Chancellor of Sambalpur University, Professor N.C. Datta of Calcutta University, Professor K.C. Pandey of Lucknow University, Professor S.K. Konar of Kalyani University, Professor D.K. Belsare of Bhopal University, Professor Professor U.C. Goswami of Gauhati University, Professor P.C. Mishra of Sambalpur University, Professor P. Natarajan of Kerala University, Professor P. S. Murthy of Bangalore University, Professor Ajit Varma of JNU, Professor A.L. Bhatia of Jaipur University, Professor A.K. Mittal of BHU, Professor S.P. Hosmani of Mysore University, Professor K. Kapoor of Udaipur University, Professor K.B. Reddy of Nagarjuna University, Professor M. Vikram Reddy of Pondicherry University, Professor B.K. Tiwari of NEHU, Professor G. Tripathi of Jodhpur University, Professor R. Ramalingam of Annamalai University, Professor Sharif U. Ahmad of Nagaland University, Professor G.K. Kulkarni of Aurangabad University, Professor S.U. Mehram of Nagpur University, Professor S.K. Battish of PAU, Professor B.M. Sharma of Manipur University, Professor B.D. Joshi of Hardwar University, Professor G.C. Pandey of Faizabad University, Professor K. C. Sharma of Ajmer University, Professor M. Raziuddin of Hazaribag University, Professor U.S. Bagde of Mumbai University, Professor Gurdeep Singh of I.S.M., Dhanbad, Dr. P.K. Goel of Karad, Professor S.P. Roy and Shri Tribhuwan Poddar of Bhagalpur University, Bhagalpur for encouragements.

I also express my deep sense of gratitude to my parents whose blessings have always prompted me to pursue academic activities deeply. I am also thankful to my wife, *Kumari Bimla* and my two lovely sons, *Kumar Pallav Shivshankaran* and *Kumar Prasun Ramakrishnan* whose natural smiles extended to me relief all through this tiresome endeavour.

Last but not the least, I am also thankful to Mr. Anil Mittal, Proprietor, Daya Publishing House, New Delhi for taking keen interest in bringing out of this book. Finally, I will always remain a debtor to all my well-wishers for their blessings, without which this book would not have come into existence.

Professor A. Kumar

Contents

Chapter 1

Role of Copepods in Aquacultural Strategies

D.K. Singh[1] & Arvind Kumar[2]

[1]Department of Zoology, A.N. College, Dumka – 814 101, Jharkhand

[2]Environmental Science Research Unit, Post Graduate Department of Zoology, S.K.M. University, Dumka, India

Freshwater fish seed production often faces a problem of an adequate food supply. Artificial feeds are widely used, but planktonic animals are very important, especially rotifers, cladocerans, and copepods. Virtually, all fish feed on plankton, especially in their early life phases. Planktivorous fish depend on small invertebrates throughout their entire lives. Copepods of the order Cyclopoida are the most important food items in freshwater aquaculture, and their nauplii are especially valuable for feeding fry (Szlauer and Szlauer, 1980). Copepods as natural food are either cultured or collected from the wild.

Copepods play very important and diverse roles in freshwater aquaculture operations, including some that are beneficial and others that are extremely adverse and may result in complete production loses. Some copepod roles are obvious for example, copepods that serve as food for small fish or copepods that are fish parasites. Others, however, are not often considered, such as copepods as micropredators of fish, their role as intermediate hosts of fish parasites, and their role as hosts and vectors of human diseases. While progress has been made in better understanding the biology of copepods further research, especially including studies of the taxonomy of different copepod species.

Free-living copepods, and zooplankton in general, may be harvested from the wild. Zooplankton may be collected from specific depths in lakes using a custom-made mechanical device (Szlauer *et al.*, 1978) that operates by propelling water through a huge plankton net using an out-board boat engine. Sendlak (1980) surveyed the possibilities of acquiring different planktonic animals mainly cladocerans and copepods, from Irisko Lake, Poland. According to his studies annual lake productivity in Poland ranged from 6.25 to 62 g m^{-3} of water (possibly even 90 g m^{-3} in eutrophic lakes). The second major method of acquiring copepods relies on harvesting plankton from small rivers flowing from lakes. Such outlet rivers contain large amounts of zooplankton. The diel and annual plankton flow from Plon Lake, Poland by the River Plonia has been estimated to be 372 and 735000 kg, respectively, while the same river removed 65.0 and 25251 kg respectively from Miedwie Lake, Poland (Szauer, 1976,1977,1983/84). In general Europe, several attempts have been made to actually use copepods and other components of zooplankton for feeding fish in aquaculture (Anwand, 1978, Szlauer and Szleauer, 1980,1982). The results were promising but better infrastructure and funding are needed.

Copepods can also methods for marine copepods are well advanced (Ogle, 1979, Ohno and Okamura, 1988, Payne and Rippingale, 2001), but relatively few attempts have been made to culture freshwater species. One example may be a method of mass culture of *Paracyclops fimbriatus* developed recently by Szlauer (1995) using observations made from a mass occurrence (13000 individuals L^{-1}) during experiments on municipal sewage sludge.

Taiwanese scientists have been successful at rearing brackish-water copepods. Prof. Shin-Hon Cheng has developed a method of culturing Apocyclops royi on a semi-industrial scale (Cheng *et al.*, 1999, 2000). Some 20 metric tones of different brackish-water marine copepods are exported each year from Taiwan to Japan. There are very few cases of using copepods as fish feed on an industrial scale. One of the most recent examples is an American company (Argent) that is marketing a product that they claim is made of cyclopoid copepods. According to the supplier, these copepods are a "selectively bred, biologically engineered microorganism cultured in a pristine Arctic lake". These copepods are reported to contain the highest known concentration of the fundamental and critical biological pigment, estaxanthene, accounting for their striking orange coloration. Complimenting this pigmentation are previously unknown levels of highly unsaturated fatty acids (HUFAS), and especially those of the famed "omega-3" family.

The majority of lernaeids have undergone extensive morphological adaptations hiding their close affinity with their freshwater cousins of the genus Cyclops. Fish parasites within this family belong to 14 valid genera (Ho, 1998) with 180 nominal species (The World of Copepods 2002) and about 110 valid species (Ho, 1998). Representatives of the genus *Lernaea* have been studied more intensively than any other freshwater copepod group due to its economic importance. Lernaeids are probably the best known copepod parasites and examples appear in most invertebrate zoology textbooks. There are 105 nominal *Lemaea* species (The world of Copepods 2002) of which only about 37 are valid (Kabata, 1979). Roddubnaja (1973) complicated understanding of the species concept of lernaeids by producing different phenotypes, resembling different described species of *Lernaea* from a single material specimen. *Lernaea* species occur on all continents, with the majority in Africa. The only cosmopolitan species in *Lernaeae cyprinacea*, which can infect a variety of freshwater fishes. Originally, *L. cyprinacea* was not present in South America and Australia, but it was accidentally introduced there with cyprinids.

Female *Lernaea* are highly metamorphosed vermiform ectoparasites (not mesoparasites) without segmentation that reach lengths of 12-16 mm (plus an additional 6 mm of egg sacs). The head, equipped with "antlers" that anchor the parasite in the subdermal tissues of a host fish, earned the parasite its vernacular name of "anchor worm". The rest of the body and egg sacs protrude into the water. This particular way of attachment is very pathogenic by its nature (Dzidziul, 1973; Khalifah and Post 1976; Kabata, 1985; Shariff and Roberts, 1989). Initially the skin and muscles adjacent to the head become hyperemic, swollen, and susceptible to secondary infections. The parasite's attachment evokes severe aute inflammation. The host's connective tissue reacts to the parasite, forming a thick fibrotic capsule around the anchor. *Lernaea* spp. Can cause severe fin damage. Wounds caused by the parasite's implantation occasionally develop into fistulae, penetrating the visceral cavity, including the heart and sometimes resulting in peritonitis and death (Kabata, 1985). Considering the invasive method of attachement and severity of the associated damage, it is surprising that *Lernaea* induced fish kills are not common (Shariff and Roberts, 1989).

Feeding and the gut structure of *Lernaea* were described by Sabatini *et al.* (1988). The life cycle comprises 3 nauplius stages, 5 copepodid stages, and adults (Grabda, 1963a). Copepodids settle on fish, mature, and copulate; and then males die and females undergo transformation while attached permanently to the host. *Lernaea* spp. Are warm-water parasites and according to Schaperclaus (1979) in some areas of the United States, no fewer than 10 generations can appear in the course of a year. The history of fish mortalities caused by *Lernaea* spp. Goes back to 1880, when in one of the lakes of the Masurian Lake District (presently in Poland), lernaeosis almost wiped out an entire population of the crucian carp (Benecke, cited by Kocylowski and Miaczynski, 1960). According to Kocytowski and Miaczynski (1960) a mass mortality of crucian carp was also reported by Kozikowska in Karasiowe Lake, Poland and by Grabda in Wilczak Hatchery, Poland. The copepod intensities reached 40 parasites per fish. An interesting account of site selection per fish. An interesting account of site selection by adult *L. cyprinacea* was published by Dorovskikh (1996). In recent years, the numbers of *Lernaceae* have drastically declined in Central Europe. In Poland, there has been no published record of *Lernaea* within the last 30 years although more than 280 papers dealing with fish parasites were published during that period (Piasecki and Wolinski, unpubl, data).

In North America, *Lernaea* spp. infect a number of cultured freshwater fishes. In summer in waters with temperatures ranging from 25 to 28 °C, this parasite finds excelled conditions for reproduction utilizing a number of fish cultured for food and ornamental purposes. An interesting case of unusual fish mortality caused by *L. cyprinacea* was published by Goodwin (1999). During June and July of 1998, at least 3 Arkansas fish farms polyculturing bighead carp (*Hypophthalmichtys nobilis*) with channel catfish (*Ictalurus punctatus*) suffered major losses of channel catfish associated with massive infections by *L. cyprinacea*. The catfish had few adult *Lernaea* attached to their skin, but there were from 8-50 copepodids on the surface of each catfish gill filament. The copepods were grazing on the gill tissue, and their feeding activity was associated with gill damage including epithelial hyperplasia, telargiectasis, and hemorrhage. Catfish skin was also covered by copepodids.. Bighead carp in the same ponds were reported to have had humerous adult copepods on their skin but did not die during the epizootic. It is possible that the filter-feeding apparatus of the carp captured the copepodids thus preventing heavy infection of the gill filaments (Goodwin, 1999). Lernaea copepodids have not been implicated previously in fish losses resulting from parasite damage to gills. The loss of catfish in these cases is likely to have been due to their being polycultured with the bighead carp, a species that is a common host for adult larnaeds. There is no treatment for *Lemaea* infections that is legal for use with food fish. Most farmers have quite polyculturing these 2 fish species.

Lernaeids also cause problems for small bait-fish and ornamental fish species. Arkansas produces millions of dollars worth of minnows each year. These fish are marketed when they are only single parasite in a critical location is enough to kill a minnow. In goldfish and koi, lerneids cause some mild infections, but disfigure the fish making them unsuitable for sale. The primary method for control of this species in non-food fish is the pesticide Dimilin (diflubenzuron).

Lernaea cyprinacea was introduced into South Africa in the beginning of the 20th century via the importation of the common carp, *Cyprinus carpio*. Since then, the copepod has spread very quickly, is now a common parasite infection all farmed species in Brazil, and is also very common in wild fish in all main drainage basins throughout the country. Gabrielli and Orsi (2000) studied 8 fish species on 53 fish farms of the state of Parana for the presence of lernaeids. All fish species were infected, but 100 per cent prevalence was found only in *C. carpio, leporinus Lacrocephalus,* and *Prochilodus lineatus*. Wild fishes in rivers were infected at low intensities. In Brazil, lernaeids cause mortality in several species of farmed fish. Another issue is the presence of parasites in a growing number of "fish-and-pay" enterprises. At fish-and-pay facilities, anglers pay for the privilege of fishing in well-stocked ponds (usually earthen ponds). The demand for this kind of leisure activity has led to the construction of thousands of fish-and-pay sites in Brazil. The presence of fish parasites, especially lernaeids in fish-and-pay ponds, and the uncontrolled movement of fish throughout the country pose serious health threats to the fish farm industry (Pavanelli *et al.*, 2000). *Lernaea cyprinacea* has also been recorded on cultured fishes on Caribbean islands (Fajer *et al.*, 1985).

Another area relatively recently conquered by lernaeids is Australia (Hall, 1983; Rowland and Ingram, 1991; Langdon, 1992). Heavy infections with *Lernaea* sp. resulted in mortality of golden perch (*Macquaria ambigua*). Murray cod (Maccullochella peeli), and silver perch (*Bidyanus didyanus*) broodstock held in ponds (Callinan, 1988), there was a heavy infection of *Lernaea* sp. Morray cod broodfish in 2 ponds at a research station following the use of common carp as a forage fish. Replacement of carp by goldfish (*Carassius auratus*) as a forage fish seemed to stop the problem (Rowland and Ingram, 1991). This suggests that carp is an important host to *Lernaea* sp. and may be responsible for the spread and increased prevalence of infections in native fish (including those cultured in ponds) (Rowland and Ingram, 1991). Carp were introduced to Australia more than a century ago and have since reached such a substantial biomass that they are considered a pest not unlike rabbits are on land. Carp are blamed for spreading, not only lernaeids, but also the pathogenic tapeworm, *Bothriocephalus* sp. and other disease agents.

In Africa, rapid increases in *Lernaea* burdens have been observed in association with growing environmental stress in some areas. High prevalence of anchor worm: were recorded by Oldewage (1993) in tilapia of Lake Victoria.

The second major branch of freshwater cyclopoids parasitic on fishes is represented by the genus *Lamproglena* spp. These copepods typically are gill dwellers, and as such they have the potential to cause fish losses in aquaculture *Lamproglena* comprises more than 40 nominal species (Piasecki, 1993a). They occur in Africa (Mark and Avenant-Oldewage, 1996; Ibraheem and Izawa, 2000), Asia (Kuang and Qian, 1985; Kumari *et al.*, 1989) and Europe (Cakic *et al.*, 1998; Galli *et al.*, 2001). To date, no fish kills caused by *Lamproglena* spp. have been reported.

The family Ergasilidae (Poecilostomatoida) comprises 24 valid genera (Amado *et al.*, 1995) with 249 nominal species (The World of Copepods 2002). The overwhelming majority of species occurs in freshwater environments. The morphology of ergasilids largely resembles that of free-living cyclopoids, but some may be extensively transformed, e.g., *Mugilicola*. The best known ergasilids are representative of the genus *Ergasilus*, which contains 153 nominal species (The World of Copepods 2002) and more than 80 valid species (Kabata, 1985). The best known species is *Ergasilus sieboldi*, which is 1.7 mm long and attaches to fish gills using its 2nd antennae. The antennae, transformed into powerful hooks, hold the gill filaments tightly and can cause tissue damage and obstruct blood flow. Parasites feeding on epithelial cells stimulate hypertrophy and consequently a coalescence of secondary gill lamellae. This in turn drastically reduces the surface available for gas exchange. Lesions on gills are often attacked by secondary pathogens such as bacteria and fungi. Feeding of *E. sieboldi* was described in detail by Einszporn (1965a,b). This particular species attaches to the outside of the gill allowing some of its congeners to explore the space between the gill filaments. In cases of extremely heavy infections of whitefish, the parasite attaches not only to gill filaments but also to the fins (Kozikowska, 1975). The life cycle comprises 6 nauplius stages, 5 copepodid stages, and adults. Males die after copulation, while females remained attached to the fish host (Abdelhalim *et al.*, 1991).

In Central Europe, the Ist spring generation of *E. sieboldi* becomes sexually mature in mid-June. Their eggs hatch and the copepodids attack fish. They mature and produce a 2nd generation in September. Sometimes a 3rd generation follows by the end of the session. One female can produce 200 off-springs. Theoretically, the ensuing 2nd generation can comprise 40000 descendants and in the 3rd, as many as 8 million (Schaperclaus, 1992) *Ergasilus* sieboldi is not host-specific and can infect a majority of freshwater fishes; however the tench. *Tinca tinca*, appears to be the most susceptible. This fact is attributed to the sluggishness of this fish, which may make it more vulnerable to copepod attack. Other less-infected fishes include pike (*Esox lucius*), bream (*Abramis brama*), whitefish (*Coregonus lavaretus*), vendace (*Coregonus albula*), carp (*Cyprinus carpio*), and roach (*Rutilus tutilus*).

Schaperclaus (1954) described a case of a single, 36 cm-long tench that harbored some 3600 specimens of *E. sieboldi* on its gills. This heavily infected fish had a condition factor of only 0.88. Similarly, intensive infection of paled (*Coregonus peled*) was reported by Abrosov and Bauer (1959,1961) from Pskov Lake, Russia. Heinemann (1934) found 5431 specimens of *E. sieboldi* on a single tench that died of asphyxia. Severe infection with *E. sieboldi* can result in heavy losses in the yield of tench. In lake Scharmuzel, Germany, the yield of tench dropped from 5000 kg before the appearance of *E. siebeldi* to 350 kg after its unwanted introduction. In 2 other small German lakes, the yield of tench dropped from 31-47 to 16.5 kg ha^{-1} after the invasion of this parasite and with copepod prevalences of only 50 %. In Lake Grimiz, Germany, the yield of tench between 1926 and 1931 declined from 4583 to 111 kg (Schaperclaus, 1992), again a reduction apparently related to *Elgasilus* infections.

The second most common host of *E. sieboldi* is the pike. Pike deaths attributed to *E. sieboldi* infections were described from 2 Pomeranian lakes in Germany by Lehmann (1924) and Rumphorst (1924). Deaths of whitefish (Coregonus wartmanni) in Zugersee, Germany were described by Baumann (1913). A very extensive account on aspects of *E. sieboldi* occurrence in different hosts was published by Grabda (1963a). *Ergasilus sieboldi* causes mass fish-kills not only of tench but also of perch, Perca fluviatilis in Wlodawa Lake, Poland (Kocylowski, 1954), whitefish in Cugersk Lake in the USSR, and bream in a number of lakes (Markevic, 1956).

In Southeast Asia, *Ergasilus* sp. occurs on the gills of cultured fishes including *Osteophilus hasselti, O. gouramy* (in Indonesia), *Ctenopharyngodon idella* (in Malaysia), and on *Oxyelotris marmoratus* (in Thailand). The parasite has the potential to adversely affect aquaculture in this region, but so far, no major outbreak attributed to *Ergasilus* sp. has been reported (Kabata, 1985). Another ergasilid, Sinergasilus major, may have importance in Central Asian aquaculture, including the former Soviet Republics and China regarding the host *Ctenopharyngodon idella* (cf. Bauer and Babaev, 1964). *Sinergasilus ieni* is another important parasite, affecting *Hypophthalmichthys molitrix* (cf. Musselius, 1973). In Japan, Nakajima and Egusa (1973) found *Pseudoergasilus zacconis* on gills of cultured ayu, *Plecoglossus altivelis*. In spite of this record, Egusa (1992) saw no serious threat by ergasilids to Japanese freshwater aquaculture.

Traditional treatment of ergasilosis included the use of organophosphate compounds, pesticides, or a mixture of copper sulfate and terric sulfate. It has been observed that well-developed aquatic vegetation helps reduce infection levels. It apparently limits movements of small bream that otherwise spread this parasite all over the lake (Schaperclaus, 1992).

Another important group of freshwater parasites is the family lernaeopodidae (Siphonostomatoida). The freshwater branch of this family is represented by the 7 genera of *Salmincola, Achtheres, Coregonicola, Basanistes, Tracheliastes, Pseudotracheliastes* and *Cauloxenus* (cf. Kabata, 1979). This group is characterized by their relatively large size, comparable to that of *Lernaea*. Their larnaeopodid attachment mechanism is unique, however, and seems to inflict less damage than that associated with *Lernaea* or *Ergasilus* spp. The body of a lernaeopodid female can be divided into 3 major parts: the large genital trunk (responsible for reproduction), the elongate and movable cephalothorax surmounted with mouth appendages, and finally a pair of more less powerful "arms", which are transformed 2nd maxillae. The maxillae are fused to a mushroom-shaped anchoring structure known as a bulla. The bulla, a product of the frontal gland, is attached to the host. From an aquaculture perspective, the most important lernaeopodid genus is Salmincola with 16 valid species (Kabata, 1969). The vernacular name for *Salmincola* spp. is the gill maggot and the genus has a circumpolar distribution in the northern hemisphere. Gill maggots may occur in high intensities and create serious problems for fish kept at high densities in cages and other aquaculture facilities (Kabata, 1970; Vaughan and Coble, 1975; Kabata and Cousens, 1977; Sutherland and Wittrock, 1985). In natural fish populations, the prevalence and intensity of infection are usually low and have little impact on fish (Black, 1982; Black *et al.*, 1983; Bowen and Stedman, 1990; Amundsen *et al.*, 1997). The cycle of the gill maggot, *S. californiensis*, was described by Kabata and Cousens (1973). It consists of 6 stages: the copepodid (free-swimming infective stage), 4 chalimus stages (which attach to a host by the frontal filament), and adults (the adult female attaches to a host a bulla, while the male remains mobile).

The best known species of gill maggot is *S. californiensis*. It was originally confined to salmonid fishes of the genus Oncorhynchus in streams emptying into the northern Pacific Ocean, but has recently moved eastwards in the US, e.g., to Missouri and Arkansas (Hoffman, 1984) and subsequently to Iowa, New Jersey, and West Virginia (Sutherland and Wittrock, 1985). Sutherland and Wittrock (1985) reported the parasite from an aquaculture site in Iowa, stating a prevalence of 83 % and a mean intensity of 4.6 female copepods. Infections were associated with hyperplasia of gill filaments caused by female copepods and atrophy of growth inhibition of affected gill filaments. The above authors also noticed a harmful effect of chalimus larvae, such as hyperplasia and sometimes fusion of the basal elements of adjacent gill filaments. Gall *et al.* (1972) found evidence of the influence of *S. californiensis* on the reproductive performance of a domesticated strain of rainbow trout. Johnson and Heindel (2001) reported problems with high numbers of S. californiensis in broodfish of Chinook almon, *O. tshawytscha*, and they described a method for their manual removal.

Another significant species from the same genus is *Salmincola edwardsii*, a parasite specific to Arctic charr, *Salvelinus alpinus*, and brook trout *Salvelinus fontinalis*. This parasite infected health problems in fishes cultured in Quebec and New Brunswick, Canada (Parissa Irani-Bunin, pers. Comm..). Salmincola salmoneus is specific for Atlantic salmon, Salmo salar. Heavy infestations with this parasite of salmon broodstock kept in hatcheries in the Maritime provinces of Canada were reported by McGladdery and Johnston (1988).

Salmincola carpionis was reported from central Japan infecting cultured brook trout and whitespotted charr, Salvelinus leucomaenis (cf. Nagasawa *et al.*, 1997,1998). *Salmincola stellatus* were recently reported in Hookaido, Japan in the buccal cavity of cultured taimen, hucho perryi, with intensities ranging from 10 to 50 individuals per fish (Nagasawa *et al.*, 1994).

The closest relatives of *Salmincola* spp. are representatives of Achtheres. The latter genus differs from the former one in having a prominent genital process. According to Kabata (1969), there are 6 valid species of this genus. *Achtheres percarum* which occurs in Europe on perch (*Perca fluviatilis*) and zander (*Sander lucioperca*) can be pathogenic to fish. Because of its attachment to gill filaments, it can trigger epithelial hyperplasia which fuses gill lamellae, and cause partial or total loss of gill filaments. Prevalences often reach 100 %, and the number of Achtheres females can reach 40 per fish (Kozikowska *et al.*, 1957). The highest observed number of all stages of the parasite in zander was 117 (Piasecki, 1993b) and 81 individuals in perch (Piasecki and Woloszyn, 1991). The file cycle of *A. percarum* includes 7 stages: nauplius, copepodid, 4 chalimus stages and adults (Piasecki and Kuzminska, 1996).

Basanistes huhonis is another "cousin" of *Salmincola*. It looks similar to *Salmincola* spp. But is covered with 12 prominent humps or outgrowths. It infects *Danube salmon, Hucho hucho* and broodstock may host very high numbers of this parasite (Ivaska, 1951; Witkowski and Blachuta, 1980).

The least known freshwater lernaeopodid genus is *Tracheliastes*, and the most-pathogenic species within this genus is *T. maculates*. This copepod has a very slim appearance, it is slightly longer than Lernaea, and its bulba attaches permanently to the scales of common bream. Its presence is initially associated with focal inflammation about its feeding area, with pronounced congestion of the skin and extravasations of blood. With heavy infections, diffuse inflammation of the skin can be observed and even extensive perforation of the scales to which the parasite's bullae are attached. There are 2 documented cases of extreme infection levels of this parasite on bream. The 1st was Jamno lake, Poland and was associated with fish mortalities. The intensity reached 56 individuals per fish (Grabda and Grabda, 1958). The other mass infection occurred in Pierzchaly lake, Poland with an intensity of 61 individuals per fish (Piasecki, 1991). The life cycle of *T. maculates* consists of 7 stages: premolt nauplius, infective copepodid, 4 chalimus stages, and adults (Piasecki, 1989).

Sea lice of fish lice of the family Caligidae (Siphonostomatoida), athogenic to marine fishes cultured in cages, have only 1 freshwater representative, *Caligus lacustris*. This copepod has caused only 1 known outbreak in cultured salmonids (Rokicki, 1987) but may have the potential to negatively affect freshwater aquaculture operations.

Copepods as Fish Parasites and Vectors of Fish Diseases

Another unfavourable impact of copepods on aquaculture facilities is that they may serve as intermediate hosts of important fish parasites such as tapeworms or nematodes. The presence of parasites may lead to fish mortalities or adversely affect the market value of the fish or fish products when parasites are present in fish muscle. Some important human parasites utilize copepods and fishes as intermediate hosts. A number of parasitic diseases can affect humans, when an infected copepod is accidentally ingested with water.

One of the better known and the most spectacular parasites are representatives of the genus *Ligula*. They commonly occur in Europe, Asia and North American in temperate zone. The final hosts are piscivorous birds. The first intermediate hosts are planktonic copepods of the genera *Cyclops, Eucyclops, Megacyclops, Acanthocyclops,* and *Eudiaptomus*. The 2nd intermediate hosts for these parasites are fishes, mainly of the family Cyprinidae. This tapewater occurs in the peritoneal cavity of fish in the form of larvae known as plerocecoids. Plerocercoids are usually between 10 cm and 1 m long (Dubinina, 1966). Infection leads to cachexia, impairment of fish growth, arrested reproduction, and death (Brylinski, 1972). Bream infect themselves in their 1st years of life when they feed on plankton. The infection is fatal. All large, market-size bream are those fortunate individuals which avoided infection in theis earlylife. Once they grow up and change the planktonic copepods in their diet to benthic macrofauna, their chances of becoming infected with *Ligulas* spp. decrease. Large pleorocercoids can occur in the muscle of fishes (Jezewski and Karbowiak, 2002). In the US, ligulosis is a sporadic problem in fat-head minnows, but given the small size of the host and the large size of the parasite, outbreaks can be quite severe.

Another interesting tapeworm transmitted by planktonic copepods is Triaenophorus nodulosus (or *T. crassus*). Its scolex is armed with 4 characteristic anchors. The final hosts are predacious fishes such as pike. The 1st intermediate hosts are copepods representing *Cyclops, Eucyclops, Mesocyclops, Paracyclops, Acanthocyclops, Orthocyclops, Diaptomus,* and *Eudiaptomus*. In the intestine of the final host, the adult tapeworms are no longer than 30 cm and are not pathogenic. However, plerocercoids occurring in fishes serving as the 2nd intermediate host can be pathogenic, setting in the liver where they become encapsulated. The presence of such plerocercoids in fish fry can lead to abdominal distension, other serious lesions, and death. Mortalities of fry associated with *Triaenophorus* spp. Infections have been recorded at may European aquaculture facilities and lakes. In kongigsee (Bauern, Germany), this parasite was a factor that strongly limited the size of the local population of Arctic charr (Shaperclaus, 1992).

The Asian tapeworm, *Bothriocephalus acheilognathi*, occurs in grass carp (*Ctenopharyngodon idellus*) and carp, not only in Asia but also the Europe and North America. This tapeworm has only 1 intermediate host, which are various species of free living cyclopoids. *Bothriocephalus acheilognathi* is less than 20 cm long, and in Europe it can be pathogenic, occurring in 1-2-moold fish in high numbers (e.g. 20-40 parasites per fish). Intensities in 2-yr-old fish varied from to 362 parasites per fish. Mortalities have been reported, even in older fish, especially after overwintering (Klenov and Vasilkov, 1972). The Asian tapeworm was introduced into the US more than 20 yr ago. It is a major problem in grass carp and golden shiners. Infections severe enough to cause loss of fish are rare, but there are many regulations regarding interstate transport that prohibit the movement of infected fish. The State of Utah requires that all grass carp coming into the state to be treated with the antihelmintic, praziquantel, before transport.

Other important cestodes include representatives of *Proteocephalus*. These species have 4 characteristic circular suckers on their scolices and they only use cyclopoids as intermediate hosts. *Proteocephalus exiguous* infects coregonids, and densities of this species and its congeners may be high and can lead to local destruction of the intestinal mucosa. In the US, a closely related tapeworm, *Corralobothrium* sp., is commonly seen in cultured channel catfish but is of no commercial importance. The broad fish tapeworm, *Diphyllobothrium* sp., is potentially pathogenic to humans and can be very abundant in fishes in some areas in North America. In the area of along the border of Maine and the Province of New Brunswick, Canada (e.g., Spednic Lake) in the mid-1990s, the viscera and muscle of fishes were so heavily infected that the fish were useless for human consumption and were thus unmarketable (Piasecki unpubl. Data).

The most interesting parasite in eel management in Europe has been the nematode *Anguilicola crassus*. The species was accidentally introduced in Western Europe in the 1970s along with imported Japanese eels, and by the end of 20th century, it managed to spread with eels to most of Europe. This nematode's only intermediate hosts are planktonic copepods. Small fish serve as paratenic hosts, and the adult nematode inhabits the swimbladder of eels. In the estuary of the Odra River (Szczecin, Lagoon) in Poland, almost 70 % of eels were infected (Garbacik-Wesolowska *et al.*, 1994), and intensities can be very high. In recent years, the entire swimbladder is sometimes found to be filled with these uglylooking nematodes, and infections may affect the marketability of the eels.

Parasitic copepods are potentially capable of transmitting viruses and bacteria that cause important fish diseases. Nylvund *et al.* (1991,1993) discussed the possibility of transmission by sea lice of the furunculosis bacterium, *Aeromonas salmonicida*, and the virus that causes infectious salmon anemia (ISA). Sea lice are considered marine copepods, but freshwater fish-lice (Caligus lacustris) can also become abundant in aquaculture cages (Rokicki, 1987) and may be disease vectors. *Salmincola* is reported to transmit IHN virus between fishes (Mulcahy *et al.*, 1990), but the actual transmission mechanism is unclear because *Salmincola* adult females are immobile.

Hosts and Vectors of Human Diseases

The broad fish tapeworm, *Diphyllobothrium latum*, is probably one of the better-known fish parasites because humans are its final host. This parasite uses 2 intermediate hosts, the 1st a copepod, the 2nd a fish. Plerocercoids in fish muscles can be accidentally eaten by humans when the fish is consumed raw or undercooked. Many papers have described infections of people with adult broad fish tapeworms (e.g. Dick *et al.*, 1991), but only a few represented cases where humans had become intermediate hosts for this parasite. A good example is so-called "sparganosis" recorded in some Far Eastern countries. It occurs when humans replace fish as the 2nd intermediate host of broad fish tapeworms by drinking water carrying *Diphyllobothrium*-laden copepods (e.g. Kim and Lee, 2001).

The most important human disease linked to copepods is cholera. The association of *Vibrio cholerae*, the causative agent of cholera. It is now well documented that *V. cholerae* is autochthonous to aquatic environments and closely associated with copepods. Seasonal cholera outbreaks are associated with algal blooms, which in turn provide enough food to cause rapid growth of planktonic copepods. The *V. cholerae* associated with copepods also rapidly reproduces.

References

Abdelhalim, Al., Lewis, J. W. and Boxshall, G. A. 1991. The life-cycle of Ergasilus sieboldi Nordmann: Poecilo-stomatoida) parasitic on British freshwater fish. J. Nat. Hist. 25: 559 – 582.

Abrosov, V. N. and Bauer, O. N. 1959. Ergasilose des peleds (Coregonus peled) in den Seen des Pskowsker Gebiets IZV. Gosud. Naucno-issled. Inst. Ozern. Rech. Rybn. Hozaist. 49: 213 – 216.

Abrosov, V. N. and Bauer, O. N. 1961. Die Peladmarane- ein neues Object der Fischwirtschaft. Ztschr. F. Fisherei. 10: 299 – 307.

Amedo MAPM, J. S. and Ho, GEF Rocha. 1995. Phylogeny and biogeography of the Ergasilidae (Copepoda, Poecilostomatoide), with reconsideration of the taxonomic status of the Vaigemidae. Contr. Zool. 65: 233 – 243.

Amundsen, P. A., Kristoffersen, R., Knudsen, R. and Klemetsen, A. 1997. Infection of *Salmincola edwardsii* (Copepoda:Lemaeopodidae) in an age-structured population of Arctic charr, a long-term study. J. Fish. Biol. 51: 1033 – 1046.

Black, G. A. 1982. Gills as an attachment site for *Salmincola edwardsii* (Copepoda:Lemaeopodidae). J. Parasit. 68: 1172 – 1173.

Black, G. A., Montgomery, W. L. and Whoriskey, F. G. 1983. Abundance and distribution of *Salmincola edwardsii* (Copepoda) on anadromous brook trout, *Salvelinus fontinalis* (Mitchill) in the Moisie River system, Quebec. J. Fish Biol. 22: 567 – 575.

Bowen, C. A. and Stedman, R. M. 1990. Host-parasite relationship and geographic distribution of *Salmincola corpulentus* (Copepoda : Lemaeopodidae) on bloater (Coregonus hoyi) stocks in Lake Huron. Can. J. Zool. 68: 1988 – 1994.

Cakic, P., Petrovic, Z., Kataranovski, D. and Fister, S. 1998. Detection of the parasitic copepods *Lamproglena pulchella* Nordmann (1932) and Larnaea (Sic.) *cyprinacea* Linnaeus (1758) on the gills of fish from Yugoslav waters. Acta. Vet. Belgrade. 48: 131 – 138.

Cheng, S. H., Chen, H. C., Su, M. S. and Ho, J. S. 1999. Effects of temperature and salinity on the maturation in *Apocyclops royi* (Cyclopidae, Cyclopoida). The 7th International Conference on Copepoda, Curitiba, Brazil 25-31 July, 1999. Programme and Abstracts Curitiba, Brazil. The World Association of Copepodologists. Pp. 80.

Cheng, S. H., Chen, H. C., Chang, S. L., Chen, T. I. And Liao, I. C. 2001. Study on the optiman density of mass culture in copepod *Apocyclops royi*. The 6th Asian Fisheries Forum. 25-30 Nov., 2001, Kaoshiung, Taiwan: Asian Fisheries Society. Pp. 58

Dick, T. A., Dixon, B. and Choudhury, A. 1991. *Diphyllobothrium, Anisakis* and other fish-borne parasitic zoonoses. SE Asian J. Trop. Med. Public Health. 22: 150 – 152.

Dorovskikh, G. N. 1996. Location *of Lernaea cyprinacea* (Copepoda: Lernaeidae) on the body of crucian. Parazitologia. 30: 540 – 544.

Egusa, S. 1992. Infection diseases of fish. Rotterdam-Brookfield, AA Balkema.

Fajer, E., Prieto, A. and Vinjoy, M. 1985. *Lernaea cyprinacea* (Copepoda:Lemaeidae) en peces en cultivo intensive en Cuba. Revis. Salud Anim. 7: 411 – 414.

Gabrielli, M. A. and Orsi, M. L. 2000. Dispersao de *Lemaea cyprinacea* (Linnaeus) (Crustacea: Copepoda) na regiao norte do estado do Parana, brasil. Review Brasil. Zool. Curitiba. 17: 395 – 399.

Galli, P., Crosa, G., Bertoglio, S., Mariniello, L., Ortis, M. and D'Amelio, S. 2001. Populations of *Lamproglena pulchella* von Nordmann, 1832 (Copepoda: Eudactylinidae) in cyprinid fish in rivers with different population levels. J. Appl. Ichthyol. 17: 93 – 96.

Garbacik-Wesolowska, A., Szkudlarek, A. and Sobecka, E. 1994. Nematode *Anguillicola crassus* in eels from the Szczecin Lagoon and Lake Letowskie. Bull. Sea. Fish. Inst. 2: 34 – 37.

Goodwin, A. E. 1999. Massive *Lemaea cyprinacea* infestations damaging the gills of channel catfish polycultured with bighead carp. J. Aquat. Animal Health. 11: 406 – 408.

Hall, D. N. 1983. Occurrence of the copepod parasite *Lernaea cyprinacea* L., on the Australian grayling *Prototroctes maraena* Gunther. Proc. R. Soc. Victoria. 95: 273 – 274.

Hartig, J. H., Jude, D. J. and Evans, M. S. 1982. Cyclopoid predation on Lake Michigan fish larvae. Can. J. Fish. Aquat. Sci. 39: 1563 – 1568.

Ho, J. S. 1998. Cladistics of the Lernaeidae (Cyclopoida), a major family of freshwater fish parasites. J. Marine Syst. 15: 177 – 183.

Hoffman, G. L. 1984. *Salmincola californiensis* continues the march eastward. Am. Fish. Newslett. 12: 4.

Huq, A. and Colwell, R. R. 1996. Environmental factors associated with emergence of disease with special reference to cholera. East Mediter. Health J. 2: 37 – 45.

Ibraheem, M. H. and Izawa, K. 2000. On the morphology of *Lamproglena monody* Capart, a parasitic Copepod on the gills of tilapia in Egypt. Zool. Middle East. 21: 103 – 108.

Johnson, K. A. and Heindel, J. A. Efficacy of manual removal and ivermectin gavage for control of *Salmincola californiensis* (Wilson) infestation of Chinook salmon, Oncorhynchus tshawytscha (Walbaum), captive brood-stcks. J. Fish Dis. 24: 197 – 203.

Kabata, Z. 1979. Parasitic Copepoda of British fishes, London. Ray Society.

Kabata, Z. 1985. Parasites and diseases of fish cultured in the tropics. London and Philadelphia, P.A.: Taylor and Francis.

Kabata, Z. and Cousens, B. 1977. Host-parasite relationship between sockeye salmon, *Oncorhynchus nerka*, and *Salmincola californiensis* (Copepoda: Lernaeopodidae). J. Fish Res. Bd. Can. 34: 191 – 202.

Khalifah, K. A. and Post, G. 1976. Histopathological effect of *Lernaea cyprinacea* (a copepod parasite) on fish. Prog. Fish-Cult. 38: 110 – 113.

Kim, J. R. and Lee, J. M. 2001. A case intramuscular sparganosis in the sartorius muscle. J. Korean Med. Sci. 16: 378 – 380.

Klenov, A. P. and Vasil'kov, G. V. 1972. Seasonal and growth dynamics of bothriocephalosis in grass carp in the Mironovskij fish pond. Bjull Vses. Inst. Gel'mintol. Im. K.I. Skrjabina. Moskva. 8: 25 – 27.

Kozikowska, Z. 1975. Ergasilus sieboldi Nordm., adaptations of the parasite to the morphology and biology of hosts and additional data on the occurrence of other species of parasitic crustaceans in the area of Warmia and Masuria. Acta. Univ. Wratislav. No. 301. Prace Zool. 7: 3 – 40.

Kuang, P. R. and Qian, J. H. 1985. Three new species of the genus Lamproglena from China (Copepoda: Cyclopoida, Lernaeidae). Acta. Zootax. Sin. 10: 363 – 369.

Kumari, P., Khera, S. and Gupta, N. K. 1989. On six new species of the genus Lamproglena Nordmann (Copepoda: Eudactylinidae), ectoparasitic on freshwater fishes of India. Res. Bull. Punjab Univ. Sci. 40: 9 – 23.

Langdon, J. S. 1992. Major protozoan and metazoan parasitic diseases of Australian finish. In Dl Bryden ed. Fin Fish Workshop, Post Graduate Committee in Veterinary Science. University of Sydney, Proceedings, Sydney. Australia, University of Sydney. 182: 1 – 28.

Mamcarz, A. 1990. Conditions for growth of paled larvae, *Coregonus peled* (Gmel.) during net-pen rearing. Acta. Acad. Agric. Techn. Olst. Protectio Aquarum Piscatoria. Supl. 17: 3 – 57.

Marten, G. G., Bordes, E. S. and Nguyen, M. 1994. Use of cyclopoid copepods for mosquito control. Hydrobiologia. 292: 491 – 496.

Marx, H. M. and Avenant-Oldewage, A. 1996. Redescription of *Lamproglena clariae* Fryer, 1956. (Copepoda, Lernaeidae), with notes on its occurrence and distribution Crustaceana. 69: 509 – 523.

McGladdery, S,. E. and Johnston, C. E. 1988. Egg development and control of the parasite *Salmincola salmoneus*, on Atlantic salmon kelts (*Salmo salar*) exposed to four different regimes of temperature and photoperiod. Aquaculture. 68: 193 – 202.

Mulcahy, D., Klaybor, D. and Batts, W. N. 1990. Isolation of infectious hematopoietic necrosis virus from a leech (*Piscicola salmositics*) and a copepod (*Salmincola* sp.), ectoparasites of sockeye salmon *Oncorhynchus nerka*. Dis. Aquat. Organ. 8: 29 – 34.

Muroga, K., Takahashi, S. and Yamanoi, H. 1979. Non-cholera *Vibrio* isolated from diseased ayu. Bull Jpn. Soc. Sci. Fish. 45: 829 – 834.

Nagasawa, K., Ikuta, K. and Kitamura, S. 1997. Distribution of *Salmincola carpionis* (Copepoda: Lernaeopodidae) in the buccal cavity of salmonids. Bull. Nat. Res. Inst. Aquacult. 26: 35 – 39.

Nagasawa, K., Ikuta, K., Nakamura, H., Shikama, T. and Kitamura, S. 1998. Occurrence and effects of the parasitic copepod *Salmincola carpionis* on salmonids in the Nikko District, Central Japan. In: (eds. H-U Dahms, T. Glatzel, H. J. Hirche, S. Schiel, H. K. Schminke). Proceedings of the 6th International Conference on Copepoda. J. Mar. Syst. 15: 269 – 272.

Nagasawa, K., Watanabe, J. R., Kimura, S. and Hara, A. 1994. Infection of *Salmincola stellatus* (Copepoda: Lernaeopodidae) on Sakhalin taimen Hucho perryi reared in Hokkaido. Bull. Faculty Fish. Hokkaido Univ. 45: 109 – 112.

Nylund, A., Bjorknes, B. and Walace, C. 1991. *Lepeophtheirus salmonis* – a possible vector in the spread of diseases on salmonids. Bull Eur. Assoc. Fish Path. 11: 213 – 216.

Nylund, A., Walace, C. and Hovland, T. 1993. The possible role of *Lepeophtheirus salmonis* (Kroyer) in the transmission of infectious salmon anemia. In: (Eds. G. A. Boxshall, D. Dafaye). Pathogens of wild and farmed fish: sea lice, Chichester, UK: Ellis Horwood. Pp. 367 – 373.

Ogle, J. 1979. Adaptation of a brown water culture technique to the mass culture of the copepoda Acartia tonsa. Gulf. Res. Rep. 6: 291 – 292.

Ochno, A. and Okamura, Y. 1988. Propagation of the calanoid copepod, *Acartia tsuensis* in outdoor tanks. Aquacutlure. 70: 39 – 51.

Oldwage,, W. H. 1993. The past, present and future role of piscine parasitic copepods in Africa. Fifth International Conference on Copepoda. University of Maryl and Baltimore County, 6-12 June, 1993. Baltimore, USA. TheWorld Association of Copepodologists. Pp. 73.

Pavanelli, G. C., Eiras, J. C., Takemoto, R. M., Ranzani-Paiva, M. J. and Magalhaes, A. R. M. 2000. Sanidade de peixes, ras, crustaceae e moluscos. In: (eds. W. C. Valenti, C. R. Poli, J. E. Pereira, J. R. Borghetti). Aquaculture no Brasil, Bases para um desenvolvimento sustentavel. Brasilia: Conselho Nacional de Desenvolvimento Cientifico e Technologico. Pp. 197 – 245.

Payne, M. F. and Rippingale, R. J. 2001. Intensive cultivation of the calanoid copepod *Gladioferens imparipes*. Aquaculture. 201: 329 – 342.

Piasecki, W. 1991. Tracheliastosis of bream, Abramis brama (L.) in Pierzchalsie Lake near Braniewo. Materialy XVI Zjazdu Polskiego Towarzystwa Parazytologiczengo, Poznari, 12-13 September, 1991. Poznan, Poland Polskie Towarzstwo Parazytologiczne. Pp. 28.

Piasecki, W. 1993a. Comparative morphology of the three species of Lamproglena (Copepoda, Cyclopoida, lernaeidae) described by von Nordmann, based on re-examination of the types. Mitt. Zool. Mus. Berlin. 69: 307 – 315.

Piasecki, W. 1993b. Crustacean parasites from three fish species from the Odra River estuary, Poland. Acta. Ichthyol. Piscal. 23: 77 – 87.

Piasecki, W. and Kuzminska, E. 1996. Development stages of Achtheres percarum von Nordmann, 1832 (Lernaeopodidae) parasitic on *perch-Perch* fluviatilis L. Sixth International Conference on Copepoda. Oldenburg Bremerhaven, Germany, 29 July-3 Aug., 1996. Oldenburg, Germany: The World Association of Copepodologists. Pp. 89.

Piasecki, W. G. 2000. Attacks of cyclopoid Acanthocyclops robustus (Sars) on newly hatched cyprinids. Electr. J. Polish Agric. Univ. Fish. 3:

Rowland, S. J.and Ingram, B. A. 1991. Diseases of Australian native freshwater fishes. Sydney NSW Fisheries. Pp. 33.

Sabatini, M. A., Marini, M., Fratello, B., Gelosini, C., Mola, L. and Benedetti, I. 1988. The digestive apparatus of the parasitic copepoda Lemaea cyptinacea L. during development and in the adults. Acta Embryol. Morph. Exp. 8: 389 – 409.

Schaperclaus, W. 1992. Fish diseases (English translation of "Fischkrankheiten", 5th corrected and substantially enlarged edition, 1996). Rotterdam. AA Balkema.

Sendak, H. 1990. Mechanical catch of zooplankton in a lake. Ph. D. thesis, Agricultural University of Szczecin, Poland.

Shariff, M. and Roberts, R. J. 1989. The experimental histopathology of *Lernaea polymorpha* Yu, 1938 infection in naïve Aristichthys nobilis (Richardson) a comparative with the lesion in naturally infected clinically resistant fish. J. Fish Dis. 12: 405 – 414.

Chapter 2

Vermiculture of *Eudrilus eugeniae* and *Eisenia fetida* in Different Organic Wastes

M. Senthamil Chelvi and M.R. Rajan

Department of Biology, Gandhigram Rural Institute Deemed University, Gandhigram – 624 302, India

ABSTRACT

The present study deals with the vermiculture of *Eudrilus eugeniae* and *Eisenia fetida* in different organic wastes for a period of 40 days. Five experimental vermibeds were prepared by using different organic wastes such as black gram husk, red gram husk, pressmud, goat droppings and cowdung. Among the different organic wastes the biomass of *E. eugeniae* and *E. fetida* were higher in black gram husk and red gram husk respectively.

Keywords: *Vermiculture, Eudrilus eugeniae, Eisenia fetida, Organic wastes.*

Introduction

The use of earthworms to convert organic materials into valuable manure is called vermicomposting several researchers have used various organic materials like cattle dung, poultry manure, horse manure, rabbit droppings, for vermiculture and vermicomposting (Edwards, 1983; Bano and Kale, 1986; Balasubramanian and Bai, 1995 and Ravinder Reddy and Sreenivasa Raju, 1998). In the present study different types of organic wastes like black gram husk, red gram husk, goat droppings, press mud and cow dung were subjected to vermicomposting by using epigenic earthworms *Eudrilus eugeniae* and *Eisenia fetida*.

Materials and Methods

Organic wastes such as black gram husk, red gram husk, goat droppings and cowdung were the raw materials used for the preparation of vermibed. Cowdung was collected from the University Dairy farm. Black gram and red gram husk were purchased from the local market in Batlagundu, Madurai, India. Goat droppings were collected from Anna Nagar near Gandhigram. Pressmud was collected from sugar factory, Madura sugras, Pandiyarajapuram, Madurai.

Cowdung and other wastes were dried under shade and powdered. Wastes were mixed with cowdung in the ratio of 1 : 1 and added required amount of water to hold 70–80 per cent moisture. All the wastes were predecomposed for 10 days. Vermibeds prepared by cowdung serve as control. Vermibeds were prepared in round containers (12 cm dia) and were maintained in triplicates. Earthworms *Eudrilus eugeniae* and *Eisenia fetida* were collected from the mass culture tanks in the Department of Biology, GRI, Gandhigram. Equal weight of 5 worms were introduced in each vermibed. Moisture level of the beds were maintained at 70–80 per cent by sprinkling water regularly and care was taken not to soak the bed with excess of water. The vermibed were monitored everyday and the vermicasting were periodically removed and separately stored. Growth and reproduction of worms were recorded in the control and experimental vermibeds. The growth performance of the worms were recorded by observing the weight gained by the worms and the reproduction by the number of cocoons produced at 10, 20, 30 and 40th day. The number and weight of hatchlings were also noted.

Results and Discussion

Table 2.1 shows the biomass of adult and young worms (Juveniles) of *Eudrilus eugeniae* in different organic wastes. Biomass was increased in the vermibeds containing cowdung, black gram husk, red gram husk and pressmud (*i.e.* 11.816, 9.887, 8.006 and 7.059 respectively). Mba (1978) also observed influence of different mulch treatment on the growth rate and activity of earthworm *Eudrilus eugeniae.* The young worms have considerably increased in vermibeds with cowdung and black gram husk only (2.558 and 3.334). Among the different organic wastes, the biomass of adult and young worms of *Eudrilus eugeniae* were increased in black gram husk. The vermibed prepared with goat droppings showed negative result in the growth and reproduction of worms. This result showed the unsuitability of the wastes and it may be due to the presence of some unwanted materials in the goat droppings.

Table 2.2 shows the biomass of adult and young worms (Juveniles) of *Eisenia fetida* in different organic wastes. The growth of the worms have considerably increased in all experimental vermibeds. Similar growth of *E. fetida* was studied by Reinceke and Vijoen (1990). In the present study the reproduction and production of young worms have considerably increased in different vermibeds (Table 2.2). Among the different organic wastes the biomass of adult worms were increased in pressmud.

The number of cocoons produced by *Eudrilus eugeniae* and *Eisenia fetida* in different organic wastes were shown in Table 2.3. The vermibed prepared with black gram husk, cowdung and red gram husk showed higher number of cocoons production of Eudrilus species (*i.e.* 13, 10 and 6) Reinceke and Hallatt (1989) reported similar result in *Perionyx excavatus.* The vermibed prepared with pressmud showed very less number of cocoons. Among the different organic wastes used in vermibed preparation the cocoons of *Eisenia fetida* were more and same in black gram and red gram husk. From the present study, it was concluded that the growth and reproduction of *Eudilus euginiae* and *Eisenia fetida* were higher in black gram husk mixed with cowdung as substrate.

Table 2.1: Biomass of Adult and Young Worms of *Eudrilus eugeniae* in Different Organic Wastes Under Different Intervals (Days)

Substrates	*Initial*	*10th Day*		*20th Day*		*30th Day*		*40th Day*	
		Adult	*Juveniles*	*Adult*	*Juveniles*	*Adult*	*Juveniles*	*Adult*	*Juveniles*
Cowdung	2.785±0.003	5.276±0.254	–	6.962±0.355	–	7.761±0.868	0.070±0.001	9.887±0.343	2.558±0.241
Pressmud	2.786±0.002	5.265±0.834	–	6.302±0.399	–	6.615±0.355	–	7.059±0.393	–
Black gram husk	2.788±0.001	6.301±0.430	–	7.638±0.108	–	10.135±1.051	0.045±0.538	11.816±1.341	3.334±0.181
Red gram husk	2.784±0.001	4.381±1.088	–	6.686±0.684	–	7.463±1.122	–	8.006±1.043	–
Goat droppings	2.786±0003	2.055±0.093	–	1.782±0.187	–	1.443±0.189	–	–	–

Table 2.2: Biomass of Adult and Young Worms of *Eisenia fetida* in Different Organic Wastes Under Different Intervals (Days)

Substrates	*Initial*	*10th Day*		*20th Day*		*30th Day*		*40th Day*	
		Adult	*Juveniles*	*Adult*	*Juveniles*	*Adult*	*Juveniles*	*Adult*	*Juveniles*
Cowdung	1.178±0.004	3.018±0.067	–	3.403±0.191	–	3.444±0.088	0.295±0.337	3.777±0.091	0.704±0248
Pressmud	1.177±0.002	2.968±0.163	–	3.070±0.108	–	3.276±0.020	1.265±0.138	3.162±0.164	1.451±0.231
Black gram husk	1.177±0.005	3.001±0.113	–	3.483±0.104	–	4.018±0.241	0.079±0.004	3.869±1.397	1.424±0.195
Red gram husk	1.178±0.004	3.018±0.067	–	3.260±0.163	–	3.753±0.109	0.393±0.011	4.254±0.034	0.540±0.030
Goal droppings	1.178±0.005	2.641±0.020	–	3.263±0.062	–	3.663±0.010	–	4.005±0.081	–

Table 2.3: Number of Cocoons of *Eudrilus eugeniae* and *Eisenia fetida* in Different Organic Wastes

Earthworms Species	*Organic Wastes*	*Initial*	*10th Day*	*20th Day*	*30th Day*	*40th Day*
Eudrilus eugeniae	Cowdung	–	–	14	14	10
	Pressmud	–	–	1	2	6
	Black gram husk	–	–	12	12	13
	Red gram husk	–	–	2	2	2
	Goat droppings	–	–	–	–	–
Eisenia fetida	Cowdung	–	–	20	20	31
	Pressmud	–	–	13	13	19
	Black gram husk	–	–	10	10	28
	Red gram husk	–	–	16	16	28
	Goat droppings	–	–	–	–	–

Acknowledgement

Authors are thankful to Department of Biology, Gandhigram Rural Institute-Deemed University, Gandhigram, India for offering facilities to carry out this study.

References

Balasubramanian, P.R. and Bai, R.K., 1995. Recycling of cattle dung, biogas plant effluent and water hyacinth in vermiculture. *Bioresource Technology*, 52(1): 85–87.

Bano, K. and Kale, R.D., 1986. Conversion of poultry manure in to protein. In: *Verms and Vermicomposting*, (Eds) Dash, M.C., Senapathi, B.K. and Mishra, P.C, pp. 125–150.

Edwards, C.A., 1983. Earthworm ecology in cultivated soils. In: *Earthworm Ecology: From Darwin to Vermiculture*, (Ed) Satchell, J.E. Chapman and Hall, London, pp. 123–138.

Mba, C.C., 1983. Utilization of *Eudrilus eugeniae* for disposal of cassava peel. In: *Earthworm Ecology: From Darwin to Vermiculture*, (Ed.) Satchell, J.E. Chapman and Hall, London, pp. 315-321.

Reinceke, A.J. and Hallatt, L.S., 1989. The growth and cocoon production of *Perionyx excavatus*. *Biol. Fert. Soils*, 8: 303–306.

Reinceke, A.J. and Vijoen, S.A., 1990. The influence of feeding pattern on growth and reproduction of the vermicomposting earthworm *Eisenia fetida*. *Biol. Fert. Soils*, 10: 1–4.

Ravinder Raddy, V. and Sreenivasa Raju, 1998. Population, growth and biomass production of Earthworm *Eudrilus eugeniae* in farmyard manure. *Journal of Res., ANGRAU*, 26(1): 17–19.

Chapter 3

Antibacterial Activity of Citrus Limon Fruit Juice Against Pathogenic Bacteria

***Pradeep M. Tumane*[1], *Bharat J. Wadher*[1], *Aqueel Khan*[2] *and Ashok V. Gomashe*[3]**

[1]P.G. Department of Microbiology, [2]P.G. Department of Biochemistry, Rashtrasant Tukdoji Maharaj, Nagpur University, L.I.T. Campus Premises, Nagpur – 440 033

[3]Shri Shivaji Science College, Department of Microbiology, Nagpur

ABSTRACT

Antibacterial activity of the citrus limon fruit juice was screened by cup-plate diffusion method, against clinical isolates of drug resistant strains of *Salmonella typhimurium, Salmonella paratyphi, Shigella flexneri, Proteus mirabilis* and *Proteus vulgaris*. Standard strains of *Proteus mirabilis* (NCIM-2300) and *Proteus vulgaris* (NCIM-2027).

Almost all isolates were found to be resistant to one or more antibiotics and chemotherapeutic agents. Fruit juice of citrus limon was found to be effective against all clinical isolates and standard strains.

Keywords: *Antibacterial properties, Citrus limon, Resistant bacteria.*

Introduction

Traditional system of Indian medicines has the greater possibility of offering new drugs in a variety of pathophysiological states, if the source of medicinal plants to be investigated is based on its specific mention in traditional system texts. It is well known fact that antibiotic resistance in pathogenic bacteria is increasing day by day and has posed a significant public health problems, particularly 'Third World'. Action to mitigate the problem include development of new antimicrobials, better infection control and greater conservation of existing drugs (Livermore *et al.*, 1999).

India has a rich tradition in the use of medicinal plant to develop drugs from plants. Nowadays herbal drugs are prescribed widely even when their biologically active compounds are unknown because of their effectiveness, minimal side effects in clinical experience and relatively low cost (Valiathan, 1998).

Citrus limon is the fruit belonging to family Rutaceae, a small tree which cultivated in Uttar Pradesh, Maharashtra, Tamil Nadu, Mysore, Rajasthan, Bihar and other parts of India.

Thus the present investigation reveals that the citrus limon fruit juice have potential bactericidal effect against the pathogenic bacteria.

Materials and Methods

The fresh fruit of citrus limon were bought from Nagpur region and stored in refrigerator at 4 C. The fruits of citrus limon were washed with sterile distilled water and then with 90 per cent alcohol. The juice was collected in a sterile screw cap test tube and used as a extract.

Antibacterial Testing

In vitro antibacterial activity was carried out against 24 hours cultures of clinical isolates and standard strains. Inoculum of standard and test organisms were prepared by inoculating a loopful of organisms in a 5 ml Hisitests broth (HI-MEDIA) and incubated at 37 C for 6–8 hrs; till a moderate turbidity was developed. The turbidity was match with 0.5 McFarland standard (Finegold and Baron, 1986), 0.1 ml of this suspension was thoroughly mixed with 25 ml of liquid Hisitest Agar (HI-MEDIA), poured in presterilized petriplates and set aside. After solidify, the seeded agar was punched out with flamed cork borer in order to make 3 cups (10 mm diameter), at a spaced out position in the petriplate. All the three cups were filled with 0.1 ml (100 µl) of citrus limon fruit juice. These agar plates were set a side at room temperature for one hour for diffusion and then incubated at 37 C for 18–24 hrs. After incubation the zone of inhibition was measured in mm diameter and the mean value of triplicate was recorded.

Organisms

The standard strains of organisms used in this study were; *Proteus mirabilis* (NCIM-2300); *Proteus vulgaris* (NCIM-2027) and the clinical isolates of *Salmonella typhimurium; Salmonella paratyphi; Shigella flexneri; Proteus vulgaris* and *Proteus mirabilis.* These clinical isolates were obtained from M.I.M.M.S.R. Medical College, Latur and Government Medical College, Nagpur. All the bacterial strains maintained on Trypticase soya-agar slopes (HI-MEDIA) in refrigerator.

Antibacterial Susceptibility Testing

Antibiotic sensitivity of clinical isolates was studied by Kirby-Bauer disc diffusion method. (Bauer *et al.*, 1966). A sterile cotton swab dipped into the inoculum tube and rotated firmly against the upper inside wall of the tube to express excess fluid. Then the swab was used to streak the entire agar surface of the petriplate three times turning the plate 60 C between each streaking. Four to five antibiotic discs were placed on each plate with enough spacing. All the plates were later on incubated at 37 C for 18–24 hours. After incubation plates were examined for zone of inhibition. Zones were measured and recorded as sensitive or resistant as per the zone size interpretative chart (HI-MEDIA).

Antibiotic and Chemotherapeutic Discs

Commercially available standard antibiotic and chemotherapeutic discs were obtained from HI-MEDIA, Mumbai. The abbreviation used for the antibiotics and chemotherapeutic agents are given in brackets. The antibiotic disc used were, Amikacin-30mcq(AK); Amoxyclav-30mcq(Ac); Ampicillin-10mcq(A); Cephaloridin-30mcq(Cr); Cephotaxime-30mcq(Ce); Chloramphenicol-30mcq(C); Ciprofloxacin-5mcq(Cf); Colistin-10mcq(C1); Co-Trimoxazole-25mcq(Co); Gentamicin-10mcq(G); Kanamycin-30mcq (K); Nalidixic acid-30mcq(Na); Neomycin-30mcq(N); Nitrofurantoin-300mcq(Nf); Norfloxacin-10mcq(Nx); Ofloxacin-5mcq(Of); Penicillin G-10unit (P); Streptomycin-10mcq(S); Tobramycin-10mcq(Tb); Pefloxacin-5mcq(Pf); Co-Trimazine-25mcq (Cm) and Cephalexin-30mcq(Cp).

Results and Discussion

The results obtained for the standard and clinical strains of pathogenic bacteria were used to study the effectiveness of citrus limon fruit juice. Five strains each of *Salmonella typhimurium, Salmonella paratyphi* and *Shigella flexneri.* Ten strains each of *Proteus mirabilis* and *Proteus vulgaris* which were found to be resistant to one or more antibiotics and chemotherapeutic agents were tested against the citrus linon fruit juice (Tables 3.1 to 3.3).

Table 3.1: Antibacterial Activity of Citrus Limon Juice on *Proteus* species

Sl.No.	*Organisms*	*Antibiotic Resistance Pattern*	*Zone of Inhibition (in mm dia.) of Citrus Limon Juice*
	Standard Strain		
(A)	*Proteus mirabilis* NCTC	2300	32
	Clinical Isolates		
1.	*Proteus mirabilis*	Cl, S	26
2.	*Proteus mirabilis*	Ac, Ak	25
3.	*Proteus mirabilis*	Cp, Cr	27
4.	*Proteus mirabilis*	Ce, Na	29
5.	*Proteus mirabilis*	C, G	26
6.	*Proteus mirabilis*	Nf, A	29
7.	*Proteus mirabilis*	Co, A	28
8.	*Proteus mirabilis*	Ac, Ce	27
9.	*Proteus mirabilis*	G, Na	30
10.	*Proteus mirabilis*	Na, Nf	29
	Standard Strain		
(B)	*Proteus vulgaris* NCIM 2027		29
	Clinical Isolates		
1.	*Proteus vulgaris*	Ac, C, Ak	25
2.	*Proteus vulgaris*	Nf, Cf, A	24
3.	*Proteus vulgaris*	Ac, C, Na	24
4.	*Proteus vulgaris*	Ac, C, Co	23
5.	*Proteus vulgaris*	Ak, Na, S	22
6.	*Proteus vulgaris*	Ak, A, Cr	24
7.	*Proteus vulgaris*	Cr, Cl, Nf	28
8.	*Proteus vulgaris*	Ak, A	27
9.	*Proteus vulgaris*	A, Cr	28
10.	*Proteus vulgaris*	A, Cm	26

Table 3.2: Antibacterial Activity of Citrus Limon Juice on *Salmonella* Species

Sl.No.	*Organisms*	*Antibiotic Resistance Pattern*	*Zone of Inhibition (in mm dia.) of Citrus Limon Juice*
	Clinical Isolates		
1.	*Salmonella typhimurium-1*	Cr, P	19
2.	*Salmonella typhimurium-9*	Nx, Cf, C	19
3.	*Salmonella typhimurium-7*	Cm, Na	20
4.	*Salmonella typhimurium-3*	A, Cl, Cm	22
5.	*Salmonella typhimurium-4*	G, C, Ak	24
	Clinical Isolates		
1.	*Salmonella paratyphi-12*	Ac, Ce	26
2.	*Salmonella paratyphi-14*	Cr, Na	24
3.	*Salmonella paratyphi-19*	Ac, Co, G	25
4.	*Salmonella paratyphi-21*	Co, G	25
5.	*Salmonella paratyphi-27*	Ac, Cp	26

Table 3.3: Antibacterial Activity of Citrus Limon Juice on *Shigella flexneri*

Sl.No.	*Organisms*	*Antibiotic Resistance Pattern*	*Zone of Inhibition (in mm dia.) of Citrus Limon Juice*
	Clinical Isolates		
1.	*Shigella flexneri-3*	P, K	28
2.	*Shigella flexneri-4*	P, K	26
3.	*Shigella flexneri-8*	Of, P	27
4.	*Shigella flexneri-5*	P, Tb	26
5.	*Shigella flexneri-1*	Ac, Ce	27

Also standard strains of *Proteus mirabilis* (NCIM-2300) and *Proteus vulgaris* (NCIM-2027) were used against citrus limon fruit juice to compare the results with clinical isolates (Tables 3.1 to 3.3).

Citrus limon fruit juice inhibited the growth of standard strains and clinical isolates of the pathogenic bacteria tested. Also it is clear that the multiple drug resistant strains are also found to be equally sensitive to fruit juice of 'Citrus limon'.

In recent past there have been increased concern over increases in level of antibiotic resistance in common pathogens towards routinely used antibiotics. The appearance of multidrug resistance (MDR) has been attributed to improper use of antibiotics and patients not taking complete course of treatment as prescribed. This has been prevalent because of lack of knowledge among masses due to illiteracy (Ganesh Chand *et al.*, 2003).

Among different mechanism of resistance development the resistance of bacteria can be inherent or it can be gained by taking up the resistance genes from other bacterial cell in vicinity. The indiscriminate use of antibiotics give rise to resistant strains of bacteria, which transfer the character to susceptible bacteria, hence making them resistant (Maiden, 1998).

According to the Pentti (1999) there are two ways of fighting the development and spread of resistant bacteria. The first is to reduce the use of antimicrobial agents to decrease selection of resistant bacteria. The second major way to tackle resistance is by improving hygienic measures to prevent the spread of resistant bacteria.

It is also possible to control the drug resistance problem by using a natural, non-toxic, easily available and broad spectrum antimicrobial agents. In case the broad spectrum antimicrobial agent so developed and if found that no resistance is appearing in bacteria, then it is possible to control the disease effectively.

In the present investigation, it is observed that the fruit juice of citrus limon was found to be effective against almost all bacterial tested including antibiotic resistant strains.

Fresh citrus limon fruit yield about 30 per cent of juice which contains about 6.5 to 8 per cent of citric acid and also vitamin C (Ascorbic acid). The juice itself gradually diminishing in acidity when kept; the citric acid gradually decomposed into glucose and carbonic acid. (Wallis, 1985). According to Nagoba *et al.*, (1998, 1999, 2000, 2002) from this laboratory has reported the therapeutic use of citric acid against more than 50 cases of *Pseudomonas aeruginosa* infections. The *Pseudomonas aeruginosa* strains isolated were found to be resistant to atleast nine antibiotics, well inhibited by 2 per cent citric acid. It was observed that the citric acid treatment is simple, reliable, non-toxic, effective and economical (Nagoba *et al.*, 1998).

Gundidzo *et al.* (1993) reported the antimicrobial property of Delbergia melanoxylon extract. They reported that the citric acid extract exhibited strongest antimicrobial activity.

Osato *et al.*, 1993, has reported a antimicrobial and antioxidant activity of unripe papaya, against *Bacillus subtilis, Enterobacter cloacea, Escherichia coli, Salmonella typhi, Staphylococcus aureus, Proteus vulgaris, Pseudomonas aeruginosa* and *Klebsiella pneumoniae.* They reported Vitamin C, malic acid, citric acid and glucose are some of the possible antioxidative components in papaya.

It is undoubtedly clear that citrus limon juice contains major amount of citric acid, which is responsible for the antimicrobial activity against majority of pathogens tested.

Conclusion

Based on the available literature and the findings of this investigation, it is amply clear that the juice of citrus limon is highly effective against all the pathogenic bacteria tested, including drug resistant strains. The active compound identified is a citric acid *i.e.* powerful antiseptic.

Acknowledgement

The authors is grateful to Prof. Dr. Sudhir U. Meshram, Head, P.G. Department of Microbiology Rashtrasant Tukdoji Maharaj Nagpur University, Nagpur and Dr. S.R. Deshmukh, Dept. of Microbiology, Govt. Medical College, Latur.

References

Bauchner, H., 1997. Parent's impact on antibiotic use. *APUA Newsletter*, 15: 1–3.

Bauer, A.W., W.M. Kirby, J.C. Sherris and M. Turek, 1966. Antibiotic susceptibility testing by standardized single disc method. *Am. J. Clin. Pathol.*, 45: 493.

Finegold, S.M. and Baron, E.J., 1986. In: *Diagnostic Microbiology*, (Eds.) Finegold and Baron, E.J., 7th Edition. The C.V. Mosby Co., St. Louis, p. 176.

Ganesh Chand, R.K. Agrawal and Vijendra Mishra, 2003. *In vitro* susceptibility patterns among common bacterial clinical isolates in Bundelkhand region. *Asian J. of Microbiol. Biotech. Env. Sc.*, 5(4): 491–494.

Gundidzo, M and N. Gaza, 1993. Antimicrobial activity of *Dalbergia melanoxylon* extracts. *J. Ethanopharmacol.*, 40(2): 127–130.

Livermore, D.M., A.P. Macgowen and M.C.J. Wale, 1999. Surveillance of antimicrobial resistance. Selection from *Brit. Med. J.* (South Asia Ed.), 14: 954–955.

Macfarlane, J., W. Holmes, R. Macfarlane and N. Britten, 1997. Influence of patient's expectations on antibiotic management of acute lower respiratory tract illness in general practice: Questionnaire study. *Brit. Med. J.*, 315: 1211–1214.

Nagoba, B.S., R.L. Gandhi, B.J. Wadher, S.R. Deshmukh and S.P. Gandhi, 1998. Citric acid treatment of a severe electric burn complicated by multiple antibiotic resistant Pseudomonas aeruginosa: Case report. *BURNS* (London), 24: 481–483.

Nagoba B.S., S.R. Deshmukh, B.J. Wadher, L. Mahabaleshwar, R.C. Gandhi, P.B. Kulkarni, V.A. Mane and J.S. Deshmukh, 1998. Treatment of superficial Pseudomonal infections: An effective and economical approach. *J. Hosp. Infect.* (London), 40: 155–157.

Nagoba, B.S., A.B. Pathan and B.J. Wadher, 1999. Citric acid treatment of superficial nosocomial infectious: A new era in antimicrobial chemotherapy. *Kuwait Medical Journal*, 31(1): 72–74.

Nagoba, B.S., P.B. Kulkarni, B.J. Wadher, U.P. Kulkarni and L. Mahabaleswar, 2000. Citric acid treatment of Diabetic foot: A simple and effective approach. *J. Asso. Phy.*, India, 48: 739–741.

Nagoba, B.S., B.J. Wadher and A.G. Chandorkar, 2002. Citric acid treatment of non-healing ulcers in Leprosy patients. *Brit. J. Dermato.*, 146: 1101.

Osato, J.A., L.A. Santiago, G.M. Remo, M.S. Cuadra and A. Mori, 1993. Antimicrobial and antioxidant activities of unripe papaya. *Life Sci.*, 53(17): 1383–1389.

Pentti, H., 1999. Control of antimicrobial resistance: Time for action. Selection from *Brit. Med. J.* (South Asia Ed.), 14(11): 953–954.

Stuart. B.L., 1999. Antimicrobial resistance: Bacteria on the defence. Selection from *Brit. Med. J.* (South Asia Ed.), 14(11): 952–953.

Valiathan, M.S., 1998. Healing plants. *Curr. Sci.*, 75: 1122–1126.

Wallis, T.E., 1985. *Textbook of Pharmacognosy*, 5th edn. (First Indian Edn.). CBS Publishers and Distributors, New Delhi, pp. 275–277.

W.H.O. Drug Information. 1993. *Essential Drugs*, 7(2): 68–73.

Chapter 4

Antibacterial Activity of Plant Extracts on Different Human Pathogens

Mamta Rawat

Department of Zoology, J.N.V. University, Jodhpur – 342 005, Rajasthan

ABSTRACT

Because of the appearance of bacterial resistant to antimicrobial agents, more effort is being made to find alternative antimicrobial component, so, the natural products are more preferred to the synthetic ones. In this series, the antibacterial activity of plant extract was evaluated against gram-negative organisms, *Escherichia coli, Klebsiella pneumoniae, Pseudomonas aeuroginosa, Proteus* spp. and Gram-positive Cocci, *Staphylococcus aureus.* The extracts from following plants were used *Nerium indicum, Acacia nilotica, Zizipus glabrata, Ocimum americanum, Ficus bengalensis,* and *Mimosa hamata.* The extracts were prepared ether and distill water. The test was performed using nutrient agar. The results showed that methanol and acetone extracts were most effective and aqueous extract were the least.

To sum up, these extract showed different inhibitory effect against all tested strains.

Keywords: *Antibacterial activity, Pathogens, Plant extracts.*

Introduction

Inspite of production of a number of new antibodies during last three decades by pharmacological industries, which are resident to the drugs by microorganisms to drugs are being utilized as therapeutic agents. (Cohen, 1992 and Nascimento *et al.*, 2000).

Several microorganism-derived antibiotics are currently in use to treat a variety of infectious human disease. Most of them have a limited antibacterial species of the pathogen; some even lead to serious side effects. Therefore actions must be taken to reduce this problem, for *e.g.*, to control the use of antibiotic, develop research to better understand the genetic mechanism of research and its continue studies to develop new drugs, either synthetic or natural. The ultimate goal is to offer appropriate and efficient antimicrobial drugs to the patient. Efforts are thus being directed to identify antibiotic sources other than traditional microorganisms.

For a long period of time, plants have been a valuable source of natural products for maintaining human health, especially in the last decade, with more intensive studies for natural therapies. India has a rich tradition in use of medicinal plants to develop drugs from plants. The use of plant compound for pharmaceutical purposes has gradually increased here. Says WHO, medicinal plants would be the best source to obtain a variety of drugs. Medicinal and aromatic plants constitute a major source of natural organic compound and serve as the immediate source of drugs (Dobhal, 1994). About 80 per cent of individuals from developed countries use traditional medicine, which has compound derived from medicinal plants. The ayurvedic system of medicine has descriptions of thousands of plants for the management of curable diseases (Tripathy, 1996). Therefore, such plants should be investigated to better understand their properties, safety and efficiency.

In this present investigation various solvent extracts from various plants were screened for their antibacterial activity against selected human pathogens.

Materials and Methods

Plant Material

Leaves of different plants were collected from the different localities; Rajasthan, India and plant were identified.

Preparation of Extract

For the evaluation of the antibacterial effect of plants against selected human pathogens, acetone, methanol, petroleum ether and aqueous of leaf of each plant were extracted. For extraction, fresh leaves were collected, washed, air-dried individually in shade, powdered separately and stored at temperature range of 15–20°C. The materials were extracted with above solvents using soxhlet apparatus. The extracts were concentrated in a rotary vacuum evaporator.

Application of Extract

The test organisms were obtained from the institute of microbial technology (IMTECH), Chandigarh, India and maintained on nutrient agar. The Kirby-Bauer disk diffusion method (Bauer *et al.*, 1966) was used for testing antibacterial activity. The media (25 ml) inoculated with suspension of each test organisms was poured into sterilized petty dishes and left to gel at room temperature. Whatman's no. 1 filters paper discs (5 mm diameter) were impregnated with 25 mg of each solvent extract individually. The filter paper discs were placed equidistantly on inoculated media and diffusion of solution was allowed to occur for 30 min at room temperature. After incubation at 37 C for 24 hrs the diameter of zones of inhibition in the bacterial growth were measured to detect and quantify the antibacterial activity. Microorganisms, which showed a zone of inhibition > 8 mm were sensitive.

Results and Discussion

The data pertaining to the antimicrobial potential of plant extracts in acetone, methanol, petroleum, ether and distil water are presented in Table 4.1–4.4 respectively.

Table 4.1: Antibacterial Activity Caused by the Plant Extract (in Acetone) through Agar Diffusion Method

Microorganism	*Escherichia coli*	*Klebsiella pneumoniae*	*Pseudomonas aeuroginosa*	*Proteus Species*	*Staphylococcus aureus*
Nerium indicum (Kaner)	–	+	–	–	+
Acacia nilotica (Babul)	–	–	+	–	+
Zizipus nummularia (Ber)	–	+	–	–	+
Ocimum Americana (Jungli tulsi)	+	+	+	+	–
Ficus bengalensis (Bergad)	–	–	–	–	–
Mimosa hamata (Alaili)	+	–	+	+	–

(+) Susceptibility (inhibition zone ≥ 7 mm)

(–) Absence of susceptibility.

Table 4.2: Antibacterial Activity Caused by the Plant Extract (in Methanol) through Agar Diffusion Method

Microorganism	*Escherichia coli*	*Klebsiella pneumoniae*	*Pseudomonas aeuroginosa*	*Proteus Species*	*Staphylococcus aureus*
Nerium indicum (Kaner)	–	–	–	–	–
Acacia nilotica (Babul)	–	+	+	+	–
Zizipus nummularia (Ber)	–	–	+	–	–
Ocimum Americana (Jungli tulsi)	+	+	–	–	+
Ficus bengalensis (Bergad)	–	–	–	–	–
Mimosa hamata (Alaili)	+	+	–	–	–

(+) Susceptibility (inhibition zone ≥ 7 mm)

(–) Absence of susceptibility.

Table 4.3: Antibacterial Activity Caused by the Plant Extract (in Petroleum Ether) through Agar Diffusion Method

Microorganism	*Escherichia coli*	*Klebsiella pneumoniae*	*Pseudomonas aeuroginosa*	*Proteus Species*	*Staphylococcus aureus*
Nerium indicum (Kaner)	+	+	+	–	–
Acacia nilotica (Babul)	–	–	–	–	–
Zizipus nummularia (Ber)	–	–	–	–	–
Ocimum Americana (Jungli tulsi)	–	+	+	–	–
Ficus bengalensis (Bergad)	–	–	–	–	–
Mimosa hamata (Alaili)	–	–	–	+	–

(+) Susceptibility (inhibition zone ≥ 7 mm)

(–) Absence of susceptibility.

Table 4.4: Antibacterial Activity Caused by the Plant Extract (in Petroleum Ether) through Agar Diffusion Method

Microorganism	*Escherichia coli*	*Klebsiella pneumoniae*	*Pseudomonas aeuroginosa*	*Proteus Species*	*Staphylococcus aureus*
Nerium indicum (Kaner)	–	+	+	–	–
Acacia nilotica (Babul)	–	–	–	–	–
Zizipus nummularia (Ber)	–	–	–	–	–
Ocimum Americana (Jungli tulsi)	+	+	–	–	–
Ficus bengalensis (Bergad)	–	–	–	–	–
Mimosa hamata (Alaili)	–	+	+	–	–

(+) Susceptibility (inhibition zone ≥ 7 mm)

(–) Absence of susceptibility.

The extracts from various plants species (*Nerium indicum, Acacia nilotica, Zizipus rummularia, Ocimum americana, Ficus bengalensis* and *Mimosa hamata*) presented; antimicrobial activity to at least one of the tested microorganisms. The extracts from *Ocimum americana* (Jungli tulsi) presented highest activities *i.e.* they were able to inhibit highest percentage of microorganism.

Whereas the extract from *Ficus bengalensis* did not show any antimicrobial activity. Rajendhran *et al.* (1998) reported that *E. coli* was resistant to acetone extract of *Ocimum* which is similar to present findings. The acetone fraction of plant extract was found to be most effective of all and distill water extract to be the least.

K. pneumoniae, the common nosocomial pathogens is resistant to many antibiotics such as ceftazidine and gentamicin, the infections caused by them not only require expensive antibiotic treatment but also increase the morbidity and mortality in patients. This bacterial strain was observed most sensitive against all solvents of plant extract.

E. coli, resistant to ampicillin, sulphonamides, tetracyclines and trimethoprim, was observed most effective to *Ocimum americanum* followed by *Mimosa hamata* and *Nerium indicum* respectively.

The variable susceptibility of microorganism to plant extracts was reported by several authors. Ramasamy *et al.* (2001) studies the effects of plants namely *Anisometes indica, A. malabreca, Blumea lacera* and *Melio azadirachta* on different human pathogens. Apart from this, different work on various indigenous plants of various regions of India have been recently reported by Darokar *et al.* (1998), Rajendaran *et al.* (1998), Dushyent and Bohra (2000).

From this, it is obvious that all plant extracts that were selected are having antibacterial activity with difference in their magnitudes. But the effective biomolecules which act as antibacterial substance have to be identified, isolated and subjected to extensive scientific and pharmacological screening that can be used as sources of new drugs.

Finally the study expresses that plant extracts have great potential as antimicrobial compounds against microorganisms. Thus, they can be used in the treatment of infectious disease caused by resistant microbes. Also the present study support the use of alcoholic and acetone extract of different plant extracts.

We are trying to extend the study of inhibitory effect of these plants extract via their chemical compositions. Also the antibiotic resistant species of bacteria, will be practiced against synergistic effect from the association of antibiotic with plant extract.

This effect will enable use of respective antibiotic when it is no longer effective by itself during therapeutic treatment.

References

Bauer, A.W., Sherris, T.M. and Kirly, W.H.M., 1966. Antibiotic susceptibility testing by a standardized single disk method. *American Journal of Clinical Pathology*, 45: 493.

Cohen, M.L., 1992. Epidemiology of drug resistance: Implications for a post-antimicrobial era. *Science*, 257: 1050–1055.

Darokar, M.P., Mathur, A., Dwivedi, S., Bhalla, R., Khaniya, S.P.S. and Kumar, S., 1993. Detection of antibacterial activity in the floral petals of some higher plants. *Current Science*, 76: 187–189.

Dobhal, R., 1994. Medicinal and aromatic plants. *Current Science*, 71: 397–398.

Dushyent, G. and Bohra, A., 2000. Toxic effect of various plant part extracts on the causal organism of typhoid fever. *Current Science*, 78: 780–781.

Nasciment G.G.F., Locatelli, J., Freitas, P.C., Silva, G.L., 2000. Antibacterial activity of plant extracts and phytochemicals on antibiotic-resistant bacteria. *Brazilian Journal of Microbiology*, 31(4): 285–299.

Rajendhran, J., Mani, M.A. and Navaneetha Kannan, K., 1998. Antibacterial activity of some selected medicinal plants. *Geobios*, 25: 280–282.

Ramasamy, S., Chales Manoharan, A. and Rajendran, J., 2001. Antibacterial effect of various plants extracts on the human pathogens. *Indian Journal of Environment and Ecoplanning*, 5(2): 317–319.

Tripathy, Y.B., 1996. Research strategy and ayurveda. *Current Science*, 73: 418.

Chapter 5

Interaction Between Engineering, Astrology and Vastu: A Holistic Approach

Gurkirpal Singh[1] ***and Sarbjit Singh Sooch***[2]

[1]*Assistant Professor, College of Architecture, Institute of Engineering and Technology Bhaddal, Ropar, Punjab*

[2]*Research Engineer, Department of Civil Engineering, Punjab Agriculture University, Ludhiana – 141 004, Punjab*

ABSTRACT

Vastu is a science which has its logic in astrology and fulfils the requirements (and concepts) of present day engineering.

Keywords: *Engineering, Astrology, Vastu.*

Introduction

In Vedic period vastu was devised for regulating standards in shelter construction (temples, palaces, houses, etc.) and town planning. Vastu can be referred to as the old form of engineering of house building and town planning, which has got its basis in astrology. It can be referred to as the building byelaws of ancient Indian civilization which effectively helped in regulating development of settlements.

Materials and Methods

Interaction between Vastu astrology and engineering has been discussed and considered for designing buildings *viz.*, houses, shops, industries.

Results and Discussions

For the best design of a building different sciences are being compared / co-related with each other as following.

A plot consists of eight directions according to shasters. Different planets govern different direction and have specific influences on the inhabitants. By understanding the directions, one is able to balance the different forces harmoniously. Vastu deals with interaction between personal space and energy created by a structure in relation to the cosmic space and cosmic energy. Vastu is the holistic application of energy fields of space matter to produce a synchronized field to produce a synergistic effect (Gayatri G.D.).

Vastushastra is helpful to the masses due to its multi-disciplinary and holistic approach to building, transcending the physical act of building.

In vastu, we design buildings and interior with the basic five elements *viz.*, earth (in the form of elements of structure), water (by the help of water bodies), air, (by ventilation), fire (sunlight and other forms of energy), space which give rise to fields of energy systems which are advantageous to human body.

Astrological norms conform the cosmic influence on a person according to the position of the stars at the time of his birth. Vastu when fine tuned with it can prevent and minimize negative effects of the vastu and related forces affecting the concerned occupants of structure (Raman B.V.)

Astrology and Vastu definitely have a direct relationship. For instance the planet Jupiter rules the north east while the planet Venus rules south east and similarly every planet signifies a particular direction. Thus the puja room is in the northeast direction and the kitchen is placed in southeast direction. In order to implement vastu in religion dominating society. It has been linked to religion. People applied it without reasoning the science behind it, as such the reason behind it was lost.

Scientific Validation

With the help of lecher antenna in Germany the various radiations existing in an environment– their source, intensity, directions and effects on residing people has been studied.

The lecher antenna helps us relate the energy axis of the human body with negative and positive radiations within a space. With the help of antenna, we can match the building radiations to the human radiations and achieve a positive response by talking corrective measures to neutralize the negative radiations present due to faults in an existing structure or environment.

The cell in the body has a wavelength of 20 to 22 centimeters and all radiation having wave lengths less than 20 centimeters, disturb and affect the functioning of the various cells in the human body, leading to over a period of time, to all kinds of health problems.

Vastu and Engineering

Though vastu has got its logic in astrology but it is relevant and it fulfils all the concepts of modern day advanced engineering.

For Example

1. Vastu advises main openings towards northeast since in tropical climate rain blows from southwest so opening towards northeast is favorable since it wont be receiving direct rain.
2. Similarly, vastu prefers sites sloping towards north since sites sloping towards north receive much less solar radiation as compared to sites sloping towards south.

3. Also vastu advises to sit facing north or east because maximum glare during the day comes from south and west which effects eye site and health.
4. For living areas north and east are recommended from point of view of coolness and comfort while bedrooms which are used during night time rather than during hot daytime are recommended in south or west (Bangalore, N.R.)
5. The sun emits two types of rays: ultra violet and infrared. These rays have various effects on human body and environment. The sun emits ultra violet rays in the morning hours, which have a good effect on human body and environment surrounding him. These rays kill bacteria. This is the (scientific) reason, why the kitchen is preferred to be located on the eastern side. It is always a good practice to focus our eyes towards the morning sun as the rays improve our power of vision. This is the reason why we orient the kitchen platform towards the eastern side so that while working in early hours; the ultra violet rays of the sun will enter the eyes.
6. On the other hand sun emits infrared rays during evening hours, but these are harmful. Water bodies placed at the south-west corners are therefore not desirable because sun travels through the southern sector, as such infrared rays are reflected due to water bodies placed at south-west corner. These reflected rays enter the house thus adversely effect the human body.

All the above stated examples highlights the common areas, enveloping areas between vastu and engineering to arrive at a new understanding of vastu–the–holistic view. All the above stated examples are in tune with engineering (modern day) principles/recommendations. Thus after so many centuries of its inception vastu is still going strong today.

Conclusion

Vastu though shrouded in mystery and symbolism, follows scientific principles acceptable to modern world and has wide applications.

References

Gayatri Devi Vasudev. *Astrology and Hoax of Scientific Temper.*

Gayatri Devi Vasudev. *Vastu, Astrology and Architecture.*

Niranjan Babu, 2005. Habitation: Blending Vastu and Astrology. *The Astrological Magazine,* Bangalore, 94(1): 60–64.

Raman, B.V., 1995. Vastu and its relevance to modern times: Relevance of Vastu. In: *Proceedings 1st All India Symposium on Vastu,* Bangalore, June 3-4, 1995.

Chapter 6

Biodiesel: As an Alternative Fuel

***Rashmi Shrivastava*[1] *and Praveen Jain*[2]**
[1]*Lecturer, University Institute of Technology, Bhopal*
[2]*Regional Director, M.P. Bhoj (Open) University, Bhopal*

ABSTRACT

Larger part of modern research is focussed on cleaner and greener technologies and eco-friendly products. Fast depletion of conventional energy sources is also an issue of great concern. This thrust has led to BIODIESEL: An effective alternative source.

This article discusses and presents a way to manage biowaste and producing automobile fuel from it that is by all means a better product. Biodiesel can be manufactured from natural renewable sources such new and used vegetable oil and animal fats. Biodiesel is biodegradable, safer diesel replacement fuel which easily blends with conventional diesel fuel and reduces an emissions of unburned hydrocarbons. Thus biodiesel is a stronger, more eco-friendly and economic substitute for conventional diesel for future.

Introduction

Biodiesel (fatty acid alkyl esters) is a cleaner-burning diesel replacement fuel made from natural, renewable sources such as new and used vegetable oils and animal fats. Just like petroleum diesel, biodiesel operates in combustion-ignition engines. Blends of up to 20 per cent biodiesel (mixed with petroleum diesel fuels) can be used in nearly all diesel equipment and are compatible with most storage and distribution equipment. These low level blends Bio-20 (20 per cent and less) don't require any engine modifications and can provide the same payload capacity and range as diesel.

Higher blends, even pure biodiesel (100 per cent biodiesel, or B-100), can be used in many engines built with little modification. Bio-100 is the name for a variety of ester-based oxygenated fuels from renewable biological sources. It can be used in compression-ignition (diesel) engines with little or no modifications. B-100 is biodegradable, non-toxic and essentially free of sulfur and aromatics.

B-100 is a methyl or ethyl ester of mainly vegetable oils. Biodiesel readily blends with diesel fuel in any per cent. The blend level is a function of economics, the desired emissions profile, material compatibility, and combustion characteristics. The focus at the moment is on a 20 vol per cent blend of biodiesel in petrodiesel.

Methyl or ethyl esters can be produced from vegetable and tree oils, animal fats, and/or used oils and tats. These oils are blended with an alcohol (usually methanol, although ethanol can be used as well) and a catalyst such as sodium hydroxide. The resulting chemical reaction (which occurs at moderate temperatures and pressures) produces an ester and glycerine. Glycerine is a valuable material used extensively in the manufacture of soaps and other consumer products.

Using Bio-100 in a conventional diesel engine substantially reduces emissions of unburned hydrocarbons, carbon monoxide, sulfates, polycyclic aromatic hydrocarbons, nitrated polycyclic aromtic hydrocarbons, and particulate matter. These reductions increase as the amount of biodiesel blended into diesel fuel increases. The best reductions are seen with Bio-100. The use of Bio-100 decreases the solid carbon fraction of particulate matter (since the oxygen in biodiesel enables more complete combustion to CO_2) and reduces the sulfate fraction (biodiesel contains less than 24 ppm sulfur), while the soluble, or hydrocarbon, fraction stays the same or increases. Therefore, Bio-100 works well with new technologies such as catalysts (which reduce the soluble fraction of diesel particulate but not the solid carbon fraction), particulate traps, and exhaust gas recirculation.

Bio-100 has physical properties very similar to conventional diesel. Emission properties, however, are better for biodiesel than for conventional diesel.

Bio-100's Physical Characteristics

Specific gravity	0.87 to 0.89
Kinematic viscosity @ 40°C	3.7 to 5.8
Cetane number	46 to 70
Higher heating value (btu/lb)	16,928 to 17,996
Sulfur, Wt%	0.0 to 0.0024
Cloud point C	– 11 to 16
Pour point C	– 15 to 13
Iodine number	60 to 135
Lower heating value (btu/lb)	15,700 to 16,735

Manufacturing Process

Bio-100 fuel can be made from new or used vegetable oils and animal fats, which are non-toxic, biodegradable, renewable resources. Fats and oils are chemically reacted with an alcohol (methanol is the usual choice) to produce chemical compounds known as fatty acid methyl esters. Bio-100 is the name given to these esters when they are intended for use as fuel. Glycerol (used in pharmaceuticals and cosmetics, among other markets) is produced as a co-product.

Bio-100 can be produced by a variety of esterification technologies. The oil and fats are filtered and preprocessed to remove water and contaminants. If free fatty acids are present, they can be removed or transformed into biodiesel using special pretreatment technologies. The pretreated oils and fats are then mixed with an alcohol (usually methanol) and a catalyst (usually sodium or potassium hydroxide). The oil molecules (triglycerides) are broken apart and reformed into esters and glycerol, which are then separated from each other and purified.

Approximately 55 per cent of the biodiesel industry can use any fat or oil feedstock, including recycled cooking grease. The other half of the industry is limited to vegetable oils, the least expensive of which is soy oil. The soy industry has been the driving force behind biodiesel commercialization because of excess production capacity, product surpluses, and declining prices. Similar issues apply to the recycled grease and animal fats industry, even though these feedstocks are less expensive compared to soy oils.

Despite the combined resources of both industries, there is only enough of the feedstocks to supply 1.9 billion gallons of biodiesel (under policies designed to encourage biodiesel use). DOE is developing a new, low cost biodiesel oil from spicy mustard seeds that could add another 5-10 billion gallons of biodiesel to the fuel supply. The mustard meal (the seed that remains once the oil is removed) is a high value pesticide that helps.

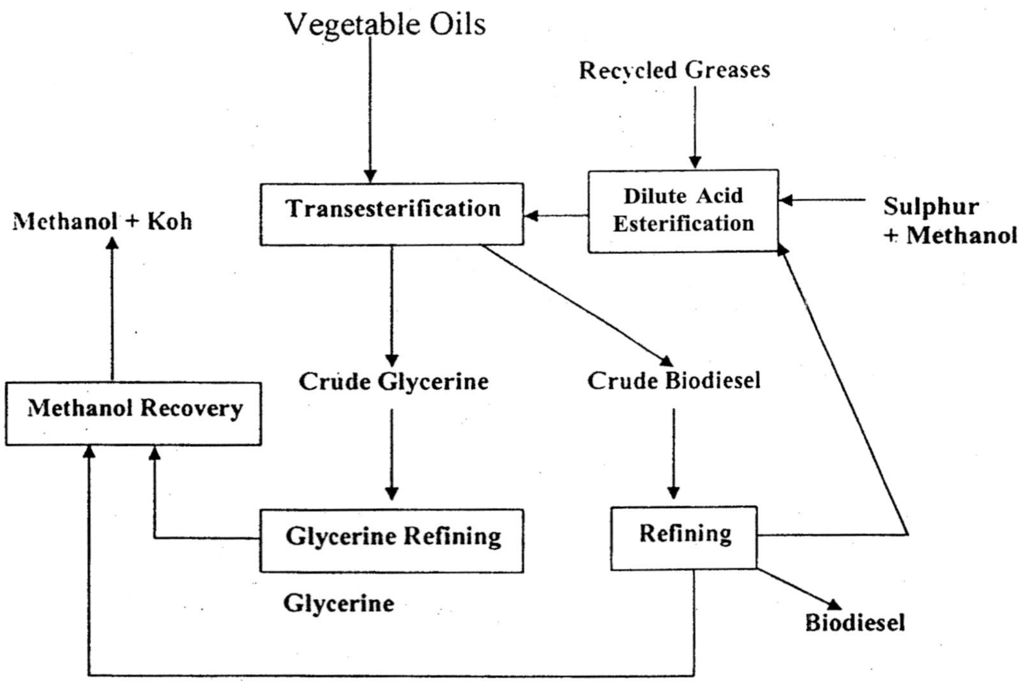

Figure 6.1: Basic Technology

Advantages

Bio-100 has a number of advantages when compared with both petroleum diesel and other alternative fuels. It needs to be emphasized, however, that the goal of the biodiesel industry is not to replace petrodiesel or to supplant the invaluable role of other alternative fuels. Each of these fuels has a place in the development of a balanced energy industry. The compression-ignition engine has certain advantages over spark-ignition engines and role of biodiesel is to contribute to the longevity and cleanliness of diesel engines. Some of them are:

1. Bio-100 is produced domestically which helps to reduce the dependence on petroleum and other fossil fuels.
2. The development of the biodiesel industry would strengthen the domestic, and particularly the rural, agricultural economy.
3. It is a renewable fuel that can be made from agricultural crops and/or other feedstocks that are considered waste, such as cooking oil and trap grease. This helps conserve resources and makes the best possible use of materials which may be perceived as having little or negative value.
4. The production of soya-based biodiesel has a positive energy balance (as much as 3 : 1), due to the high energy value of ester-based feedstocks, the low-energy requirements of the conversion process, and the nitrogen-fixing characteristic of soybeans, which reduces the need for fertilizer.
5. Biodiesel blends are competitive with other alternative fuels on a life-cycle cost basis, because of its lower associated infrastructure costs (no extensive engine or refueling modifications needed), the positive energy balance discussed above and the efficiency of the fuel on a per-Btu basis. Used in conjunction with the compression-ignition engine, which is much more efficient than spark-ignition engines, biodiesel compares quite favorably with CNG, ethanol and methanol, the other leading alternative fuels in the marketplace today.
6. It is considered readily biodegradable and non-toxic. Continued testing indicates that biodiesel degrades as fast as and is as safe as sugar in the environment and when blended with diesel accelerates the diesel's degradation in the environment. These characteristics may prove valuable in certain markets.
7. Bio 100 and biodiesel blends significantly reduce harmful tail pipe emissions as documented in tests using U.S. EFP protocols.
8. The emissions smell better and appear to help reduce nausea in those breathing the fumes.
9. Bio-100 fuel and biodiesel/petrodiesel fuel blends smell better than conventional diesel.

Properties of Bio-100

Bio-100 is much cleaner than fossil-fuel diesel ("dinodiesel"). It can be used in any diesel engine with little or no modifications–in fact diesel engines run better and last longer with biodiesel. And it can easily be made from a common waste product–used cooking oil.

1. Bio-100 fuel burns up to 75 per cent cleaner than conventional diesel fuel made from fossil fuels.
2. It substantially reduces unburned hydrocarbons, carbon monoxide and particulate matter in exhaust fumes.
3.

   Sulphur dioxide emissions are eliminated (biodiesel contains no sulphur).
4. Bio-100 based and adds no CO_2 to the atmosphere.

5. The ozone-forming potential of biodiesel emissions is nearly 50 per cent less than conventional diesel fuel.
6. Nitrous oxide (NOx) emissions may increase or decrease but can be reduced to well below conventional diesel fuel levels by adjusting engine timing.
7. Bio-100 exhaust is not offensive and doesn't cause eye irritation (it smells like French fries!)
8. Bio-100 is environmentally friendly it is renewable, "more biodegradable than sugar and less toxic than table salt" (US National Biodiesel Board).
9. Bio-100 can be used in any diesel engine.
10. Fuel economy is the same as conventional diesel fuel.
11. It is a much better lubricant than conventional diesel fuel and extends engine life–a German truck won an entry in the Guinness Book of Records by travelling more than 1.25 million km (780,000 miles) on biodiesel with its original engine.
12. Bio-100 has a high *cetane* rating, which improves engine performance.
13. Biodiesel can be produced from any fat or vegetable oil, including waste cooking oil.

Impact of Bio-100

Diesel Emissions and Cancer

According to a research the use of pure biodiesel instead of petroleum-based diesel fuel could offer a 93.6 per cent reduction in cancer risks from exhaust emissions exposure.

Greenhouse Effect

Using vegetable oils or animal fats as fuel for motor vehicles is in effect running them on solar energy. All biofuels, including ethanol, are derived from the conversion of sunlight to energy (carbohydrates) that takes place in the green leaves of plants.

Plants take up carbon dioxide (CO_2) from the atmosphere; burning plant (or animal) products in an engine releases the CO_2 uptake back into the atmosphere, to be taken up again by other plants. The CO_2 is recycled, atmospheric CO_2 levels remain constant. Thus biofuels do not increase the Greenhouse Effect–unlike fossil fuels, which release large amounts of new (or rather very old) CO_2 which has been locked away from the atmosphere.

In fact bio-100 can actually reduce CO_2 levels in the atmosphere: growing soybeans consumes nearly four limes as much CO_2 as the amount of CO_2 produced in the exhaust from soybean oil biodiesel.

Energy Efficiency

According to a comparative life-cycle study by the US Department of Energy's National Renewable Energy Laboratory, biodiesel requires only 0.31 units of fossil energy to make 1 unit of fuel.

References

Allen, C.A.W. and K.C. Wath, 1999. Comparative Analysis of the Alomization characteristics of Fifteen Biodiesel Types. ASAE Paper, 996113.

EPA, 1988. Guidance on Estimating Motor Vehicle Emission Reduction from the use of Alternative Fuels and Fuel blends. Technical Report EPA-AA-TSS-PA-87-4.

Herslof, B., O. Podluha and B. Toregard, 1979. HPLC of Triglycerides. *JAOCS*, 56: 864–866.

Isigger, A., F. Karaosmanoglu, H.A. Alcsoy, F. Hamdullahpur and O.L. Gulder, 1993. Safflower seed of Turkish origin as diesel fuel alternative. In: *Applied Biochemistry and Biotechnology*, 40/41: 89–105.

Johnson L.A. and E.G. Hammond, 1996. *Soybean Oil Ester Fual Blends*. U.S. Patent 5520708.

Chapter 7

House: The Family Temple

Gurkirpal Singh[1] *and Sarbjit Singh Sooch*[2]

[1]*Assistant Professor, College of Architecture, Institute of Engineering and Technology, Bhaddal, Ropar, Punjab*

[2]*Research Engineer, Department of Civil Engineering, Punjab Agriculture University, Ludhiana – 141 004, Punjab*

ABSTRACT

This article focuses attention towards various aspects that should be considered in detail while designing a house for a particular family need.

Keywords: *Design, House, Family.*

Introduction

House is important not only because it accommodates the basic unit of the Society but because the design of the house affects the development of the children, the well being and happiness of the family members, the interaction within the family and with other social groups in the community. The style of Living for a particular family is also an important factor in determining the design of the house. If the family values health, peace of mind, family centrism, prestige, the design of the house needs to make the provision for satisfying family living styles.

The orientation towards economy would probably lead to the construction of technologically advanced and compact house, the one which is economical to construct and maintain. The value of family centrism would encourage the planning of the family room where all members could get together to enjoy each other work and pleasures. The drive for physical health value would require the provision of adequate daylight and fresh air. It would also emphasized the provision of safety features.

The value of leisure would necessitate the provision of natural materials, continues lines and simplicity of design and furnishings. The design would provide for areas of privacy for reading or listening to music including adequate storage for records and books (Glen).

Methods and Materials

The following points should be kept in mind while designing the house:

1. Place of meeting (common activities for gathering)
   (*a*) Small gatherings/functions (small order)/food dining)/entertainment indoor/outdoor
2. Security
3. Privacy
   (*a*) Outsiders access only to limited areas
   (*b*) Children segregated from elders yet in their care, under supervision.
4. Able to provide healthy living environment
   (*a*) Light/ventilation
   (*b*) Sound insulation
   (*c*) Hygienic condition
5. Cater to needs of (cater to all walks of life, stages of life efficiently because in this house generations are to be accommodated *i.e.* it should cater to different stages in the life of man).
   (*a*) Elders
   (*b*) Active members
   (*c*) Teenagers
   (*d*) Children

Results and Discussion

Physically housing is an important factor in maintaining physical health of the people. Research has shown that there is positive relationship between poor housing and poor health. In an environment of overcrowding, there is a very less chance of training the younger children than of the older ones. The addition of each child in the family puts the strain on the family and community resources. It does not only reduce the opportunities for the additional child but also makes him dependent on the community resources (Glen). There is positive correlation between poor housing and increased mortality and morbidity. Therefore the house shapes the development of the society.

Houses are comprised of number of rooms, the activities performed in each room and the order of each activity within the house or rooms varies among families. The variation in activity patterns calls for the functional design of the house so that it meets the needs of the workers and the tasks. Here the problem is to coordinate the requirement of the work and the worker.

Housing, with in the broader context is ability to pre conceive the complex set of relationships between people, their physical environment and the influential parameters of economics, culture and tradition, is of great consequence (Mehta, 2005).

There exists a generalization of space-making that evolves structural relationships and coherence within the elements operational. The generalization is a requisite for order within the built. However, these generalization as thematic types is often interpreted, modified and accommodated with respect to the specific demand of place, people and ways of life, generating a distinct sense of place and fit (Mehta, 2005).

The house is symbolic of the existential relationship between man and God, man and nature and man and man. These relationship are expressed at the individual level of dwelling as the shrine, the courtyard and the family room. It is these relationship that infuse meaning within the built and provide a foothold to man, providing him a sense of completeness of his own world (Mehta, 2005).

The objective of considering the requirement of activity or task in designing work places in the house is to determine the conditions that result in minimum of strain on the worker and require a minimum of effort to do the activity. The words strain an effort do not only include physical strain and effort but also the effective, cognitive and temporal casts of work. The design of the house need to enhance the efficiency of the home maker when considered in terms of the ratio of input to output.

The requirements of design of a work place for performing various activities vary due to the variation in supplies, equipment and work surfaces on which to carry out processes and the number of persons present. Magrabi (1967) define activity as a function of various components such as duration of time, sequence of steps, identity of participants, the index of interactive space, the identity of the social area or area in which each item of equipment and material available to the family is utilized in the activity.

The inter-relationship among activities is another factor which needs attention in designing work places. The activities are not carried out singly in isolation. There is a constant overlapping of tasks *i.e.* the task done concurrently with the other tasks or for tasks that require more than one work place should not be considered in isolation to the effect of other factors. The requirements of all the tasks to be performed in the same place need to be coordinated (Etal, 1967).

A method for evaluating the adequacy of spatial area with respect to its use was suggested by Bassard (1951). The suggested index is based on law of family interaction in which the number of personal inter-relationships, X, is calculated from the number of persons, Y, in the family with the following formula:

$$X = \frac{Y^2 - Y}{2}$$

According to Bassard, an interactive space index may be calculated by dividing the number of square feet of floor space by the number of personal interrelationships. For total floor space, the floor space in the areas actually used or available for use in the activity should be used. Number of personal interrelationships should be calculated on the basis of those persons present within the activity area while the activity was in progress. However, sex, age, marital status, occupation and stage in the family life cycle be taken into account for modifying the index.

Conclusion

It discusses the importance of needs of a family that a house should be able to fulfill like in a temple people revere the deity, it is at a highest pedestal of importance, all activities are oriented towards it. Thus the family should be kept at highest reverence while designing a house.

There is a need to create a mode of representation that documents the living space and relationships that it accommodates, the community and the politics that it generates, the forms and the memories and aspirations it provokes (Nilanjan, 2005).

References

Bassard, J.H.S.A., 1951. Spatial Index for Family Interaction, AM. Society. Rev. 16 (April, 1951) pp. 243-246, as cited by Magrabi.

Bayer, W. Glen. *Housing and Personal Values*. New York.

Bayer, W. Glen. *Housing and Society*. New York.

Etal, E.F., 1967. Framework for studying family activity patterns. *Journal of Home Economics*, 59(2).

Magrabi, F.M., B. Paolucci, and M.E. Heifner, 1967. Framework for studying family activity patterns. *Journal of House Economics*, 59(9)714–719.

Mehta, Kajwan, 2005. Studio drawing history. *Architecture–Time, Space and People.*

Nilanjan, Bhowal, 2005. Measuring Settlement. *Architecture–Time, Space and People.*

Chapter 8

Modelling of Estimation of Gene Frequencies

G. Eswara Prasad[1] *and D. Nagarajan*[2]
[1]*S.T. Hindu College, Nagercoil*
[2]*St. Joseph's College of Engineering, Chennai – 119*

ABSTRACT

Every human being can be classified into one of four blood group O, A, B and AB. The inheritance of these blood group is controlled by three allelomorphic genes O, A and B of which O is recessive to A and B. Rao (1973) has analysed the gene frequencies of blood group. In this article to analysis the estimation of the gene frequencies and stationary distribution and comparing the findings with maximum likelihood estimates and score methods.

Keywords: *Blood group, Matured baby, Prematured baby.*

Introduction

Blood is a liquid tissue. Suspended in the watery plasma are seven type of cells and cell fragments. Red blood cells (RBCs) Patelets and five kinds of weight blood cells (WBCs) neutrophils, eosinophils, hasophils, lymphocytes and monocytes. If one takes a sample of blood, treats it with an agent to prevent clotting and spins it in a centrifuges. The red cells settle to the bottom the white cells settle on top of them forming the "buffy coat". The blood performs two major function they are transport through the body of oxygen and carbondioxide, food molecules, iron, wastes, hormones and heat and defence of the body against infection and other foreign materials. All the WBCs participate in these defence. Anaemia is a shortages of RBCs and the amount of haemoglobin in them anaemia has many causes. One of the most common is an inadequate intake of iron in the diet. Red blood cells have surface antigens that differ between people and that create the so-called blood group such as the ABO systems and the Rh systems.

For the purpose of Rh incompatibility the Rh (Rhesus) blood group operates as a one gene-two allelo systems. The gene product is identified as the Rh D antigens or D antigens. The proteins product of the genes is a transmembrane protein composed of over 400 amino acid. If a mother who is Rh negative is carrying a fatal who is Rh positive (Rr must be heterozygote) the mother has the capacity to produce immune antibodies to the Rh antigens on the fatal RBC. Fortunately the mothers systems does not come into contact with the fatal RBC until the time of birth when the placenta ruptures and some of fatal RBC enter the mother's blood stream.

Data

The study based on the data collected from 500 hospital records collected from Governmental hospital, Chennai using the frequencies A = 196, B = 88, Ab = 34, O = 182. This was collected during June, 2004 to December 2005.

Description of Model

If r, p and q are gene frequencies of O, A and B, then the expected probabilities of the six genotypes (four phenotypes) in random mating are as follows:

Table 8.1: Model Description

Phenotype	*Genotype*	*Probability*
O	OO	R^2
A	AA	$P^2 + 2pr$
	AO	
B	BB	$q^2 + 2qr$
	BO	
AB	AB	2pq

given $\overline{O}, \overline{A}, \overline{B}$ and $\overline{AB}$ are observed frequencies and estimate the gene frequencies p, q and r (Rao, 1973).

$$r' = \sqrt{\frac{\overline{O}}{N}}$$

$$p' = 1 - \sqrt{\frac{\overline{O} + \overline{B}}{N}}$$

$$q' = 1 - \sqrt{\frac{\overline{O} + \overline{A}}{N}}$$

Let D denotes the deviation, $-D = p' + q' + r' - 1$ estimate due to Bernstein are obtained as follows.

$$\tilde{r} = \left[1 + \frac{1}{2}D\right]\left[r' + \frac{1}{2}D\right]$$

$$\tilde{p} = \left[1 + \frac{1}{2}D\right][p']$$

$$\tilde{q} = \left[1 + \frac{1}{2}D\right][\tilde{q}]$$

In general, the probability and derivatives, with respect to the independent parameters P and q as follow:

Table 8.2: Model Description

Probability (π)		Derivatives $\frac{\partial \pi}{\partial p}$	
O	r^2	$-2r$	$-2r$
A	$P^2 + 2pr$	$2r$	$-2p$
B	$q^2 + 2qr$	$-2q$	$2r$
AB	$2pq$	$2q$	$2p$

The frequencies A = 196, B = 88, Ab = 34, O = 182

Approximate solution obtained by Bernstein's method are:

$P_0 = 0.26515$, $q_0 = 0.13051$, $r_0 = 0.60332$ and D = 0.00102.

Better estimation due to Bernstein are obtained as follows:

P = 0.26523, q = 0.13057, r = 0.60413 and D = 0.00102

The probability and coefficients for the calculation of efficient scores at the approximate value obtained in Table 8.3.

Table 8.3: Stationary Distribution of the Study Frequencies

	Probability (π)		$\frac{1}{\pi}\frac{\partial \pi}{\partial q}$	Observed Frequencies
O	0.36497	– 3.31057	– 3.31057	182
A	0.39090	3.09096	– 1.35727	196
B	0.17481	– 1.49533	6.9984	88
AB	0.06927	3.77363	7.6593	34
				500

From the above table the value of π show the blood group O and A are also almost same probability 0.36497, 0.3990 and the blood group B is 0.17481 and AB is 0.06927. If suggest that the blood group A is the highest probability of among four blood group.

The scores are

$$Q_p = 0.0188 \text{ and } Q_p = \frac{1}{\pi}\frac{\partial \pi}{\partial p}$$

$$Q_p = 0.10946 \text{ and } Q_q = \frac{1}{\pi}\frac{\partial \pi}{\partial q}$$

The information matrix for a single observation is

i_{pp} =9.11199, i_{pq} = 2.55548, i_{qq} = 17.13513, i_{qp} = 2.55548

$$\begin{bmatrix} 9.11199 & 2.55548 \\ 2.5548 & 17.135133 \end{bmatrix}$$

The inverse of the information matrix per single observation is

i^{pp} = 0.114432, i^{pq} = – 0.01706, i^{qq} = 0.06090,i^{qp} = – 0.01706

$$\begin{bmatrix} 0.114432 & -0.01706 \\ -0.01706 & 0.06090 \end{bmatrix}$$

The solution for correction are

δ_p = 0.00000568

δ_q = 0.000012

The corrections are very small. If the corrections are not small, the whole process has to be repeated with the second approximation. It is the information matrix need not be recalculated for each approximation. Only the new scores have to be calculated at each stage and used in conjunction with the same inverse information matrix to obtain closer approximations. When convergence is achieved, the information matrix and scores may be calculated for obtaining the last approximate values.

The maximum likelihood estimates and the variance are:

$$\tilde{p} = 0.26528, \ \tilde{q} = 0.13057, \ \tilde{r} = 0.60413$$

$$v(\tilde{p}) = 0.000228, \ v(\tilde{q}) = 0.0001218, \ v(\tilde{r}) = 0.00041898$$

reference may be made to Finney (1952), Fisher (1950) and Rao (1950)

In case of O, A, and B blood group data there is another algorithm which yields the maximum likelihood estimates.

Let p_0, q_0 and r_0 be provision estimates and compute $P_0 = \frac{p_0}{r_0}$ and $Q_0 = \frac{q_0}{r_0}$. Let us represent by P_k and Q_k the kth approximation of p/r and q/r, the $(k+1)^{th}$ approximation are found from the formula:

$$\frac{\overline{A} + \overline{AB}}{P_{KH}} = 2(\overline{O}) + \frac{\overline{A}}{2 + P_K} + \frac{2\ \overline{B}}{2 + Q_K}$$

$$\frac{\overline{A} + A\overline{B}}{P_{KH}} = 2(\overline{O}) + \frac{\overline{A}}{2 + P_K} + \frac{2\ \overline{B}}{2 + Q_K}$$

the iteration may be repeated until stable values of P and Q are obtained from which the estimates p, q and r are computed asL

$$\tilde{r} = \frac{1}{P + Q + 1} \quad \tilde{P} = \tilde{r}P, \tilde{q} = \tilde{r}Q$$

by adopting the provisional estimates

$\tilde{r} = 0.60413,\ \tilde{p} = 0.26528,\ \tilde{q} = 0.13057$

with $P_0 = 0.60332$, $q_0 = 0.21612$

The maximum likelihood estimates p, q and r

Are $\tilde{r} = 0.60332,\ \tilde{p} = 0.26515,\ \tilde{q} = 0.13051$

Which agree closely with the estimates obtained by the iterative procedure using the information matrix.

Conclusion

From the above results the maximum likelihood method and score method are almost same for the estimation of gene frequencies. The value of π show the blood group O and A are also almost same probability 0.36497, 0.3990 and the blood group B is 0.17481 and AB is 0.06927. If suggest that the blood group A is the highest probability of among four blood group.

References

Finney, D.J., 1952. *Probit Analysis*, 2nd edition. Cambridge University Press, Cambridge, Mass.

Fisher, R.A., 1950. *Statistical Methods for Research Workers*, 11th edition. Oliver and Boyd, Edinburgh.

Radhakrisha Rao, C., 1973. *Linear Statistical Inference and its Applications*. A Wiley-Interscience Publication, John Wiley & Sons Inc.

Chapter 9

Evaluation of Toxicity of Monocrotophos Against Three Vector Species of Mosquito

N. Arun Nagendran

P.G. and Research Department of Zoology and Microbiology,
Thiagarajar College (Autonomous), Madurai – 625 009

ABSTRACT

The organophosphorus pesticide, monocrotophos, was tested for its toxicity against three important vector mosquitoes *Anopheles stephensi, Aedes aegypti* and *Culex quinquefasciatus.* The results of bioassay studies on the fourth instar larvae of these mosquitoes indicate that *Aedes aegypti* was more susceptible to monocrotophos with a low LC_{50} value of 0.35 ppm followed by *Anopheles stephensi,* and *Culex quinquefasciatus.*

Keywords: *Anopheles stephensi, Aedes aegypti, Culex quinquefasciatus, Monocrotophos, Larval mortality.*

Introduction

Mosquitoes are considered as one of the major vectors because it transmits a number of diseases such as dengue fever, malarial fever, filariasis and several viruses (Ross, 1965; Service, 1983). The voracious feeding habit, high reproductive and dispersal potential and successful exploitation of the environment are the causes of rapid proliferation. Vector control is an important component of the Public Health Programmes. The main objective is to break the transmission of the parasites (Ganesh *et al.*, 2002). The application of chemical larvicides for the control of mosquito larvae as a part of control of vector borne disease is implemented by various public health authorities (Dhanraj and Thirunavukarasu, 2005). The prolonged use of insecticides resulted in the development of resistance in some vectors. The development of resistance to insecticides by arthropods represents a major concern of programmes of pest control in agriculture and vector control for arthropod borne diseases. Approximately 40 per cent of the 506 medically important arthropods have been reported to exhibit insecticide resistance of which 50 per cent are mosquitoes (WHO, 1992). The development of resistance in vectors is mainly due to over use of insecticides. Hence, the present work aims in determining the optimum quantity of monocrotophos against three important vector species, *Anopheles stephensi, Aedes aegypti* and *Culex quinquefasciatus* to restrict the development of insecticidal resistance.

Materials and Methods

The larvae of *Anopheles stephensi, Aedes aegypti* and *Culex quinquefasciatus* were obtained from Centre for Research in Medical Entomology (CRME), Madurai as egg rafts. The egg rafts were inoculated in sterilized tap water. The temperature was maintained as 29–31 C, which is optimal for normal growth. The larvae hatch with overnight incubation. The larvae were fed with yeast tablets and dog biscuits in the ratio of 3 : 2. Water is changed for once in two days. The early fourth instar larvae were selected for the study. The insecticide monocrotophos was procured from commercial sellers.

Larval bioassay was performed on early fourth instar larvae according to the standard procedure of WHO (1992). The tests were carried out by exposing 20 early fourth instar larvae in glass jars containing 249 ml of tap water and 1 ml of different concentrations of the insecticide.

Results and Discussion

Despite their toxicity to non target organisms and development of resistance, chemical control methods using synthetic insecticides had been favoured so far because of their speedy action. In the present investigation, an organophosphate insecticide monocrotophos was tested for its efficacy on fourth instar larvae of three medically important mosquitoes, *Anopheles stephensi, Aedes aegypti* and *Culex quinquefasciatus* were subjected to different concentrations of monocrotophos as given in Table 9.1. The concentration of pesticide, per cent mortality of larvae, LC_{50} value, regression equation, 95 per cent confidential limits, standard error and slope for the effect of monocrotophos on *Anopheles stephensi, Aedes aegypti* and *Culex quinquefasciatus* were given in Table 9.1.

The susceptibility of larvae of *Anopheles stephensi* to fenthion has been reported Chockalingam *et al.* (1991). Similar to their observation, in the present study larval susceptibility of *Anopheles stephensi* was noticed to another organophosphate compound, monocrotophos. Fifty per cent larval mortality was obtained with 0.44 ppm.

The toxicity studies of monocrotophos on the fourth instar larvae of *Aedes aegypti* indicate that the insecticide was toxic to the mosquito larvae. The LC_{50} value as per the probit analysis was 0.3135 ppm. In support to the present observation, the susceptibility of *Aedes aegypti* larvae to organophosphate insecticide has already been reported by Gjullin and Issak (1957).

Culex quinquefasciatus had 50 per cent mortality at 0.5851 ppm. Curtis and Pasteur (1981) and Villani *et al.* (1983) have reported the resistance developed by *Culex quinquefasciatus* against organophosphates. However, recently, Rao *et al.* (2004) have reported that *Culex quinquefasciatus* is highly susceptible to an organophosphate compound fenthion. Their studies revealed that 0.33 ppm concentration of fenthion restricted only 40 per cent egg emergence and no emergence was observed at 0.55 ppm.

Baktharatchagan and David (1991) have reported that *Culex quinquefasciatus* was more susceptible than *Aedes aegypti* and *Anopheles stephensi* against the insecticide cypermethrin. In contrast, in the

Table 9.1: Toxicity of Monocrotophos to the Fourth Instar Larvae of *Anopheles stephensi*, *Aedes aegypti* and *Culex quinquefasciatus*

Species	*Concentration of Monocrotophos (in ppm)*	*% of Mortality*	*LC_{50} (in ppm)*	*Regression Equation*	*S.E.*	*95% Confidential Limit*		*Slope*
						Upper Limit	*Lower Limit*	
Anopheles stephensi	0.35	33.33	0.4055	Y = 7.856323 + 7.286905 X	1.821	0.425	0.386	1.369
	0.44	56.66						
	0.50	76.66						
	0.55	83.33						
Aedes aegyptii	0.30	46.66	0.3135	Y = 6.08834 + 2.160784 X	1.306	0.376	0.260	2.886
	0.35	53.33						
	0.40	63.33						
	0.55	68.33						
Culex quinquefasciatus	0.45	23.33	0.5851	Y = 488248 + 6.395243 X	3.576	0.735	0.465	1.430
	0.50	35.00						
	0.66	51.66						
	0.70	78.33						

results of the present study indicate that the fourth instar larvae of *Aedes aegypti* was more susceptible followed by *Anopheles stephensi* and *Culex quinquefasciatus*. However, it is concluded that the use of insecticides at optimal concentration will be effective in controlling dreadful disease causing vector mosquitoes, *Anopheles stephensi, Aedes aegypti* and *Culex quinquefasciatus* with a minimum impact on other non target organisms and the environment.

Acknowledgement

The author acknowledges the University Grants Commission for the financial assistance provided to carry out this study under UGC-MRP/219/04 (SERO) and the Director, Centre for Research in Medical Entomology, ICMR, Madurai for the provision on mosquito eggs.

References

Baktahratchagan, R. and David, B.V., 1991. Evaluation of Trebon (ethofenprox) against *Cx. quinquefasciatus An. stephensi* and *Ae. aegypti* and on non target organisms. *Ind. J. Malario.*, 28: 249–253.

Chockalingam, S., Kuppusamy, A., Punithavathy, G. and Manoharan, T., 1991. Synergistic effect of insecticide with plant extracts against mosquito vectors. *J. Appl. Zool. Res.*, 2(2): 92–95.

Curtis, C.F. and Pasteur, N., 1981. Organophosphate resistance in vector populations of the complex of *Culex pipens*. L. *Bull. Ent. Res.*, 71: 153–171.

Dhanraj, B. and Thirunavakkarasu, 2005. Evaluation of sumilarv 0.5 per cent GC (pyripyroxifen) against *Cx. quinquefasciatus* mosquito larvae. *Pestology*.

Ganesh, K.N., Vijayan, V.A. and Urmila, J., 2002. Role of esterases and mono oxygenase in deltamethrin resistance in *Anopheles stephensi* at Mysore. *Ind. J. Expt. Biol.*, 40: 583–588.

Gjullin, C.M. and Isaak, I.W., 1957. Present status of mosquito resistance to insecticide in the San Joaquin valley in California. *Mosquito News*, 17(2): 67–70.

Rao, T.R., 1984. *The Anophelenes of India*. Malarial Research Centre (ICMR) report, New Delhi.

Ross, H.H., 1965. *A Textbook of Entomology*. pp. 377.

Service, M.W., 1983. Management of vectors in Youdeowei. In: *Pest and Vector Management in Tropics*, (Ed) M.W. Sevice. p. 265–280.

Villani, F., White, G.B., Curtis, L. and Mites, S.J., 1983. Inheritance and activity of some esterases associated organophosphate resistance in mosquito complex of *Culex pipens* L. *Bull. Ent. Res.*, 73: 153–170.

World Health Organization, 1992. Instruction for determining the susceptibility of resistance of mosquito larvae of insecticides. 5th report of WHO expert committee on Vector Biology and Control. WHO. Tech. Rep. Ser. 818.

Chapter 10

Evaluation of Greengram Germplasm for Tolerance to Pre-harvest Sprouting

K.L. Narasimha Rao, Ch. Mallikarjuna Rao and Y. Koteswara Rao
Regional Agricultural Research Station, Lam, Guntur – 34

ABSTRACT

Screening of forty greengram germplasm for different physiological and yield attributes was carried out in augmented block design at RARS, Lam during *Kharif* 2004. The mean, range and coefficient of variation for different characters were worked out. Data pertaining to number of pods sprouted per plant, number of seeds sprouted per plant, imbibition (per cent), wall thickness and yield per plant were recorded. The ANOVA indicated significant differences among checks and test entries studied. The association studies indicated that yield is negatively correlated with all the other physiological characters.

***Keywords**: Greengram, Pre-harvest sprouting (Weather damage), Correlation.*

Introduction

Greengram [*Vigna radiata* (L.) Wilczek], also known as mungbean is one the most important pulse crops of India. In India pulses form major sources of dietary protein (20–30 per cent) protein, which is 2 to 2.5 times higher than that of cereals (Ritu *et al.*, 2005). Greengram is an excellent source of easily digestible protein of low flatulence and is cultivated during *Kharif, Rabi* (in peninsular India) and during summer as a sandwich crop under irrigated condition. Greengram production remained low and static around 450 kg/ha due to biotic, abiotic stresses and low input technology followed by the farmers. Of the various constraints in greengram production, weather damage (pre-harvest sprouting) is one important abiotic factor which is coming in the way of adoption of recommended technologies. Greengram is a rainy season crop and comes to harvest when the monsoons are still active. Rainfall during the pre-harvest period provide moisture which is absorbed by dry pods and seeds. Increasing water content leads to testa expansion and an increased seed respiration rate. Subsequent drying during the periods between rainfall causes testa shrinkage. The consequences of this cyclic wetting and drying are testa discoloration and cracking, reduced seed weight and vigour. Weather damaged seed is more prone to fungal infection. The most vulnerable period is between physiologic maturity and harvest resulting in losses in yield and quality. Thus, evaluation of greengram germplasm for tolerance to pre-harvest sprouting has become necessary.

Materials and Methods

Forty greengram genotypes along with four local checks (LGG 460, LGG 450, LGG 410 and LGG 407) were grown during *Kharif* 2004 at Regional Agricultural Research Station, Lam, Guntur in augmented block design. These forty genotypes include breeding lines, cultivars and commercial varieties grown all over India supplied by ICAR under ICAR ad-hoc scheme on greengram. Recommended cultural practices and plant protection measures were followed to raise the crop. Data were recorded on number of pods sprouted per plant, number of seeds sprouted per plant, imbibition (per cent), pod wall thickness and yield per plant. Seeds were collected from 10 randomly selected plants in each line at harvest and dried to 10 per cent moisture. Fully matured, unbroken and disease free seeds were selected for sprouting. Sprouting was carried out in dark by the method described by Whyte (1973) and Anon (1974) at 27 and 19 C day/night temperatures, respectively for 72 hr. Imbibition per cent was estimated by the method of Levari (1960) for a period of 18 hr. Pod wall thickness was measured using screw gauge. The augmented design analysis was carried out using IRRISTAT 4.4 software and correlation as method suggested by Falconer (1964).

Results and Discussion

The mean, range and coefficient of variation for different characters of test and check entries are given in Table 10.1. The mean of checks is found be higher than that of the test entries except for wall thickness which is on par indicating that the checks are more suitable compared to test entries. The local check used LGG 450 is tolerant to pre-harvest sprouting with pod wall thickness of 0.131 mm and yield of 6.31 g/plant. The other check LGG 460 recorded high yield of 7.28 g/plant. So, the genotypes with medium to low pod wall thickness and high yield when compared to local checks should be selected. The genotype M 524 recorded pod wall thickness of 0.146 mm and yield of 7.28 g/plant while the genotype M 365 recorded 0.142 mm and 7.92 g/plant respectively. The genotype M 1643 recorded 0.139 mm with highest yield of 10.69 g/plant. The genotype M 247 showed 0.115 mm thickness and yield of 9.36 g/plant while M 352 recorded 0.111 mm pod wall thickness and 9.31 g/plant yield respectively.

Table 10.1: Mean and Range of Different Characters of Greengram

Character	*Test Entries*		*Checks*		*SE±*	*CV (%)*	*LSD (%)*
	Range	*Mean*	*Range*	*Mean*			
Pods sprouted	35.14–69.46	50.11	52.17–68.18	50.66	2.242	5.80	4.52
Seeds sprouted	21.53–43.16	34.97	37.96–54.17	35.78	1.683	5.01	3.39
Imbibition	21.19–91.11	51.90	55.41–70.63	52.03	2.274	5.59	4.49
Pod wall thickness	0.102–0.146	0.112	0.113–0.135	0.112	0.002	3.09	0.01
Yield	2.14–10.69	5.33	6.01–7.28	5.35	0.795	15.90	1.61

The ANOVA (Table 10.2) showed significant differences for the characters studied at 5 per cent and 1 per cent levels indicating that the material had lot of variation among them. The association among different physiological characters and yield is given in Table 10.3. The yield is found to be negatively correlated with all the characters studied, while wall thickness is highly significantly correlated with number of pods sprouted and imbibition. The imbibition is highly positively correlated with number of pods sprouted. The pod wall thickness is significantly positively correlated with number of seeds sprouted indicating that genotypes with increase in pod wall thickness lead to more sprouted seed as a result of more imbibition rate. Hence, genotypes with medium to low pod wall thickness need to be selected for tolerance to pre-harvest sprouting.

Table 10.2: Analysis of Variance (Mean Sum of Squares) for Different Characters of Greengram

Source	*D.F*	*Pods Sprouted*	*Seeds Sprouted*	*Imbibition*	*Pod Wall Thickness*	*Yield*
Blocks	4	4.152	9.337*	3.645	0.00006**	2.625**
Treatment	3	18.928*	9.978*	11.036*	0.0002**	1.489*
Error M.S	12	5.271	2.892	3.465	0.00001	0.291
Total	**19**					

*Significance at 5 per cent level.

** Significance at 1 per cent level.

Table 10.3: Correlations Among Different Characters of Greengram

Character	*Pods Sprouted*	*Seeds Sprouted*	*Imbibition*	*Pod Wall Thickness*	*Yield*
Pods sprouted	1.000	0.180	0.680**	0.636**	– 0.327*
Seeds sprouted		1.000	0.300*	0.320*	– 0.153
Imbibition			1.000	0.859**	– 0.406**
Pod wall thickness				1.000	– 0.265
Yield					1.000

* and ** denote significance at 5 per cent and 1 per cent levels, respectively.

The physiological studies relating to pre-harvest sprouting need to be stressed for further research. More emphasis should be on thin pod wall thickness, wax content of pod wall and their genetics.

References

Anon., 1974. Sprouts in your kitchen, Menlopark, CA, U.S.A., Lane Magazine, Book Co., pp. 64.

Falconer, D.S., 1964. *Introduction to Quantitative Genetics*, 2nd edn., Edinburgh, Oliver and Boyd., pp. 312–329.

Levari, R., 1960. Ph.D Thesis, Jerusalem (in Hebrew). Quoted in Mayer, A.M. and Poljakoff-Mayber, A., 1963. The Germination of Seeds, New York, Pergamon press, pp.72.

Ritu Saxena, R. Singh, P.K. and Ravi Saxena, R., 2005. Multivariate analysis in mungbean. *Indian Journal of Pulses Research*, 18(1): 26–29.

Satyanarayana, A., 1999. Impact of new technologies: Case studies on greengram/blackgram. Brain storming meeting on strategies for increasing pulse production held at New Delhi, 26-27th March, 1999.

Whyte, K.C., 1973. *The Complete Book of Sprouting*. San Francisco, CA, U.S.A., Tronbador Press.

Chapter 11

Embedded Type Third Order Slope Rotatable Designs Over All Directions

G.V.S.R. Anjaneyulu*, K. Sri Suvarna Lakshmi and V.L. Narasimham
Acharya Nagarjuna University, Nagarjuna Nagar, Guntur – 522 002, A.P.

ABSTRACT

In this article an attempt is made to construct Third Order Slope Rotatable Designs Over All Directions in four dimensions using embedding techniques.

Keywords: *Slope rotatable designs, Third order slope rotatable designs, Third order slope rotatable designs over all directions, Variance-sum third order slope rotatable designs.*

Introduction

As the estimation of slope of response surface occurs frequently in practical situations, many authors diverted their attention from Rotatability to Slope Rotatability. The study of the slope of response surface was initiated by Atkinson (1970). Slope Rotatability is a Design criteria for factors with continuous quantitative levels where it is assumed that the slope of a response function fitted to a set of quantitative factor levels is the response of interest. Slope Rotatable Central Composite Designs for Second Order models were discussed by Hader and Park (1978). Slope rotatability over all directions is a useful concept when the slope of a second order response is to be studied. Slope Rotatable Over All Directions designs ensure that knowledge of the slope is acquired symmetrically, whatever direction later becomes of more interest as the data are analyzed. Anjaneyulu *et al.* (1997) established that

Corresponding Author: E-mail: gvsr_anjaneyulu@rediffmail.com.

second order slope rotatable designs over all directions is also a Variance-Sum Second order slope rotatable design. Anjaneyulu *et al.* (1995, 2000) introduced and constructed third order slope rotatable design over an axial direction. Slope rotatability is potentially useful in cases where rates of reaction (in chemical experiments), rates of deterioration (for example, in a food product) are measured. The requirements of slope rotatability are sensible when it is not known in which direction in the x-space useful changes in slopes are to be anticipated. Then equal variances (exactly or approximately) for a given distance from the center of the design are desirable.

Gardiner *et al.* (1959) introduced third order rotatable designs. Das and Narasimham (1962) constructed some second order rotatable designs and third order rotatable designs. Anjaneyulu *et al.* (1993) constructed second order slope rotatable design using Draper and Herzberg type embedding techniques. Anjaneyulu *et al.* (2004) introduced that any Variance-Sum Third Order Slope Rotatable Design is a Third Order Slope Rotatable Design Over All Directions introduced by Park and Lee (1995). Huda (1982) constructed some third-order rotatable designs in three dimensions. Huda (1983) constructed two third order rotatable designs in four dimensions. When, these designs are used, the results of the experiments performed according to the two-dimensional designs need not be discarded *i.e.*, the designs may be performed 'Sequentially' in the factors by starting with the experiments involving two factors rather than all three factors and thus can results in saving of resources. In view of the applications of such design in agronomical and industrial experiments we construct third order slope rotatable designs in over all directions using Huda's (1983) method.

Conditions for Third Order Slope Rotatable Designs Over All Directions

The conditions for slope rotatability over all directions for the third order polynomial model with four independent variables are the following (Park and Lee (1995).

$\sum x_i^2$

1. $4V_{ii} + 6C_{i,iii} + \sum_{j \neq i} (V_{ij} + 2C_{j,iij})$ equals for all I = 1, 2, 3, 4
2. $9V_{iii} + \sum_{j \neq i} V_{iij}$ equals for all i = 1, 2, 3, 4,

   $4V_{iij} + 4V_{jji} + \sum_{k \neq i,\, j} V_{ijk} + 2C_{iik,jjk} + 6C_{jjj,iij} + 6C_{iii,jji}$ equals for all $i < j$ and the value of the second equation is two times of that of the first equation.
3. All coefficients of the terms, except for constant,

   $x_1^2, x_2^2, x_3^2, x_4^2, x_1^4, x_2^4, x_3^4, x_4^4, x_1^2, x_2^2, x_1^2, x_3^2, x_1^2, x_4^2, x_2^2, x_3^2, x_2^2, x_4^2, x_3^2, x_4^2$, equal zero.

Symmetry Conditions for Variance-sum Third Order Slope Rotatable Designs

Anjaneyulu *et al.* (2004) have derived the following conditions for Variance-sum Third Order; Slope Rotatable Designs:

(A) All sums like $\sum x_i, \sum x_i x_j, \sum x_i^2 x_j, \sum x_j^2, \sum x_i x_j x_k, \sum x_i^2 x_j x_k$

$\sum x_i^2 x_j, \sum x_i^9 x_j^9, \sum x_i x_j x_k x_l$.......... etc. are equal to zero. That is all sums of products in which at least one of the x's is with an odd power are zero.

(B) (*i*) $= N\lambda_2 =$ constant, $\forall_i$

(*ii*) $\sum x_i^4 = a\,N\lambda_4 = \text{constant}, \ \forall_i$

(*iii*) $\sum x_i^6 = b\,N\lambda_6 = \text{constant}, \ \forall_i$

(C) (*i*) $\sum x_i^2 x_j^2 = N\lambda_4 = \text{constant, for } i \neq j$

(*ii*) $= c\,N\lambda_6 = \text{constant, for } i \neq j$

(*iii*) $= N\lambda_6 = \text{constant, for } i \neq j \neq k$ ......(1)

Construction of Third Order Slope Rotatable Design Over All Directions Using Huda's Methods of Embedding

Consider the design points $\left(x_{1u}^{(i)}, x_{2u}^{(i)}\right), i = 1, 2; u = 1, 2, \ldots\ldots N^1$ where for each *i* the design points are equally spaced on the circle of radius p_i (i = 1,2)

$$\text{Let } A = \frac{N^1}{2}\left(p_1^2 + p_2^2\right)$$

$$B = \frac{N^1}{8}\left(p_1^4 + p_2^4\right)$$

$$C = \frac{N^1}{48}\left(p_1^6 + p_2^6\right) \qquad \sum x_i^2 x_j^2 x_k^2 \qquad \text{......(2)}$$

Now consider a four-dimensional point set which gives $4N^1 + 32$ points.

These design points are taken as follows:

$\left(x_{1u}^{(i)}, x_{2u}^{(i)}, 0, 0\right)\left(u = 1, \ldots\ldots\ldots N^1;\ i = 1,2\right)$

$\left(0, 0, x_{1u}^{(i)}, x_{2u}^{(i)}\right)\left(u = 1, \ldots\ldots\ldots N^1;\ i = 1,2\right)$

$(\pm p, \pm p, \pm q, 0)$

$(\pm p, \pm p, 0, q)$

$(\pm q, 0, \pm p, \pm p)$

$(0, \pm q, \pm p \pm p)$ ......(3)

from this set of points the following conditions hold good.

$$\sum_{u=1}^{N} x_{iu}^2 = A + 16p^2 + 8q^2 \ (i = 1,2,3,4)$$

$$\sum_{u=1}^{N} x_{iu}^4 = aB + 16p^2 + 8q^4 \ (i = 1,2,3,4)$$

$$\sum_{u=1}^{N} x_{iu}^2 x_{ju}^2 = B + 16p^4 \ (i = 1, j = 2 \text{ and } i = 3, j = 4)$$

$$= r16p^2q^2 \ (i = 1,2; j = 3,4)$$

$$\sum_{u=1}^{N} x_{iu}^6 = \text{bC} + 16p^6 + 8\text{q}^6 \ (i = 1,2,3,4)$$

$$\sum_{u=1}^{N} x_{iu}^4 x_{ju}^2 = \text{cC} + 16p^6 \ (i = 1, j = 2 \text{ and } i = 3, j = 4)$$

$$= 8p^4q^2 + 8p^2q^4 \ (i = 1,2; j = 3,4)$$

$$\sum_{u=1}^{N} x_{iu}^2 x_{ju}^2 x_{ku}^2 = 8p^4q^2 = \ (i \neq j \neq k; i,j,k = 1,2,3,4) \quad \text{......(4)}$$

and all other sums of powers and products upto order six are zero. The considered set of design points form a third order slope rotatable design over all directions in four dimensions if the following conditions are satisfied.

(i) $\sum x_{iu}^4 = a\sum x_{iu}^2 x_{ju}^2$

$$aB + 16\text{p}^4 + 8q^4 = a(B + 16p^4)$$

$$\text{q}^2 = \sqrt{2(\text{a} - 1)\text{p}^2}$$

(ii) $\sum x_{iu}^4 x_{ju}^2 = c\left(\sum x_{iu}^2 x_{ju}^2 x_{ku}^2\right)$

$$cC + 16p^6 = c\lfloor 8p^4q^2 \rfloor$$

$$\therefore \ C = \frac{8\,p^6 \lfloor c(\sqrt{2(\text{a} - 1)}) - 2 \rfloor}{c}$$

(iii) $\sum x_{iu}^4 = a\sum x_{iu}^2 x_{ju}^2 \ (i = 1,2; j = 3.4)$

$$aB + 16p^4 + 8q^4 = a\lfloor 16p^2q^2 \rfloor$$

$$\therefore \ B = 16p^4\left(\sqrt{2(\text{a} - 1)} - 1\right) \quad \text{......(5)}$$

the conditions holds good for any value of a, b and c

Let a = 3, b = 15, c = 3 the above conditions gives [c.f. Huda (1983)]

$\text{q}^2 = 2p^2$

$\text{B} = 16p^4$

$$\text{C} = \frac{32}{3}p^6$$

the last two requirements are satisfied if t is chosen to be such that

$$\frac{(1 + t^2)^3}{(1 + t^3)^2} = \frac{8}{N^1}$$

and $p^4 = \frac{B}{16}$

$$= \frac{\left(\frac{N^1(p_1^4 + p_2^4)}{8}\right)}{16}$$

$$= \frac{N^1}{128}(1+t)p_1^4$$

$$\therefore\ p^2 = \left[\frac{N^1}{128}(1+t)p_1^4\right]^{\frac{1}{2}}$$

$q^2 = 2p^2$

The considered design in (3) gives us a four dimensional third order slope rotatable design over all directions since it satisfied all the conditions given in (1).

Example: Third order slope rotatable design OAD in four dimensions when $N^1 = 7$ and taking a = 3, b = 15, c = 3.

Consider the design in (3) which is a third order slope rotatable design in over all directions. Then conditions in the design (2) reduces to the following:

(*i*) $q^2 = \sqrt{2(a-1)}p^2$

(*ii*) $C = \frac{32}{3}p^6$

(*iii*) $B = 16p^4$

the last two requirements are satisfied if *t* is chosen to be such that

$$\frac{(1 + t^2)^3}{(1 + t^3)^2} = \frac{8}{N^1}$$

and also let $p_2^2 = tp_1^2\ (t \geq 0)$

when $N^1 = 7$

$t^6 - 21t^4 + 16t^3 - 21t^2 + 1 = 0$

by using the Newton-Raphson Method, the solution for $0 \leq t \leq 1$ is given by $t = 0.233071$.

$\therefore\ p_2^2 = 0.233071\ p_1^2$

$B = 16p^4$

$p^2 = 0.240121\ p_1^2$

and $q^2 = 2p^2$

$= 0.480242.$

Therefore the $4N^1 + 32$ *i.e.*, 60 design points form a third order slope rotatable designs over all directions in four dimensions.

As in the case of third order rotatable design in the construction of third order slope rotatable designs over all directions, we do also have less number of design points. But construction of third order rotatable design is more restrictive and not always possible. Because, it has to fulfil the requirements such as a = 3, b = 15 and c = 3, where as in the case of this proposed design it is less restrictive. We can construct third order slope rotatable designs over all directions easily compared to third order rotatable design and our designs are general class of designs because third order rotatable design is a subset of third order slope rotatable design over all directions.

Acknowledgements

The authors are thankful to the authorities of Acharya Nagarjuna University for providing the necessary facilities in completion of their work.

References

Anjaneyulu, G.V.S.R., Dattatreya Rao, A.V. and Narasimham, V.L., 1993. Embedding in second-order-slope-rotatable designs, reports of statistical application research. *JUSE*, 40(4): 1–11.

Anjaneyulu, G.V.S.R., Dattatreya Rao, A.V. and Narasimham, V.L., 1995. Construction of third order slope rotatable designs through doubly balanced incomplete block designs. *Gujarat Statistical Review*, 21–22: 49–54.

Anjaneyulu, G.V.S.R., Dattatreya Rao, A.V. and Narasimham, V.L., 2000. Third order slope rotatable designs, *J. Ind. Soc. for Probability and Statistics*, 5: 65–74.

Anjaneyulu, G.V.S.R., Nagabhushanam, P. and Narasimham, V.L., 2004. On third order slope rotatable designs over all directions. *J. Ind. Soc. for Probability and Statistics*, 8: 92–101.

Anjaneyulu, G.V.S.R., Varma, D.N. and Narasimham, V.L., 1997. A note on second order slope rotatable designs over all directions. *Communications in Statistics: Theory and Methods*, 26(6): 1477–1479.

Atkinson, A.C., 1970. The design of experiments to estimate the slope of a response surface. *Biometrika*, 57: 319–328.

Das, M.N. and Narasimham, V.L., 1962. Construction of rotatable designs through balanced incomplete block designs. *Ann. Math. Stat.*, 33: 1421–1439.

Gardiner, D.A., Grandage, A.H.E. and Hader, R.J., 1959. Third order rotatable designs for exploring response surface. *Ann. Math. Stat.*, 30: 1082–1096.

Hader, R.J. and Park, S.H., 1978. Slope-rotatable central composite designs. *Technometrics*, 20: 413–417.

Huda, S., 1982. Some third order rotatable designs in three dimensions. *Ann. Inst. Stat. Math.*, 34: 365–371.

Huda, S., 1983. Two third order rotatable designs in four dimensions *JSPI*, 6: 241–243.

Park, S.H., 1987. A class of multi factor designs for estimating the slope of response surfaces. *Technometrics.*

Park, S.H. and Lee, M.S., 1995. "Slope-rotatability over all directions in third order response surface models. *Journal of the Korean Statistical Society*, 24(2): 519–536.

Chapter 12

Comparative Study on the Haemocytes of Some Cerambicidae

S.D. Disale[1], S.R. Wagh[2], S.M. Gaikwad[3], G.P. Bhawane[4] and G.D. Deshmukh[5]

[1]*Department of Zoology, D.B.J. College, Chiplun, Dist. Ratnagiri – 415 605*

[2]*Department of Zoology, New Arts, Commerce and Science College, Parner, Dist. Ahmednagar – 414 302*

[3]*Bhogawati Mahavidyalaya Kurukali, Dist. Kolhapur – 416 001*

[4]*Department of Zoology, Shivaji University, Kolhapur – 416 004*

[5]*Department of Zoology, Rajaram Mahavidyalaya, Kolhapur – 416 004*

ABSTRACT

Total and differential haemocyte populations were comparatively studied in both sexes of eight species of cerambicid beetles by employing standard techniques of bleeding, fixation and staining. The haemocytes observed were classified on the basis of tinctorial properties and cytoplasmic inclusions, into six types such as prohaemocytes, granulocytes, spherulocytes, plasmatocytes, oenocytes and coagulocytes.

The total haemocyte counts were found to be always more in females of all these beetles. Differential haemocyte counts revealed that prohaemocytes were more in males of *Stromatium barbatum, Xystocera globosa,* and *Xylotrechus subscutellatus* where as the granulocytes were more in females of *Batocera rubra, Cerosterna scabrator* and *Xystocera globosa.* Rest of the cell types showed somewhat similar populations in males as well as females. Out of six types granulocytes and plasmatocytes were found to contribute major populations.

Keywords: *Haemocytes, Cerambicidae, comparative study.*

Introduction

The circulating body fluid in insects is called haemolyrnph. It contains various types of cells called haemocytes. These cells were firstly classified in to four types (Cuenot, 1896). After this classification some other workers have classified these cells in different ways leading to lot of confusion. Jones (1962) revised these classifications and proposed a dynamic scheme of classification, mentioning nine types of haemocytes. Finally Price and Ratcliffe (1974) reconsidered this classification and classified the haemocytes in to six basic types, *viz.*, Prohaemocytes, Granulocytes, Spherulocytes, Plasmatocytes, Oenocytes and Coagulocytes. This classification was based on their extensive study of haemocytes in fifteen insect orders including Coleoptera.

The insect haemocytes perform variety of vital functions like phagocytosis, encapsulation nodule formation, haemolymph coagulation, tissue formation, intermediary metabolism, etc. and play important roles in life of insects (Gupta, 1979).

In present investigation, total and differential haemocyte counts were comparatively studied in both the sexes of beetles by employing standard bleeding, fixation and staining techniques (Cuenot, 1896; Jones, 1962; Price and Ratcliffe, 1974; Gupta, 1979).

Materials and Methods

The Cerambicid beetles used in present investigation were collected from their natural habitats. They were acclimatized to the laboratory conditions for two days. During this period they were fed upon their natural foods. These beetles were used for collection of haemolymph and preparation of temporary and permanent stained smears. For this purpose the acclimatized beetles were classified according to their sex. They were heat fixed and/or chemically fixed and bleed by amputation of head or thoracic appendages (Cuenot, 1896; Jones, 1962; Price and Ratcliffe, 1974; Gupta, 1979). The total haemocyte counts were taken by using Naubaurs haemocytometry (Pilat, 1935; Fisher, 1936; Jones and Tauber, 1954; Jones, 1962). The differential counts were taken by staining the smears with three different stains *viz.* Leishman, Giemsa and Pappenheims panoptic chrome (Cuenot, 1896; Jones, 1962; Price and Ratcliffe, 1974; Gupta, 1979).

Results and Discussion

Different types of haemocytes observed and the parameters studied are presented in Table 12.1. The haemocytes were identified on the basis of histomorphological peculiarities and classified in to six types as per Jones (1962); Gupta and Sutherland (1968); Price and Ratcliffe (1974) and Gupta (1979).

Among the males, total haemocyte count was found to be maximum in *Batocera rubra* (82650±240) and minimum in *Olenecamptus bilobus* (51020±124). In case of females, the total haemocyte count was found to be maximum in *Batocera rubra* (93250±135) and minimum in *Olenecamptus bilobus* (59265±175). The total haemocyte counts were always found to be higher in females.

The tinctorial properties and differential haemocyte counts revealed that there are six different types of haemocytes *viz.* Prohaemocytes, Granulocytes, Spherulocytes, Plasmatocytes, Oenocytes and Coagulocytes.

The relative percentage of Prohaemocytes among the males is found to be maximum in *Stromatium barbatum* (26.47) and minimum in *Cerosterna scabrator* (10.07). In case of females, it was found to be more in *Olenecamptus bilobus* (18.35) and less in *Xystocera globosa* (08.23).

Table 12.1: Differential and Total Haemocyte Counts

Sl.No.	Name of the Beetle	Sex	DHC (%age)						THC
			Pro	Gra	Sph	Pls	Oes	Cos	(Mean & S.D.)
1.	*Batocera rubra*	M	13.48	34.23	15.25	16.22	13.75	08.07	82650±240
		F	13.57	35.75	17.25	17.28	10.20	05.95	93250±135
2.	*Cerosterna scabrator*	M	10.07	33.34	14.93	18.75	12.15	10.76	76350±165
		F	10.40	36.41	15.03	19.07	12.71	06.36	91050±265
3.	*Coptops aedificator*	M	12.83	22.94	19.22	29.55	08.35	07.11	
		F	10.47	24.09	17.33	31.05	06.55	10.51	
4.	*Olenecamptus bilobus*	M	17.45	18.91	21.09	31.27	10.90	03.63	51020±124
		F	18.35	18.44	22.95	28.34	11.58	02.00	59265±175
5.	*Pterolophia* sp.	M	15.47	25.90	19.42	32.73	05.39	01.09	
		F	14.37	26.86	18.61	33.45	04.40	02.31	
6.	*Stromatium barbatum*	M	26.47	19.41	13.53	29.41	07.65	03.53	72500±350
		F	10.72	20.72	19.28	37.14	09.28	02.82	87000±280
7.	*Xystocera globosa*	M	17.35	27.68	11.98	35.95	04.96	02.06	
		F	08.23	31.65	16.67	36.55	05.66	01.24	
8.	*Xylotrechus subscutellatus*	M	12.64	22.23	21.45	26.65	10.21	06.82	
		F	09.33	26.69	23.57	27.48	10.22	02.71	

THC's were found to be higher in females.

The relative percentage of Granulocytes among the males is found to be maximum in *Batocera rubra* (34.23) and minimum in *Olenecamptus bilobus* (18.91). In case of females, it was found to be more in *Cerosterna scabrator* (36.41) and less in *Olenecamptus bilobus* (18.44).

The relative percentage of Spherulocytes among the males is found to be maximum in *Xylotrechus subscutellatus* (21.45) and minimum in *Xystocera globosa* (11.98). In case of females, it was found to be more in *Xylotrechus subscutellatus* (23.57) and less in *Cerosterna scabrator* (15.03).

The relative percentage of Plasmatocytes among the males is found to be maximum in *Xystocera globosa* (35.95) and minimum in *Batocera robra* (16.22). In case of females, it was found to be more in *Stromatium barbatum* (37.14) and less in *Batocera rubra* (17.28).

The relative percentage of Oenocytes among the males is found to be maximum in *Batocera robra* (13.75) and minimum in *Xystocera globosa* (04.96). In case of females, it was found to be more in *Cerosterna scabrator* (12.71) and less in *Pterolophia* sps. (04.40).

The relative percentage of Coagulocytes among the males is found to be maximum in *Cerosterna scabrator* (10.76) and minimum in *Pterolophia* sps. (01.09). In case of females, it was found to be more in *Coptops aedificator* (10.51) and less in *Xystocera globosa* (01.24).

The survey of available literature revealed that there are several reports on haemocytes, but all of them are of diverse type as far as haemocyte classification and terminology is concerned (Price and Ratcliffe, 1974; Jones, 1950; Sonawane, 1984; Disale, 1997). There may be several reasons for diversity. Utilization of different life stages of insects for study may be one of these reasons. Extensive data is now available in case of adult insect belonging to order Coleoptera (Disale, 1997). This data showed

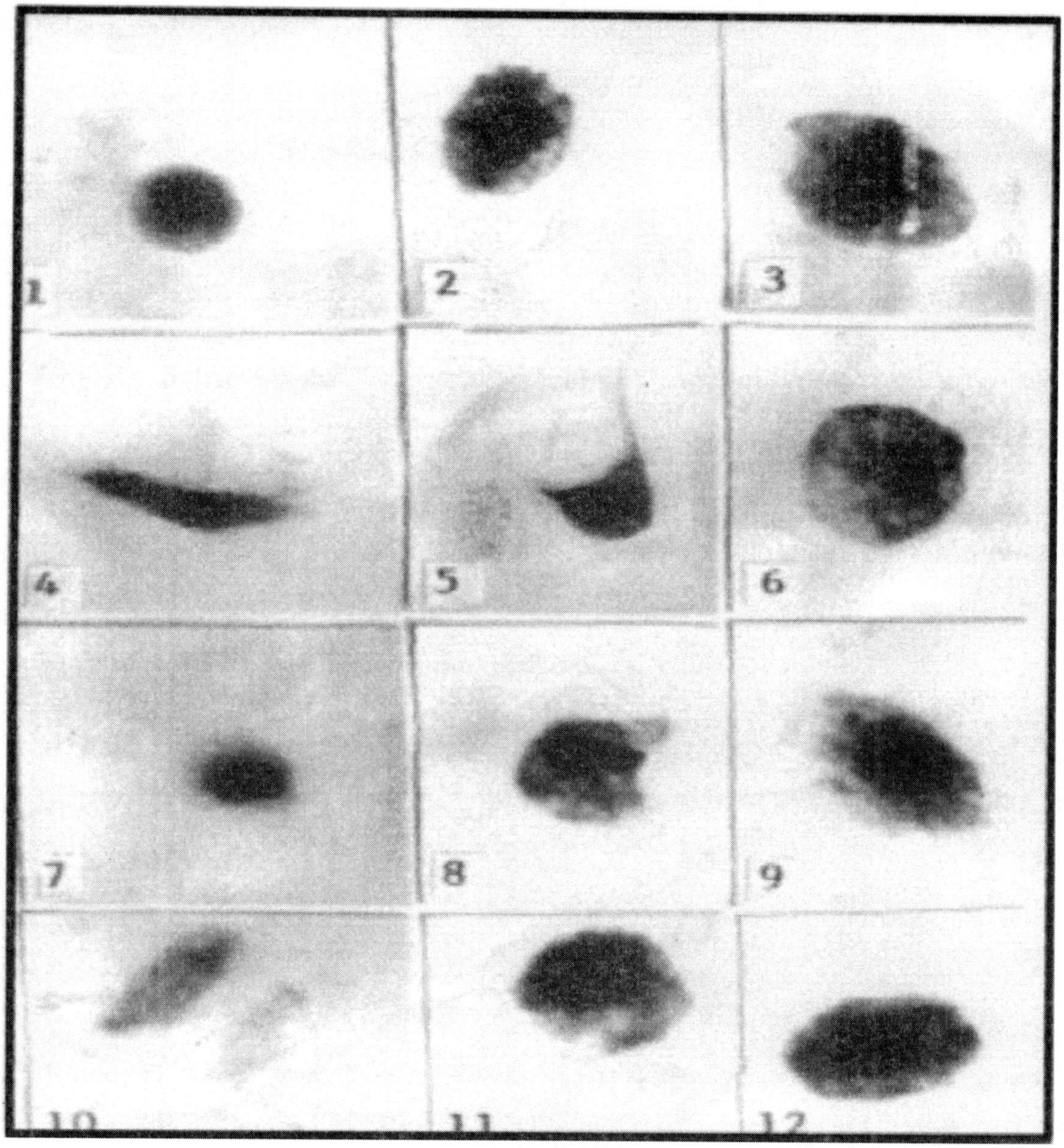

Plate 12.1: Haemocyte of Cerambicidae
(A) Prohaemocytes (Figures 1 and 7); (B) Granulocytes (Figures 2 and 8); (C) Spherulocytes (Figures 3 and 9); (D) Plasmatocytes (Figures 4 and 10); (E) Oenocytes (Figures 5 and 11); (F) Coagulocytes (Figures 6 and 12)

that there are six definite types of haemocytes in Coleopteran adults, with extremely minor exceptions.

The results obtained during present investigation indicate that the relative percentage of haemocytes is comparatively more in females than in males. Out of six types of haemocytes reported, Granulocytes contribute major population, followed by Plasmatocytes, Spherulocytes, Prohaemocytes, Oenocytes and Coagulocytes are in least amount. Similar variations in haemocyte populations among the members of a family and within a species, among two sexes were also reported in different insects (Price and Ratcliffe, 1974; Jones, 1950; Sonawane, 1984; Collins, 1963 and Disale, 1997).

References

Arnold, J.W., 1952. The haemocytes of the Mediterranean flour moth, *Ephestia* kuhniella, Zell. (Lepidoptera : Pyralididae). *Can. J. Zool.*, 30: 352–365.

Arvy, L.M., Gabe, M. and J. Lhoste, 1950. Action de quelques insecticides sur le sang du Doryophore (*C. decimlineata* Say.). *Bull. Soc. Entomol. Fr.*, 55: 122–127.

Collins, N., 1963. Les haemocytes de la larvae de *Melolontha melolontha* L. (Coleoptera : Scarabaeidae). *Rev. Pathol. Veg. Entomol. Agric. Fr.*, 42: 161–167.

Cuenot, L., 1896. Etudes Physiologique sur les Orthopteres. *Arch. Biol.*, Paris, 14: 293–341.

Disale, S.D., 1997. Coleopteran Haemocytes: A Comparative Study. *Ph.D. Thesis*, Shivaji University, Kolhapur, (India).

Fisher, R.A., 1936. The effect of a few toxic substances upon the total blood cell count in the cockroach, *Blatta orientalis* L. *Ann. Entomol. Soc. Amer.*, 29: 334–340.

Gupta, A.P. and D.J. Sutherland, 1968. Effect of sublethal doses of Chlordane on the haemocytes and mid gut epithelium of *Periplaneta americana*. *Ann. Entomol. Soc. Amer.*, 61(4): 910–918.

Gupta, A.P., 1979. Haemocyte types: their structures, synonomies, interrelationship and significance. In: *Insect Haemocytes*, (Ed.) A.P. Gupta. Cambridge University Press, Cambridge, pp. 86–127.

Jones, J.C., 1962. Current concepts concerning insect haemocytes. *Amer. Zool.*, 2: 209–246.

Jones, J.C. and O.E. Tauber, 1954. Abnormal haemocytes in mealworm (*Tenebrio molitor* L.). *Ann. Entomol. Soc. Amer.*, 47(3): 428–444.

Jones, J.C., 1950. The normal haemocyte picture of the yellow mealworm *Tenebrio molitor* Linn. *Iowa State Coll. J. Sci.*, 24: 356–361.

Mall, S. and S. Gupta, 1979. *Indian J. Entomology*, 41: 223–233.

Pilat, M., 1935. The effects of intestinal poisoning on the blood of locusts (*Locusta migratoria*). *Bull. Entomol. Res.* 26: 283–292.

Price, C.D. and N.A Ratcliffe., 1974. A reappraisal of insect haemocyte classification by the examination of blood from fifteen insect orders. *Z. Zellforsch. Microsk. Anat.*, 147(3): 313–324.

Sonawane, Y.B., 1984. Insect haemocytes: A comparative study. *M.Phil. Dissertation*, Shivaji University, Kolhapur, (India).

Taumanouff, C. and M. Lapied, 1950. Action du dichloro diphenyl trichloro ethane (DDT) sur les chenilles de la fausse teigne des ruches Galleria mellonella L. Effect de la temperature. Action sur le sang. *Acad. Agric. Fr.*, 36: 386–471.

Trehan, K.N. and H.R. Pajni, 1961. Mode of action of insecticides. *Beitr. Entomol.*, 11: 1–11.

Chapter 13

Flavobacterium ferrugineus as an Ecofriendly Biopesticide Against Larvae of *Pericallia ricini* (Arctidae : Lepidoptera)

S. Dinakaran* and S. Anbalagan

P.G. Department of Zoology, Madura College, Madurai – 625 011, Tamil Nadu, India

ABSTRACT

A pathogenic bacterial species was isolated from field collected naturally infected third and fifth instar larvae of a lepidopteran pest, *Pericallia ricini* by standard microbiological methods and taxonomically identified as *Flavobacterium ferrugineus* on the basis of twelve biochemical assays and four colony morphological characters. Reinfection study was conducted by oral administration of pure culture of the pathogenic bacterial thorough castor leaves. Percentage mortality data suggests that third instar is more susceptible to *Flavobacterium ferrugineus* than late instars. Inspection of dead larvae indicates the milky disease. This ecofriendly biopesticide is recommended for further investigations.

Keywords: *Flavobacterium ferrugineus, Pericallia ricini, Biopesticide.*

Introduction

Total eradication of agricultural pests from agro-ecosystems is impossible and scientifically unnecessary. More preciously regulation of their population density below the economic injury level is possible technically and meaningful. This solution is partially achieved by synthetic chemical

Corresponding Author: E-mail: dinkara@rediffmail.com

insecticides. However, they are equally responsible for several adverse effects such as environmental contamination, biological magnification in food chains, development of resistance in pests and harmful effects on non-target organisms. Hence he need for alternative control methods that are cost-effective and ecofriendly is greater at present. Biological control using variety of natural enemies offer an excellent alternative to chemical pesticides and it forms a major component of current Integrated Pest Management Programs (Jeyaraj, 1985; Ananthakrishnan, 1992; Narayanan, 1995). The utility of insect pathogenic microbes especially bacteria appears to be gaining acceptance and popularity in applied entomological researches (Falcon, 1971; Morris, 1982; Dinakaran and Krishnan, 2002). *Pericallia ricini* commonly called, as wooly bear is a serious polyphagous pest during its larval stages. It infests a wide range of edible crops such as potato, castor, banana, pumpkin, cotton, sesamum, moringa and brinjal (Nayar *et al.*, 1976; Chockalingam, 1979). In the present research, we isolated and taxonomically characterized a pathogenic bacterium, *F. ferrugineus* from naturally diseased larvae of *P. ricini*. Pathogenecity of pure culture was evaluated against the larval stages of same pest under laboratory conditions and the results are presented in this communication.

Materials and Methods

Naturally infected third and fifth instar larvae of *P. ricini* were collected from the castor plantations in previously autoclaved glass vials, sealed immediately and transported to laboratory in cold box. They were surface sterilized in 0.01 per cent mercuric chloride for few minutes, washed several changes in sterile water. The whole body homogenate of larvae was prepared in sterile water and streaked onto nutrient agar plates under aseptic conditions. Most consistently growing colonies were sub-cultured and maintained in agar at 37 C. Among them, a yellow colour colony was found to be more consistent in occurrence and prominent in growth. This isolate was subjected to twelve standard biochemical tests to identify at species level by following Buchannon and Gibbons (1974). Colony-morphological characters were also considered for identification. Freshly moulted third, fourth and fifth instar larvae of *P. ricini* from stock culture were starved for 5 hours and used as test larvae for pathogenecity test. As a preliminary study, bacterial suspension from pure culture incubated at 37 C for 24 hours is directly used as pesticide formulation without any quantification. The suspension was sprayed onto healthy castor leaves using an automiser and the starved larvae were then fed on the sprayed leaves. Freshly treated leaves were given daily. A control was maintained on leaves treated with only nutrient broth. The experiments were replicated thrice each comprising ten larvae. Observation on percentage mortality over control were made up to 96 hours and converted into probit mortality values.

Results and Discussion

The morphological and biochemical features of the pathogenic isolate suggest that it would be *F. ferrugineus*. Mortality data of third, fourth and fifth instar larvae of *P. ricni* fed on pure culture of this pathogen are given in Table 13.1. Third instar is more susceptible to *F. ferrugineus* when compared to fourth and fifth instar larvae. It shows 100 per cent mortality after 48 hours treatment when compared to 86–90 per cent in other two stages. Isolation and characterization of pathogenic bacteria from naturally diseased insects and then evaluation of pathogenecity of their pure culture as biopesticide is a most widely accepted protocol in microbial pesticide research (Jayaraj, 1985; Mohamed *et al.*, 1992; Dinakaran and Krishnan, 2002) as well as in understanding the aetiology of microbial diseases in productive insects (Nagendran *et al.*, 1999). For instance, Mohamed Ali *et al.* (1992) have isolated three species of bacteria (*Bacillus thuringiensis, Serratia marcescens* and *Pseudomonas aeruginosa*) from naturally infected larvae of *Hyblaea purea* and tested their pathogenecity against the same pest. Mortality of pests after treatment with bacterial pesticides varies with species and age of the larvae. First two instars of

Heliothis armigera were more susceptible to biobit, a commercial formulation of *B.thuringeiensis* than the late instars (Douressamy and Regupathy, 1993). Similarly *B. thuringiensis, S. marcescens* and *P. aeruginosa* have caused 100, 70 and 33.3 per cent mortality respectively in larvae of *H. Puera* (Mohamed Ali *et al.*, 1992).

Table 13.1: Mean, Standard Deviation and Percentage for Mortality in Larval Stages of *P. ricini* Exposed to *Flavobacterium ferrugineus*

Instar	*Exposure Duration (Hours)*					
	24		*48*		*72*	
	Mean ± S.D.	*% Mortality*	*Mean ± S.D.*	*% Mortality*	*Mean ± S.D.*	*% Mortality*
III	6.67±0.58	66.6	2.67±0.58	93.2	0.67±0.58	99.8
IV	7.67±0.58	76.7	1.67±0.58	93.2	0.67±0.58	99.9
V	7.67±0.58	76.7	1.33±0.58	89.9	1.0±0	99.9

As in the present study, oral feeding of pathogenic bacteria through leaves was found to be superior than spraying the larvae topically (Kalla *et al.*, 1997; Roychoudhary *et al.*, 1995). Therefore it could be expected that *F. ferrugineus* could be sprayed over the plant to suppress *P. ricini*. The consistent occurrence of *F. ferrugineus* in field-collected larvae of *P. ricini* suggests that this pathogen may play an important role in containing the pest in nature. So far many species of *Bacillus* such as *B. papillae, B. cereus, B. sphaericus* and *B. firmus* were reported as entomopathogenic against lepidopteran (Burges, 1981; Varma and Mohamed, 1986). *Streptococcus* sp. was reported to involve in flacherie disease of silkworm (Krishnaswami *et al.*, 1978). Inspection of dead larvae in the present study reveals milky disease in which haemolymph turns to white fluid. So far several species of *Bacillus* were reported to cause milky disease in many insect pest including larval Lepidoptera (Dutky, 1963; Steinhaus, 1964). Field trial as well as evaluation of pathogenecity to other pests are two major avenues needed research in future with reference to *F. ferrugineus* biopesticides.

Acknowledgement

This work was originally carried out at Department of Zoology, Thiagarajar College, Madurai. We are grateful to Board of Madura College, H.O.D. and Principal for support and encouragement.

References

Ananthakrishnan, T.N., 1992. *Emerging Trends in Biological Control of Phytophagous Insects*. Oxford and IBH, New Delhi, pp 239.

Buchannon, R.E. and Gibbons, N.E., 1974. *Bergy's Manual of Determinative Bacteriology*, 8th Ed, Williams and Wilkin, Baltimore.

Burges, H.D., 1981. *Microbial Control of Pests and Plant Diseases: 1970–1980*. Academic Press, New York, pp 939.

Chockalingam, S., 1979. A study on food consumption and utilization in *Pericallia ricini* (Lepidoptera : Arctidae). *Comp. Physiol. Ecol.*, 4: 78–80.

Dinakaran, S. and Krishnan, N., 2002. Bioefficacy of a bacterial pesticide (*Streptococcus rehi*) against larvae of *Pericallia ricini* (Arctidae : Lepidoptera).

Douresswamy, S. and Regupathy, A., 1992. Efficacy of *Bacillus thuringiensis* and endosulfan against larvae of *Heliothis armigera* Hubn. *J. Appl. Zool. Res.*, 4: 45–47.

Dutky, S.R., 1963. Microbial pesticides. In: *Insect Pathology: An Advanced Treatise,* (Ed.) E.A. Steinhaus. Academic Press, New York, pp. 75–116.

Falcon, L.A., 1971. Use of bacteria for microbial control. In: *Microbial Control of insects and Mites,* (Eds.) H.D. Burger and N.W. Hussey. Academic Press, London, pp. 67–96.

Jeyaraj, S., 1985. History and developmental of microbial control. In: *Microbial Control of Pest Management.* TNAU, Coimbatore, pp. 1–18.

Kalla, S., Joshi, K.C. and Pant, N.C., 1997. Effectiveness of three toxins of *Bacillus thuringiensis* against shisham defollator *Plecoptera reflexa. J. App. Zool. Res.*, 8: 115–116.

Krishnaswami, S., Narasimhanna, M.N., Suryanarayanan, S.K., and Kamaraj, S., 1978. *Silkworm Rearing Sericulture Manual II.* FAO and Oxford and IBH, New Delhi, pp. 108–109.

Morris, O.N., 1982. Bacteria as Pesticide: forest application. In: *Microbial Viral Pesticides,* (Ed.) E. Kurstar. Marcel Dekkar, New York, pp. 239–288.

Mohamed Ali, M.L., Verma, R.V. and Sudheendrakumar, V.V., 1992. Microbial pathogen associated with forest insect pests of Kerala. In: *Emerging Trends in Biological Control of Phytophagous Insects,* (Ed.) T.N. Ananthakrishnan. Oxford and IBH, New Delhi, pp. 119–128.

Nagendran, N., Thiruvalluvan M. and Krishnan, N., 1999. *Pseudomonas* sp. associated with Flacherie affected silk work. *J. Appl. Zool. Res.,* 10: 181.

Narayanan, K., 1995. Microbial control of insect pest: A comer stone in sustainable agriculture. In: *Plant-Microbe-Interaction in Sustainable Agriculture,* (Eds.) R.K. Bhel, A.L. Khurana and R.C. Dogra. CCS Publisher, New Delhi. pp. 173–182.

Nayar, K.K., Ananthakrishnan, T.N. and David, B.V., 1979. *General and Applied Entomology.* Tata Mc-Graw Hill, New Delhi, pp. 265–266.

Roychoudhary, N., Joshy, K.C. and Pandey, D.K., 1994. Effect of three variety of toxins of *Bacillus thuringiensis* against teak leaf foliator *Ectectona machaeractis* Walker (Pyralidae). *J. Appl. Zool. Res.,* 30: 57–60.

Chapter 14

Study of Important Medicinal Plants from Parner Tahasil, Dist. Ahmednagar, India

S.K. Aher, V.S. Shinde, R.B. Ghogare, A.R. Thube, Y.P. Date, S.R. Paimode and A.R. Kale

Department of Botany, New Arts Commerce and Science College, Parner, Dist. Ahmednagar – 414 302, M.S.

ABSTRACT

The present study deals with exploration of medicinally important plants from Parner Tahasil of Maharashtra. Parner is bordering the west' part of Ahmednagar District. The Tahasil has hilly as well as plain area with its distinct flora. 170 medicinally important plants belonging from 58 families have been recorded with their botanical name, vernacular name, family and plant part used in the medicines. Family Asteraceae, Cucurbitaceae, Papilionaeae were the richest in medicinal plants.

Keywords: *Medicinal plants, Parner Tahasil.*

Introduction

Plants have provided the means of curing diseases perhaps since the evolution of mankind on the earth. The methods of herbal treatment have achieved eminence especially in the Central and South Asia, China and countries surrounding the Mediterranean sea. In India the oldest record of the use of plants as medicine is mentioned in the *Rigveda*.

Maharashtra is the rich repository of medicinal wealth, which occupies an important place in vedic literature. Rich diversity of medicinal plants occurs in the Maharashtra growing naturally under diverse environmental conditions.

Parner Tahasil has an area of 1,77,676 hectares of land abundant in natural resources including immense potential for medicinal plants. The Tahasil is situated at 800 m above sea level and average rainfall is 577 m. The Tahasil has hilly as well as plain area. No scientific and systematic inventory of available resources of medicinal plants to their quantitative, qualitative parameters and their regeneration under natural conditions are available. Though a some works (Aher *et al.*, 2004) have been carried our relating to the study of few medicinal plants of the area, no exhaustive work in this valuable aspect have been so far studied.

The present study was undertaken to enlist, identify and quantity the important medicinal plants of the Parner Tahasil and gather information regarding the plant part used as medicine.

Materials and Methods

The study was carried out by adopting the methodology of Jain (1989). Ten field visits, each lasting for 2–3 days and spread over eight months (from June 2004 to January 2005) were undertaken. Information regarding medicinal plants, vernacular name and their part used were collected from local people also. Voucher specimens were collected, dried and mounted following conventional methods and preserved. Plants were identified taxonomically by using flora of Bombay Presidency (Cooke, 1901–1908) and Flora of Marathwada (Naik, 1998).

Results and Discussion

In the present study a total 170 species of medicinal plants were collected which belong to 58 different families of plant kingdom. Of these 170 species, 12 species belong to family Cucurbitaceae, 11 species to family Papilionaceae, 9 species to Asteraceae, 8 species to Apiaceae, 7 species to Caesalpiniaceae and remaining to other families. In 49 plant species leaves are used as medicines, in 41 species roots are used as medicines, in 15 species roots, in 14 species flowers where as in 9 species whole plant body is medicinally important (Table 14.1). Auti *et al.* (2004) recorded 50 medicinally important weeds from Shrirampur Tahasil, Dist. Ahmednagar. The uses, chemical constituents and the ayurvedic preparations made from many plants have also been given by Kaushik (1988) and Kaushik and Dhiman (2000).

Table 14.1

Sl.No.	Botanical Name	Ver. Name	Family	Plant Part Used
1.	*Annona squamosa*, L.	Sitaphal	Annonaceae	Seed
2.	*Annona reticulata*, L.	Ramphal	Annonaceae	Seed
3.	*Polyalthia longifollia*, L.	Ashok	Annonaceae	Leaf
4.	*Brassica compestris*, L.	Mohri	Cruciferae	Seed
5.	*Rhaphanus sativus*, L.	Mula	Cruciferae	Root
6.	*Datura strumarium*, L.	Dhotra	Solanaceae	Seed
7.	*Capsicum annum*, L.	Mirchi	Solanaceae	Seed
8.	*Nicotiana tobaccum*, L.	Tambakhu	Solanaceae	Leaf
9.	*Catharanthus roseus*, L.	Sadafuli	Apocynaceae	Flower
10.	*Nerium indicum*, L.	Kanher	Apocynaceae	Root
11.	*Eucalyptus globolus* L.	Nilgiri	Myrtaceae	Leaf

Contd...

Table 14.1–Contd...

Sl.No.	*Botanical Name*	*Ver. Name*	*Family*	*Plant Part Used*
12.	*Syzygium aromaticum*, L.	Lavang	Myrtaceae	Fruit
13.	*Cassia tora*, L.	Tarwad	Caesalpinaceae	Leaf
14.	*Syzygium cumini*, L.	Jambhul	Myrtaceae	Root, Seed
15.	*Tamarindus indica*, L.	Chinch	Caesalpinaceae	Fruit
16.	*Baulinia racemosa*, L.	Aapata	Caesalpinaceae	Leaf
17.	*Delonix resia*, L.	Gulmohar	Caesalpinaceae	Flower
18.	*Ocimum sanctum*, L.	Tulas	Labiatae	Whole plant
19.	*Ocimum bacilium*, L.	Subja	Labiatae	Whole plant
20.	*Mentha viridis*, L.	Pudina	Labiatae	Leaf
21.	*Accacia nilotica*, L.	Babhul	Mimoceae	Stem
22.	*Accacia catecheu*, L.	Kat	Mimoceae	Stem
23.	*Accacia concina*, L.	Shikakai	Mimoceae	Fruit
24.	*Citrus sinesis*, L.	Mosambi	Rutaceae	Fruit
25.	*Aegel marmalos*, Corr.	Bell	Rutaceae	Fruit
26.	*Citrus aurantifolia*, L.	Limbu	Rutaceae	Fruit
27.	*Citrus reticulata*, L.	Santri	Rutaceae	Fruit
28.	*Ferobia limoni*, L.	Kawath	Rutaceae	Fruit
29.	*Murraya konegii*, L.	Kadhipatta	Rutaceae	Leaf
30.	*Adhatoda vasica*, L.	Adulsa	Acanthaceae	Whole plant
31.	*Ricinus communis*, L.	Erand	Euphorbiaceae	Fruit
32.	*Phyllanthus emblica*, L.	Awala	Euphorbiaceae	Fruit
33.	*Barleria prionitis*, L.	Katekoranti	Acanthaceae	Fruit
34.	*Asparagus racemosa*, L.	Shatawari	Liliaceae	Leaf, Root
35.	*Allium sepa*, L.	Kanda	Liliaceae	Stem
36.	*Allium sativum*, L.	Lasun	Liliaceae	Stem
37.	*Aloe vera*, L.	Korpphad	Liliaceae	Leaf
38.	*Cocus nucifera*, L	Nariyal	Palmae	Fruit
39.	*Areca catecheu*, L.	Supari	Palmae	Fruit
40.	*Phoenix sylvestris*, L.	Sindhi	Palmae	Fruit
41.	*Sesbania sesban*, L.	Shivari	Papilionaceae	Leaf
42.	*Dalbergia sisoo*, Roxb.	Shisam	Leguminoceae	Leaf
43.	*Azadiracta indica*, L.	Kadunimb	Meliaceae	Whole plant
44.	*Lowsonia inermis*, L.	Mehandi	Lowsoniaceae	Leaf
45.	*Santalum album*, L.	Chandan	Santalaceae	Stem
46.	*Michelia champaka*, L.	Sonchafa	Magnoliaceae	Flower
47.	*Vitex nigunda*, L.	Nirgudi	Verbenaceae	Leaf
48.	*Eclipta alba*, L.	Maka	Asteraceae	Leaf

Contd...

Table 14.1–Contd...

Sl.No.	Botanical Name	Ver. Name	Family	Plant Part Used
49.	*Cynadon dactyolon,* L.	Durva	Poaceae	Whole plant
50.	*Calotropis procera,* R. Br.	Rui	Asclepidaceae	Leaf
51.	*Tridax prochumbers,* L.	Kurmudi	Asteraceae	Leaf
52.	*Lantana camara,* L.	Tantani	Verbenaceae	Leaf
53.	*Curcuma longa,* L.	Halad	Guttiferae	Stem
54.	*Arbus precatorius,* L.	Gunj	Papilionaceae	Leaf
55.	*Gyamopsis tetragonoloba,* L.	Gawar	Papilionaceae	Fruit
56.	*Trigonella foenum graeum,* L.	Methi	Papilionaceae	Seed
57.	*Ficus benegalensis,* L.	Wad	Urticaceae	Stem
58.	*Seasamum indicum,* L.	Til	Pedaliaceae	Seed
59.	*Coriandrum sativum,* L.	Kothimbir	Apiaceae	Seed
60.	*Helianthus anus,* L.	Sauryaphul	Asteraceae	Seed
61.	*Daucus carota,* L.	Gajar	Asteraceae	Root
62.	*Calocasia antiguorum,* L.	Alu	Araceae	Leaf
63.	*Melia azadirach,* L.	Bakan	Meliacesae	Leaf
64.	*Zingiber officinale,* L.	Aale	Meliacesae	Rhizome
65.	*Carus copticum,* L.	Owa	Apiaceae	Seed
66.	*Pongamia pinnate,* L.	Karanji	Papilionaceae	Seed
67.	*Boerhavia repens,* L.	Punarnava	Nyctaginaceae	Seed
68.	*Prosopis julo flora,* L.	Wedi Babhul	Mimosae	Seed
69.	*Caesalpinia bondue,* L.	Sagargota	Leguminoceae	Fruit
70.	*Spirocia oleraceae,* L.	Palak	Chenopodiaceae	Leaf
71.	*Pithecolobium duke,* L.	Vilayatichinch	Papavaraceae	Fruit
72.	*Terminalia catap,* L.	Badaam	Rosaceae	Seed
73.	*Curcuma aromaticum,* L.	Ranhalad	Scitaminae	Stem
74.	*Drimia indica,* Roxb.	Rankanda	Liliaceae	Tuber
75.	*Ficus religiosa,* L.	Pimpal	Uritaceae	Leaf
76.	*Ficus racemosa,* L.	Umbar	Uritaceae	Fruit
77.	*Guizotia abyssinica,* L.	Karle	Cucurbitaceae	Fruit
78.	*Coccinia grandis,* L.	Tondali	Apiaceae	Fruit
79.	*Bacoppa mannieri,* L.	Brahami	Umbelifevae	Leaf
80.	*Ocimum gratissimum,* L.	Ramtulas	Labiatae	Whole plant
81.	*Sapindus trifolious,* L.	Ritha	Sapindaceae	Seed
82.	*Foenicum vulgare,* L.	Badishep	Apiacae	Fruit
83.	*Panicum miliaceum,* L.	Rala	Apiacae	Seed
84.	*Hibiscus syriacus,* L.	Paidhari	Malvaceae	Whole plant

Contd...

Table 14.1–Contd...

Sl.No.	Botanical Name	Ver. Name	Family	Plant Part Used
85.	*Spilanthes paniculata*, L.	Akkalkara	Asteraceae	Flower
86.	*Terminalia belleria*, L.	Behda	Combretaceae	Leaf
87.	*Myristica beddomei*, L.	Jayphal	Myristicaceae	Fruit
88.	*Calotropis gigantea*, L.	Pandhari Rui	Asclepidaceae	Leaf
89.	*Benincasa hispida*, L.	Kohala	Cucurbitaceae	Fruit
90.	*Bryophyllum pinnatum*, L.	Panfuti	Crassulaceae	Leaf
91.	*Carica papaya*, L.	Papai	Caricaceae	Fruit
92.	*Cassia fistula*, L.	Bahava	Caesalpiniacae	Seed
93.	*Celosia argaentea*, L.	Kurdu	Amaranthaceae	Root
94.	*Clitoria ternatea*, L.	Gokarn	Papilionaceae	Leaf
95.	*Cleome gynandra*, L.	Tilwan Pandhari	Cleomaceae	Leaf
96.	*Jasminum auriculatum*, Vahl.	Jui	Oleaceae	Flower
97.	*Jasminum officinale*, Vahl.	Jai	Oleaceae	Flower
98.	*Mimosa pudica*, L.	Lajalu	Mimosae	Leaf
99.	*Nyctanthus arbortristis*, L.	Prajakta	Oleaceae	Flower
100.	*Passiflora foetida*, L.	Krishna kamal	Passifloraceae	Flower
101	*Portulacca oleracea*, L.	Ghol	Portulaccaceae	Leaf
102.	*Sida cordifolia*, L.	Chikana	Malvaceae	Leaf
103.	*Tejetus erecta*, L.	Zendu	Asteraceae	Leaf
104.	*Terminalia arjuna*, Bedd.	Arjun	Combrataceae	Stem
105.	*Zizyphus jujuba*, Lamk	Bor	Rhamnaceae	Seed
106.	*Tinospora cardifolia*, Miers	Gulvel	Menispermaceae	Leaf
107.	*Solanum melanogena*, L.	Wangi	Solanaceae	Fruit
108.	*Butea monosperma*, L.	Palas	Papilionceae	Flower
109.	*Solanum tubersum*, L.	Batata	Solanaceae	Tuber
110.	*Solanum virginianum*, L.	Bhuiringani	Asteraceae	Leaf
111.	*Regia repens, L.*	Ghatipitipapada	Acanthaceae	Leaf
112.	*Adiantum philippenes*, L.	Adiantum	Adiantaceae	Leaf
113.	*Achyranthese aspera*, L.	Aaghada	Amaranthaceae	Root
114.	*Amaranthus spinosus*, L.	Katemath	Amaranthaceae	Root
115.	*Mangifera indica*, L.	Amba	Anacardiaceae	Whole plant
116.	*Carum curvi*, L.	Shahajire	Apiaceae	Fruit
117.	*Cuminum cyminum*, L.	Jire	Apiaceae	Fruit
118.	*Tabernaemontana coronaria*, R.Br.	Tagar	Apocynaceae	Seed
119.	*Thevetia peruviana*, Pers.	Pivalakanher	Apocynaceae	Root
120.	*Borassus flabelliafer*, L.	Tad	Arecaceae	Fruit

Contd...

Table 14.1–Contd...

Sl.No.	Botanical Name	Ver. Name	Family	Plant Part Used
121.	*Gymnema sylvestre,* R.Br.	Kawali	Asclepidaceae	Root
122.	*Cartnamus tinctorious,* L.	Karadi	Asteraceae	Fruit
123.	*Echinops echinatus,* Roxb.	Utkatar	Asteraceae	Root
124.	*Cordia dichotama,* F.	Bhokar	Caesalpinaceae	Leaf
125.	*Cassia sophera,* L.	Rantakalka	Caesalpinaceae	Bark
126.	*Chenopodium album,* L.	Chandanbatva	Chenopodiaceae	Leaf
127.	*Cleome viscosa,* L.	Ran-mohari	Cleomaceae	Seed
128.	*Quispualis densiflora,* Wall.	Lal-chemeli	Combrataceae	Leaf
129.	*Evolvululous hirsutus,* Lamk.	Shankpushpi	Convolvulaceae	Leaf
130.	*Benincasa cerifera,* Savi.	Khokali	Cucurbitaceae	Seed
131.	*Cucumis colocynthis,* L.	Kadu-wrundavan	Cucurbitaceae	Fruit
132.	*Cucumis melo,* L.	Kharbuj	Cucurbitaceae	Fruit
133.	*Cucumis sativus,* L.	Kakadi	Cucurbitaceae	Fruit
134.	*Cucurbita pepo,* L.	Dangar	Cucurbitaceae	Fruit
135.	*Piplocyclos palmatus,* Jeff.	Shivalingi	Cucurbitaceae	Leaf
136.	*Lagenaria siceraria,* L.	Kadubhopala	Cucurbitaceae	Root
137.	*Luffa cylindrica,* Roxb.	Dodaka	Cucurbitaceae	Leaf
138.	*Cyperus scariosus,* R.Br.	Lawhale	Cyperaceae	Tuber
139.	*Diospyros peregrina,* L.	Temburni	Ebenaceae	Bark
140.	*Arachis hypogea,* L.	Bhuimoong	Papilionaceae	Root
141.	*Cajanus cajan,* Linn.	Tur	Papilionaceae	Seed
142.	*Cicer aurientum,* L.	Harbara	Papilionaceae	Seed
143.	*Erythrina indica,* Lamk.	Pangara	Papilionaceae	Bark
144.	*Abelmoschus esculentus,* L.	Bhendi	Malvaceae	Fruit
145.	*Abutilon indicum,* L.	Mudraa	Malvaceae	Stem
146.	*Gossypium arboreum,* L.	Kapasi	Malvaceae	Leaf, Seed
147.	*Cocculus cordifolius,* DC.	Amrutvel	Menispermaceae	Stem
148.	*Musa paradisiaca,* L.	Kel	Musaceae	Whole plant
149.	*Psidium guavajava,* L.	Peru	Myrtaceae	Fruit
150.	*Jasminum sambac,* Ait.	Mogara	Oleaceae	Flower
151.	*Cymbopogon citatus,* DC.	Gavati chaha	Poaceae	Whole plant
152.	*Sorghum bicolor,* L.	Lal-jowari	Poaceae	Seed
153.	*Clematis triloba,* Heyle ex. Roth	Ranjani	Rananculaceae	Root
154.	*Rosa damasana,* Mill.	Gulab	Rosaceae	Flower
155.	*Cyperus rotundus,* L.	Nagarmotha	Cyperaceae	Leaf
156.	*Tribulus terrestris,* L.	Sarata	Zygophyllaceae	Leaf

Contd...

Table 14.1–Contd...

Sl.No.	Botanical Name	Ver. Name	Family	Plant Part Used
157.	*Peucedanum gareolens*, L.	Balant Shepu	Apiaceae	Fruit
158.	*Hibiscus syriacus*, L.	Pandharijaswand	Malavaceae	Flower
159.	*Periploca aphylla*, L.	Ransher	Asclepidaceae	Stem
160.	*Citrulus colocynthi*, L.	Indrayan	Cucurbitaceae	Fruit
161.	*Vitiveria zizanioides*, L.	Wala	Poaceae	Flower
162.	*Boerrhavia diffusa*, L.	Getuli	Plantangenaceae	Root
163.	*Frerea indica*, L.	Makadi	Asclepidaceae	Stem
164.	*Solanum incanum*, L.	Karanda	Malvaceae	Leaf
165.	*Evolvulus alsinoides*, L.	Vishnukranta	Convolvulaceae	Leaf
166.	*Bacopa mannieri*, L.	Bramhi	Scrophulariaceae	Leaf
167.	*Abelmoschus ficulmeus*, L.	Ranbhendi	Malvaceae	Fruit
168.	*Girardinia diversifolia*, L.	Agaya	Urticaceae	Root
169.	*Datbergia latifolia*, L.	Beetee	Leguminoceae	Fruit
170.	*Tephrosia humiltonii*, L.	Unhali	Leguminoceae	Leaf

From the present study it indirectly becomes evident that:

1. Cucurbitaceae, Papilionaceae and Asteraceae families are rich in medicinal plants.
2. The density of those plants where roots or tubers are used as medicine is low.
3. Lack of scientific knowledge, has resulted in severe depletion of herbs and shrubs and severe biotic pressure of humans and grazing animals is detrimental to the regeneration of these plants.
4. Scientific studies may elaborate the prospects of growing more and more medicinal plants successively.

References

Aher, R.K., Auti, B.K., Deshmukh, R.N. and Borkar, G.B., 2004. Survey of medicinal plants from Parner (Ahmednagar District.). *Ad. Plant Sci.*, 17(1): 57–59.

Auti, B.K., Pingle, S.D., and Aher, R.K., 2004. Survey of weeds and their medicinal value from Shrirampur Tahasil (Ahmednagar District M.S.). *Ad. Plant Sci.*, 17(11): 395–401.

Cooke, T., 1901-1908. *Flora of Presidency of Bombay*, Vols. I–III. Adlard and Sons, Ltd., London.

Jain, S.K. (Ed.), 1989. *Methods and Approaches in Ethnobotany*. Society of Ethnobotanists, CDRI, Lucknow.

Kaushik, P., 1988. *Indigenous Medicinal Plants Including Microbes and Fungi*. Today and Tomorrow's Printers and Publishers, New Delhi.

Kaushik, P. and Dhiman, A.K. 2000. *Medicinal Plants and Raw Drugs of India*. Bishan Singh Mehendra Pal Singh, Dehra Dun.

Naik, V.N., 1998. *Flora of Marathwada*, Vol. I. Amrut Prakashan, Aurangabad.

Naik, V.N., 1998. *Flora of Marathwada*, Vol. II. Amrut Prakashan, Aurangabad.

Chapter 15

Some Common Medicinal Plants Used by Koch (Tribal) People of Goalpara District of Assam

A. Rahman[1], S.K. Basumatary[2] and M. Ahmed[2]

[1]*Department of Sericulture, Government of Assam, Goalpara – 783 101, Assam, India*

[2]*Department of Botany, Goalpara College, Goalpara, Assam*

ABSTRACT

A survey of some common medicinal plants has been carried out in different places of Goalpara District, which are used in various diseases and ailments. This paper deals with some chemical constituents of 37 numbers of plant species, which have been used as medicine for several common diseases among the Koch (tribal) people of Goalpara, Assam.

Keywords: *Ethnobotany, Phyto-chemical, Medicinal plants, Plant parts, Koch people, Goalpara.*

Introduction

Medicinal plants have curative properties due to the presence of various complex chemical substances of different composition, which are found as secondary plant metabolites in one or more parts of these plants. These plant metabolites, glycosides, corticosteroids, essential oils etc (Purohit, Prajapati, Sharma and Kumar, 2003).

A number of workers have drown attention towards the threatened medicinal plants of India (Jain and Shastry, 1980; Shah, 1975) and (Raghav, 2003) in the field of Chemotaxonomy and Anonymous, 1994; Jain and Borthakur, 1980; Baishya and Majumder, 1980; Basumatary *et al.*, 2004; Gogoi and Borthakur, 1991; Hajra and Baishya, 1981; Jamir, 1997 in the field of ethnobotanical research. The North Eastern Region is considered to be one of the rich floristic region for medicinal plants.

Table 15.1: List of Plant Constituents and Uses of their Plant Parts by the Koch (Tribal) People of Goalpara District

Sl.No.	Scientific Name	Family	Local Name	Chemical Principle	Plant Parts	Uses
1.	*Acorus calamus* L.	Araceae	Gondrof	Asarone, β-isomer	Rhizome	Depression and Mental disorder
2.	*Adhatoda vasica* Nees.	Acanthaceae	Taosengam	Vasicinolone, Vasicol, Peganine	Roots	Bronchitis, Bronchial, Asthma
3.	*Allium cepa* L.	Liliaceae	Piao	Arginine, Mistidine, Isoleucin	Bulb	Hemorrhoids, Dysentery, Asthma
4.	*Allium sativum* L.	Liliaceae	Rushun	Allicin, Diallyl, Disufide	Bulb	Cough, Bronchitis, Pertussis
5.	*Alocasia macrorrhiza* L. schott.	Araceae	Ola	β-sitosterol composterol	Leaves	Impetigo snake-bite
6.	*Amaranthus spinosus* L.	Amaranthaceae	Katakhunna	β-sitosterol, Stigmasterol	Leaves	Hyperdipsia, Leprosy
7.	*Argemon maxicana* L.	Papaveraceae	Sampokkata	Berberine and Protopine	Seed	Asthma and toothache
8.	*Artocarpus heterophyllus* Lam.	Moraceae	Pansung	Dihydromarin, Artocarpin, Artocarpanone	Hard wood	Fever, Boils, wound, Skin diseases
9.	*Averrhoa carambola* L.	Oxalidaceae	Kamrenga	Vitamin-C, Oxalic acid	Leaves	Scabies, Prurient
10.	*Azadirachta indica* A Juss.	Meliaceae	Neem	Tyrosine, Proline	Leaves	Leprosy, Diabetes, Skin diseases
11.	*Cannabis sativa* L.	Cannabinaceae	Bhung	Cannabinodiol, cannabitetrol	Leaves	Convulsion, Diarrhoea
12.	*Carica papaya* L.	Caricaceae	Madhu	Cryptoxanthim, Cryptoflavin	Fruits	Stomachache, Appetizer
13.	*Cassia fistula* L.	Cesalpiniaceae	Zingangpan	Crysophanol, Rhein	Fruits	Tongue sore
14.	*Centella asiatica* (L.) Urban	Apiaceae	Manik	Kaempferol, Vitamin C, Asiatic acid	Whole plant	Constipation, Jaundice, Dysentery
15.	*Cinnmonum tamata* Neem and Ebrem	Lauraceae	Tezpat	Camphora cadinene	Leaves	Cardiac disease
16.	*Cissus quadrangularis* L.	Vitaceae	Karenjora	β-silesterol, δ-amyrin	Stem	Skin disease, Leprosy
17.	*Citrus aurantium* L.	Rutaceae	Zamburi	β-sitastarol, Stigmasterol	Leaves	Urinary tract infections
18.	*Curcuma aromatica* Salisb	Zingiberaceae	Dungai	β-carcumene Xanthorhizol	Rhizomes	Bronchitis cough
19.	*Cynodon dactylon* (L) Pers.	Poaceae	Dubla bon	Phytol, Stigmasterol acetate	Whole plant	Skin disease, Vomiting, Diarrhoea
20.	*Datura stramonium* L.	Solanaceae	Dutura	Flovonoids	Leaves	Rheumatic pains
21.	*Dioscorea bulbifera* L.	Dioscoreaceae	Khandukum	Arginine, Choline, Sterols	Tubers	Diabetes, Diarrhoea
22.	*Emblica officinalis* Goerin	Euphorbiaceae	Bonparam	Vitamin-C	Fruits	Scurvy, Swollen
23.	*Eupatorium odoratum* L.	Asteraceae	Murmuripan	P-cymene, Eupatoriopirin	Roots	Antiviral, Anticancer
24.	*Holarrhena antidysenterica*	Apocynaceae	Mafapan	Holarrhine, Holarrhenine	Trunk bark	Antidysaentric, Diarrhoea

Contd...

Table 15.1–Contd...

Sl.No.	*Scientific Name*	*Family*	*Local Name*	*Chemical Principle*	*Plant Parts*	*Uses*
25.	*Leucus aspera* Sprang	Lamiaceae	Kansipa	Glucoside	Leaves	Arthralgia
26.	*Mimosa pudica* L.	Mimosaceae	Kantapan	Norepinepherine, Gentisic acid	Leaves	Haemorrhoids, Urinary
27.	*Momordica charanita* L.	Cucurbitaceae	Khanowali	Spinasterol, Stigmasterol, Traxerol	Fruit	Leprosy, Malignant ulcers
28.	*Ocimum sanctum* L.	Lamiaceae	Tulsipan	Eugenol, Carvacrol, Linalool	Whole plant	Fever, Cough
29.	*Oroxylon indicum* (L) Bents. Ex Kurz.	Bignoniaceae	Naokaful	Orixyline-A, Baicalein, Chrysin	Barks, Flower	Bronchitis, Vomiting, Neuralgia
30.	*Oxalis corniculata* L.	Oxalidaceae	Shisak	Vitexin, Isovitexin	Leaves	Induration of the breasts.
31.	*Paederia foeida* L.	Rubiaceae	Pasung	α-paederine, β-paederine	Leaves	Rheumatism
32.	*Raulvolfia serpentina* (L) Benth, ex Kurz	Apocynaceae	Chandotia	Serpantine, Serpantinine, Rauwolfinine	Leaves	Rheumatism
33.	*Ricinus communis* L.	Euphorbiaceae	Bherna	Ricinine, Kaempferol, Flycoside	Seeds	Breast tumers, Boil
34.	*Saccharum officinarum* L.	Poaceae	Kosher	Compestrol, Coumarine	Stem	Sore eyes
35.	*Solanum indicum* auct. non. L.	Solanaceae	Khangka	Astringent	Roots	Skin disease, Leprosy
36.	*Tamarindus indica* L.	Caesalpiniaceae	Tengtili	Flaronoides, Vitexin, Homoorientin	Leaves	Jaundice, Scabis tumers
37.	*Terminalia arjuna* (Roxb Ex DC) Wright & Arn.	Combretaceae	Arjun	Arjunine, Laactone, Arjunetin	Bark	Diabetes, Anaemia, Athma

Materials and Methods

During the survey period (2002–03) the medicinal plants were collected from Matia, Lakhipur, Kochdhowa etc. of Gaolpara district where the Koch (tribal) are the dominated, adopted by the techniques Jain (1964, 1967). Schultes (1964) for ethnobotanical research was followed. Local names, important medicinal plants given by medical man and elderly persons were the source for recording the information. Local name and botanical names were identified at the herbarium of Botanical Survey of India (B.S.I) Shillong, medicinal uses of roots, rhizomes, barks, bulbs, leaves, flowers, fruits and seeds are enumerated in Table 15.1.

Results and Discussion

Total 37 numbers of species of plants from 27 families (Table 15.1) were recorded which are used medically by the Koch people of Goalpara District. Some of the herbal medicines have been found to be very effective for the cure of the disease of the female and children and thus can play an important role. Large number of herbal species has been reported to prove curative effects for the common specific disease among the women and children (Sinha, 1998; Bhattacharjee, 1999). The increasing demand of medicinal plants has resulted in the rapid dwindling of these natural resources mainly for deforestation. Activities related to land degradation, primarily those addressing deforestation and desertifiaction as they relate to the local areas, are also eligible for funding. Along with the United Nation Development Programme and the United Nation Environment Programme the World Bank is an implementing agency Global Environment Facility (GEF) (Purohit, Prajapati, Sharma and Kumar, 2003).

Acknowledgement

We are thankful to Dr. P.J. Hindique, Reader Department of Biotechnology Gauhati University Guwahati and Dr. S. Phukan, Deputy Director, Botanical Survey of India, Shillong for herbarium and library facilities while writing this paper. We thank Mr.Manob Koch, Gojen Koch, Rajib Koch, for encouragement and help provided during the study period.

References

Anonymous, 1994. *A Handbook of Folklore Material of North East India.* ABILAC, Guwahati.

Baishya, C.L. and Majumder, R., 1980. Folklore claims from the Brahmputra Valley, Assam. *Enthomedicine,* 6: 139–145.

Basumatary, S.K., Ahmed, M. and Deka, S.P., 2004. Some medicinal plant leaves used by Boro (Tribal) people of Golpara district, Assam. *Natural Product Radience,* 3(2): 88–90.

Bhattacharjee, S.K., 1999. *Handbook of Medicinal Plants.* Pointer Publishers.

Chandal, K.P.S., Shukla, G. and Neelam Sarma, 1996. *Biodiversity in Medicinal and Aromatic Plants in India: Conservation and Utilization.* NBPGR, New Delhi, pp. 1–239.

Gogoi, P. and Barthakur, S.K., 1991. Plants Religio-culture belief of Tai-Khumtis of Assam (India). *Enthobotany,* 3(182): 89–95.

Hajra, P.K. and Baishya, A.K., 1981. Ethnobotanical notes on the Miris (Mishings) of Assam Plains. In: *Glimpse of India Ethnobotany,* (Ed.) S.K. Jain. pp. 161–169.

Jain, S.K. and Barthakur, S.K., 1980. Ethnobotany of the Mikirs of India. *Elon Bot.,* 34: 264–272.

Jain, S.K. and Sastry, A.R.K., 1980. Threatened plants of India: A state of the art Report, Botanical Survey of India, Howrah.

Jamir, N.S., 1997. Ethnobotany of Naga Tribes in Nagaland: 1-Medicinal herbs. *Ethnobotany*, 9: 101–104.

Prajapati, N.D., Purohit, S.S., Sharma, A.K. and Kumar, T., 2003. *A Handbook of Medicinal Plants*. Agrobios (India).

Raghav, C.S., 2003. Medicinal plants on the touchstone of science. *Science Reporter*, pp. 52–53.

Schultes, R.E., 1964. The role ethnobotanical search for new medicinal plants. *Lioydia*, 25(4): 257–266.

Shah, N.C., 1975 Prospect of botanical drug from hill district of Uttar Pradesh. *Indian Drugs*, 12(11): 17–20.

Sinha, R.K., 1996. *Ethnobotany: The Rennaissance of the Traditional Herbal Medicine*. Ina Shree Publishers, Jaipur.

Chapter 16

Quantification of Organic Matter Fractions in Sewage Biosolid Composts

T. Chitdeshwari
Department of Soil Science and Agricultural Chemistry, Tamil Nadu Agricultural University, Coimbatore – 641 003, Tamil Nadu

ABSTRACT

Composting experiment was conducted to evaluate the suitability of sewage biosolid composts for land application with two structuring agents *viz.*, Coir pith (CP) and Green leaf manure (GLM). The results indicated that, the coirpith compost had higher values for the various organic matter fractions which might be due to the high lignin content of the material. The increase in fulvic and humic acid contents was noticed with increasing ratios of sewage sludge addition. The humic acid and Fulvic acid ratio (HA/FA ratio) for 2 : 1 ratio of coir pith compost was 1.80 while for GLM it was 1.97 which showed their suitability for land application.

***Keywords**: Sewage biosolid composts, Organic matter fractions.*

Introduction

The mounting production of sewage biosolid due to population explosion and industrial activity of the country has resulted in the accumulation of waste products which inturn leads to health hazards and environmental problems. The high organic matter content and useful plant nutrients particularly N, P and trace elements in the biosolid made it as a substitute for chemical fertilisers in crop production which was a useful alternative to dumping and incineration. Its use in agriculture will enrich the soil with organic matter and plant nutrients, which in turn will have a beneficial effect on growth and crop yield. Hence to know the suitability of the sewage biosolid compost application to the soil, a composting experiment was carried out with two different structuring agents.

Materials and Methods

Composting experiment was conducted with sewage biosolid and two structuring agents *viz.*, Coirpith and Green leaf manure (Glyricidia) at different ratios (1.0 : 0.5, 1.0 : 1.0, 1.0 : 1.50 and 1.0 : 2.00 ratio of structuring agents and sewage sludge). The moisture content of the compost was maintained at 60 per cent water holding capacity and turned regularly. The aerobic composting was carried out for 60 days and the samples were drawn. The samples were analysed for various fractions of organic matter *viz.*, fulvic acid, humic acid and their ratio since they are the most important chemical criterion in defining the compost maturity. The procedure outlined by Stevenson (1965) was employed for quantifying the organic matter fractions. The results were furnished in Tables 16.1 and 16.2.

Table 16.1: Effect of Sewage Biosolids Composting on the Changes in the Fractions of Organic Matter on 60th Day (%)

Ratios	*Humic Acid (% of OM)*				*Fulvic Acid (% of OM)*			
	S_1	S_2	S_3	*Mean*	S_1	S_2	S_3	*Mean*
0.5 : 1.0	1.20	0.466	0.624	0.572	0.554	1.39	1.02	1.20
1.0 : 1.0	1.12	0.541	0.899	0.677	0.706	1.72	1.44	1.43
1.5 : 1.0	1.06	0.486	1.18	0.805	0.824	2.07	1.69	1.61
2.0 : 1.0	1.13	0.513	1.27	0.968	0.917	2.28	1.91	1.77
Mean	**1.13**	**1.502**	**0.993**	**0.756**	**0.750**	**1.87**	**1.52**	**1.50**
		R	S	R × S	R	S	R × S	
	SE_d	0.10	0.08	0.17	0.06	0.06	0.11	
	CD (0.05)	0.20	0.17	0.35	0.13	0.11	0.23	

Table 16.2: Effect of Sewage Biosolids Composting on the Changes in the Fractions of Organic Matter on 60th Day (%)

Ratios	*OM (%)*				*HA/FA Ratio*			
	S_1	S_2	S_3	*Mean*	S_1	S_2	S_3	*Mean*
0.5 : 1.0	19.4	28.0	21.9	23.1	2.58	2.24	1.79	2.20
1.0 : 1.0	19.4	31.3	24.5	25.1	2.07	1.93	2.40	2.13
1.5 : 1.0	18.1	.32.9	26.1	25.7	2.18	1.88	2.09	2.05
2.0 : 1.0	18.8	33.7	27.7	26.7	2.22	1.80	1.97	2.00
Mean	**18.9**	**31.5**	**25.1**	**25.2**	**2.26**	**1.96**	**2.06**	**2.09**
			R		S		R × S	
		SE_d	1.37		1.41		2.37	
		CD (0.05)	NS		2.81		NS	

Results and Discussion

The humic substances *viz.,* humic acid and fulvic acid contents are most important chemical criterian in defining the compost maturity as reported by Bernal *et al.* (1998). The humic and fulvic acids of the composts collected at 60th day revealed the presence of high content of humic acid and low contents of fulvic acid. The humic and fulvic acid extracted with composts revealed that, the coir pith compost had higher values due to the high lignin content of the material. The increase in fulvic and humic acid contents with higher ratios could be attributed to the higher quantity of sludge addition. Besides, the increasing microbial load of the sludge compost materials might have influenced the humification process.

The values obtained for humic acid, fulvic acid and humic acid/fulvic acid ratios for composts prepared with two structuring agents in different ratios were critically examined. Increasingly the humic acid and fulvic acid contents were high in the coir pith based sludge compost than the GLM compost. Whereas the humic acid/fulvic acid ratios were high in the latter except for 0.5 : 1.0 ratio. However, it is worth to state at this point that HA/FA ratio for 2 : 1 ratio of coir pith compost was 1.80 while for GLM it was 1.97 which were very much near to the proposed humic acid/fulvic acid (1.90) by Jimenez and Garcia (1992).

References

Bernal, M.P., C. Paredes, M.A. Sanchez-Monedero and J. Cegarra, 1998. Maturity and stability parameters of composts prepared with a wide range of organic wastes. *Bioresource Tech.,* 63: 91–99.

Jimenez, I.E. and P.V. Garcia, 1992. Determination of maturity indices for city refuse composts. *Agric. Eco. Systems Environ.,* 38: 331–343.

Stevenson, F.J., 1965. Gross chemical fractionation of organic matter. In: *Methods of Soil Analysis Part II,* (Eds.) C.A. Black, D.D. Evans, J.L. White, L.E. Ensminger and F.E. Clark. American Society of Agronomy, Madison, Wisconsin, pp. 1409–1421.

Chapter 17

Influence of Vermicompost on the Growth and the Secondary Metabolites of Selected Medicinal Plants

D. Srinivasa Rao[1], *Indira A. Jeyaraaj*[2], *R. Jeyaraaj*[3] *and M. Lakshmi Prabha*[1]
[1]*Ph.D. Scholar in Biochemistry,* [2]*Reader in Biochemistry,* [3]*Reader in Zoology*
Kongunadu Arts and Science College, Coimbatore – 641 029, Tamil Nadu, India

ABSTRACT

Vermicompost is extremely homogenous and potential organic manure for plant growth. Our study was aimed to find out the effect of vermicompost at different concentrations (25 per cent, 50 per cent, 75 per cent, 90 per cent and 100 per cent) on plant growth and the content of medicinally important secondary metabolites like total phenols, tannins, alkaloids, flavonoids and steroids in *Ocimum sanctum*, Linn; *Coleus aromaticus*, Benth; *Glycyrrhiza glabra*, Linn; and *Alpinia galanga*, Sw; Among the various concentration of vermicompost taken, vermicompost at the concentration of 75 per cent recorded the maximum growth. 100 per cent vermicompost did not support the growth of plant. However, the concentration of secondary metabolites like total phenols, tannins, alkaloids, flavonoids and steroids were higher in 90 per cent vermicompost.

Keywords: *Vermicompost, Plant growth, Secondary metabolites.*

Introduction

Vermicompost is rich source of macro and micronutrients, vitamins, minerals, thus it improves soil structure, physico-chemical properties and act excellent soil amendments (Slater and Hopp,

1947). Earthworms have beneficial physical, biological and chemical effects on soils and can increases plant growth and crop yields (Hiranmai Yadav and Vijayakumari, 2004). These beneficial effects have been attributed to improvements in soil properties and structure, to greater availability of mineral nutrients to plants and to biological active metabolites acting as plant growth regulators (Gajalakshmi and Abbasi, 2004). Earthworms fragment organic waste substrates, stimulates microbial activity greatly and increases rates of mineralization, rapidly converting the wastes in to humus-like substances with a finer structure than composts but possessing a greater and more diverse microbial activity (Ehler's 1975). There have been suggestions that the dramatic increases microbial activity in organic matter by earthworms could result in production of significant quantities of plant growth regulators such as IAA, gibberellins and cytokines (Edwards and Neuhauser, 1988). Vermicompost affects the amounts of primary and secondary metabolites in plants (Julka and Paliwal, 2005). The use of plants as medicine is as old as human civilization. Medicinal plants have their values in the secondary metabolites present in various plant tissues. Secondary metabolites are biosynthetically derived from primary metabolites and they do not have no obvious role in plants primary metabolism, secondary metabolites often play ecologically significant roles in how plants deal with their environment and therefore important in their ultimate survival (Brattsten, 1979). The biological forms of these secondary metabolites make them so useful as pharmaceutical. But these secondary metabolites are produced in much smaller quantities and various costly techniques like tissue culture etc, are used to improve the synthesis of secondary metabolites in medicinal plants. The focus of this study is to determine how far the cheapest, eco-friendly vermicompost application influences the secondary metabolite synthesis in selected medicinal plants like *Ocimum sanctum*, Linn; *Coleus aromaticus*, Benth; *Glycyrrhiza* Linn; and *Alpinia galanga*, Sw.

Materials and Methods

Preparation of Vermicompost

Earthworm *Eudrilus eugineae* was collected from Aarthi Farms, Kondegoundampalayam Village, Pollachi Taluk, Coimbatore, Tamilnadu, India. The earthworms were cultured at premises of Kongunadu Arts and Science College, Coimbatore, Tamilnadu, India for six months and were used for the decomposition of campus and hostel wastes, the resultant vermicompost was used for the subsequent studies. The different concentrations of vermicompost (25 per cent, 50 per cent, 75 per cent and 90 per cent) were prepared and applied to the selected medicinal plants viz, *Ocimum sanctum*, Linn; *Coleus aromaticus*, Benth; *Glycyrrhiza glabra*, Linn; and *Alpinia galanga*, Sw and find the effect on the growth and secondary metabolite synthesis.

Measurement of the Growth Parameters and the Secondary Metabolites Content

The growth of selected plants were measured by various growth parameters namely number of leaves, internodes, leaf length and width. The secondary metabolites *viz.*, total phenols, tannins, flavonoids, alkaloids and steroids were assayed according to methods described as by Zakaria (1991).

Statistical Analysis

Values were represented as mean ± SD. Data were analyzed using one-way analysis of variance (ANOVA) and group means were compared using DMRT. P values 1 per cent, 5 per cent were considered significant.

Results and Discussion

Medicinal plants are highly esteemed for millennia as a rich source of therapeutic agents for prevention and treatment of various ailments. The secondary metabolites such as total phenols, tannins,

Table 17.1: Growth and Content of Secondary Metabolites in *Osicum sanctum* on Varying Concentration of Vermicompost

Composition of Vermicompost	*Growth Parameters*				*Secondary Metabolites*				
	No. Leaves	*Internodes*	*Leaf Length (cm)*	*Leaf Width (cm)*	*Alkaloids (µg/g)*	*Steroids (µg/g)*	*Flavanoids (µg/g)*	*Tannins (µg/g)*	*Total Phenols (µg/g)*
Control	17c	22e	2.76±0.25a	1.30±0.20c	2.86±0.11e	0.158±0.008e	0.314±0.009e	0.339±0.009e	0.165±0.007e
25%	21bc	27d	3.06±0.11a	1.30±0.10bc	3.99±0.02d	0.277±0.066d	0.473±0.008d	0.555±0.014d	0.221±0.012d
50%	25b	32c	3.25±0.19a	1.53±0.05abc	4.81±0.09c	0.335±0.007c	0.683±0.020c	0.687±0.012c	0.363±0.007c
75%	33a	48c	3.35±0.25a	1.63±0.05ab	5.77±0.16b	0.478±0.009b	0.835±0.005b	0.779±0.012b	0.442±0.011b
90%	36a	52b	3.53±0.30a	1.80±0.10a	6.63±0.92a	0.683±0.013a	0.975±0.006a	0.883±0.011a	0.684±0.007a

Data represent an average of 3 replicates in two repetitive experiments.

Control: Red loamy soil

Treatment: Vermicompost supplemented to control at 25 per cent, 50 per cent, 75 per cent and 90 per cent.

Statistics: Mean ± SD, ANOVA (DMRT)

***P < 0.001, **P < 0.05, significant levels.

Table 17.2: Growth and Content of Secondary Metabolites in *Glycyrrhiza glabra* on Varying Concentration of Vermicompost

Composition of Vermicompost	*Growth Parameters*				*Secondary Metabolites*				
	No. Leaves	*Internodes*	*Leaf Length (cm)*	*Leaf Width (cm)*	*Alkaloids (µg/g)*	*Steroids (µg/g)*	*Flavanoids (µg/g)*	*Tannins (µg/g)*	*Total Phenols (µg/g)*
Control	71c	27b	21.4±1.15e	0.90±0.10e	1.38±0.53e	0.213±0.001e	0.539±0.009e	0.174±0.006e	0.278±0.007e
25%	75c	28b	25.3±0.66d	1.63±0.15d	1.86±0.14d	0.414±0.116d	0.675±0.006d	0.255±0.001d	0.310±0.001d
50%	77c	29b	28.7±0.25c	2.12±0.11c	2.58±0.11c	0.480±0.008c	0.745±0.005c	0.352±0.009c	0.513±0.001c
75%	85b	32a	31.9±l.80b	2.16±0.23b	3.81±0.09b	0.571±0.050b	0.815±0.007b	0.441±0.016b	0.678±0.006b
90%	116a	32a	38.2±l.05a	3.53±0.25a	5.45±0.06a	0.870±0.006a	0.989±0.006a	0.597±0.021a	0.822±0.002a

Data represent an average of 3 replicates in two repetitive experiments.

Control: Red loamy soil

Treatment: Vermicompost supplemented to control at 25 per centm 50 per cent, 75 per cent and 90 per cent.

Statistics: Mean ± SD, ANOVA (DMRT)

***P < 0.001, **P < 0.05, significant levels.

Table 17.3: Growth and Content of Secondary Metabolites in *Coleus aromaticus* on Varying Concentration of Vermicompost

Composition of Vermicompost	Growth Parameters				Secondary Metabolites				
	No. Leaves	Internodes	Leaf Length (cm)	Leaf Width (cm)	Alkaloids (μg/g)	Steroids (μg/g)	Flavanoids (μg/g)	Tannins (μg/g)	Total Phenols (μg/g)
Control	9c	9d	3.8±0.10a	3.8±0.10e	1.66±0.090e	0.167±0.006e	0.319±0.005e	0.283±0.007e	0.117±0.006e
25%	14bc	12c	4.53±0.05ab	4.43±0.15d	2.08±0.010d	0.326±0.021d	0.532±0.008d	0.366±0.010d	0.280±0.006d
50%	17b	14c	4.76±0.15ab	4.80±0.14c	2.86±0.06c	0.522±0.009c	0.665±0.013c	0.453±0.016c	0.360±0.007c
75%	19a	28b	5.30±0.10ab	5.50±0.20b	3.50±0.18b	0.681±0.009b	0.731±0.010b	0.572±0.012b	0.525±0.006b
90%	27a	31a	5.60±0.20ab	6.30±0.20a	4.81±0.15a	0.740±0.008a	0.854±0.007a	0.766±0.016a	0.733±0.007a

Data represent an average of 3 replicates in two repetitive experiments.

Control: Red loamy soil

Treatment: Vermicompost supplemented to control at 25 per cent, 50 per cent, 75 per cent and 90 per cent.

Statistics: Mean ± SD, ANOVA (DMRT)

$^{***}P < 0.001$, $^{**}P < 0.05$, significant levels.

Table 17.4: Growth and Content of Secondary Metabolites in *Alpinia galanga* on Varying Concentration of Vermicompost

Composition of Vermicompost	Growth Parameters				Secondary Metabolites				
	No. Leaves	Internodes	Leaf Length (cm)	Leaf Width (cm)	Alkaloids (μg/g)	Steroids (μg/g)	Flavanoids (μg/g)	Tannins (μg/g)	Total Phenols (μg/g)
Control	1a	1a	33.5±1.36d	1.10±0.20d	1.92±0.04e	0.147±0.006e	0.439±0.004e	0.133±0.027e	0.083±0.007e
25%	3a	1a	38.4±0.43c	1.73±0.10c	2.64±0.08d	0.279±0.006d	0.565±0.013d	0.274±0.011d	0.184±0.006d
50%	3a	1a	39.9±0.36b	2.26±0.37b	3.04±0.06c	0.361±0.008c	0.623±0.011c	0.386±0.007c	0.134±0.005c
75%	7a	1a	43.2±0.66a	2.53±0.05a	5.51±0.17b	0.523±0.009b	0.723±0.005b	0.553±0.008b	0.501±0.003b
90%	2a	2a	43.9±1.05a	2.81±0.10a	6.73±0.19a	0.681±0.003a	0.819±0.003a	0.663±0.005a	0.690±0.008a

Data represent an average of 3 replicates in two repetitive experiments.

Control: Red loamy soil

Treatment: Vermicompost supplemented to control at 25 per cent, 50 per cent, 75 per cent and 90 per cent.

Statistics: Mean ± SD, ANOVA (DMRT)

$^{***}P < 0.001$, $^{**}P < 0.05$, significant levels.

steroids, and alkaloids make the plants as useful as pharmaceutical. But these compounds are produced in low amounts. In this study, the cheaply available vermicompost was tested for its influence on growth parameters and secondary metabolites of vermicompost treated medicinal plants like *Ocimum sanctum*, Linn; *Coleus aromaticus*, Benth; *Glycyrrhiza glabra*, Linn; and *Alpinia galanga*, Sw and the result are depicted in the Tables 17.1–17.4 respectively.

The various growth parameters like number of leaves, internodes, leaf length and width increased with increasing concentration of vermicompost from 25 per cent to 90 per cent. The maximum number of leaves, internodes, leaf length and width was observed with 90 per cent vermicompost application on 45th day plant action in all medicinal plants studied and is significantly increased when compared to control and other treatments. Similarly, 90 per cent vermicompost application showed maximum levels of secondary metabolites synthesis.

The greater amounts of easily available form of macro and micronutrients in the vermicompost results better growth and yield of the plant (Lee 1985). Besides the major nutrients vermicompost is a rich source of trace elements *viz.*, calcium, magnesium, zinc, copper, iron, and manganese (Lakshmi Prabha *et al.*, 2005). It helps in providing balanced nutrients to encourage vigorous plant growth. Vermicompost increases the microbial activity in organic matter, which result in production of significant quantities of plant growth regulators such as Indole acetic acid, gibberllins and cytokinins by microorganisms. These growth hormones might enhance the synthesis of secondary metabolites of selected medicinal plants *Ocimum sanctum*, Linn; *Coleus aromaticus*, Benth; *Glycyrrhiza glabra*, Linn; and *Alpinia galanga*, Sw. Our study is as in accordance with findings Kale and Bano (1986) who found that plant growth as shoot weight, root weight, root and shoot length were increased by the application of vermicompost in better way than chemical fertilizer.

References

Brattsten, L.B., 1979. Biochemical defense mechanisms in herbivores against plant alleochemicals. In: *Herbirous: Their Interaction with Secondary Plant Metabolism*, (Eds.) G.A. Rosenthal and D.H. Jansen. Academic Press, London, pp. 200–270.

Edwards, C.A. and Neuhauser, E.F., 1988. *Earthworms in Waste and Environmental Management*. SPB Academic Publishing, The Hague, Netherlands.

Ehler's, W., 1975. Observations on Earthworm channels and infiltration on tilled and untilled loess soil. *Soil Science*, 119: 242–249.

Gajalakshmi, S. and Abbasi, S.A., 2004. Earthworm and vermicomposting. *Indian J. Biotechnology*, 3: 486–494.

Hiranmai Yadav and Vijayakumari, 2004. Impact of vermicompost on biochemical characters of chilli (*Capsium Annuum*). *J. Ecotoxicol. Environ. Monit.*, 14(1): 51–56.

Julka and Paliwal, 2005. In: *Proceedings Vermitechonology Transfer to NSS Programme Officers*, (Eds.) R. Jeyaraaj and Indira A. Jeyaraaj. Rohini Press, Coimbatore, Tamil Nadu, India, p. 5–21.

Kale, R.D. and Bano, M.R., 1986. In: *Proc. Nat. Semi. on Organic Waste Utilization*, pp. 151–160.

LakshmiPraba, M., Indira, A. Jeyaraaj and Jeyaraaj, R., 2005. Macro and micronutrient changes in vermicomposting of vegetable wastes using *Eudrilus eugeniae. South Asian Journal of Socio Political Studies*, p. 129.

Lee, K.E., 1985. *Earthworms: Their Ecology and Relationships with Land Use*. Academic Press, Sydney, pp. 411.

Slater, C.S. and Hopp, 1947. Relation of fall protection to earthworm populations and soil physical conditions. In: *Proceedings Soil Science Society of America*, 12: 508–511.

Zakaria, 1991. Isolation and characterization of active compounds from medicinal plants. *Asia Pacific Journal of Pharmacology*, 6(1): 15–20.

Chapter 18

Effect of Organic-Inorganic Fertilizers on Groundnut-Wheat Cropping Sequence

B.M. Kamble, M.S. Shirke and B.A. Chougule

Agricultural Research Station, K. Digraj., Dist. Sangli (M.S.)

ABSTRACT

A field experiment was conducted during the year 2002–03 in medium black soils (pH 8.13). The effect of inorganic fertilizer in terms of yield in groundnut wheat cropping sequence was found significant with targeted yield equation. The treatment with organic manure as per the STCR equation with FYM @ 5 t/ha and biofertilizer, PSB (T_7) gave the highest pod yield (23.75 qt/ha) and grain yield (42.35 qt/ha) of groundnut and wheat respectively, while the total groundnut equivalent yield of the system was 61.99 qt/ha. The yield of groundnut and wheat was higher to the tune of 18.75 and 5.87 per cent respectively over the target. The treatments T_7 (T_5 +FYM + BF) was found economically profitable and best over others with highest net (Rs. 63034/ha) returns and B : C ratio 1 : 2.77.

Keywords: *Cropping sequence, Biofertilizer, FYM, Fertilizer, Targeted yield, Wheat, Groundnut.*

Introduction

Groundnut is one of the important oilseed crop included in cropping system, similarly wheat is a main cereal grown in rabi season under irrigated condition. The importance of highly intensive crop is well recognized to meet the growing demands of ever increasing population. It is necessary to choose appropriate yield targets and fertilizer use practices so that twin objective of high yield and maintenance of soil fertility over seasons could be achieved (Santhi and Selva Kumari, 1999). A cropping system which is not only productive and profitable but also time saving and maintains soil fertility, is of greater importance in present condition. The determination of the fertilizer dose for cropping system is complex due to factors like soil, nutrient fixation and residual effects. To encourage the rational use of fertilizer, it is essential that the cultivators are made aware of the profitability of the fertilizer application under sequence cropping with the use of fertilizer prescription equations.

Soil testing is widely accepted all over the world as quick and effective means of evaluating fertility status of soil (Pal, 1993). Information on nutrient requirement, plant density and tillage practices is very limited particularly when the nutrients are supplied through organic and inorganic sources, sustainability of higher yield could be achieved and maintained only through integrated input management (Dudhatra *et al.*, 2002). Therefore there is need for developing an efficient nutrient management system, with use of organic manures, inorganic fertilizers and biofertilizers to maintain soil fertility for better crop production. The efficient and judicial use of organic, inorganic and biofertilizsers help in increasing yields in groundnut (Singh *et al.*, 1990). Keeping the above points in view; an experiment on fertilizer application based on soil test value was undertaken in present investigation.

Material and Methods

A field experiment was conducted on groundnut-wheat cropping sequence based on soil test crop response approach conducted during 2002–03 in medium deep black soils at Agricultural Research Station, K. Digraj, Sangli (M.S.). The experimental soils having pH-8.13, EC-0.21 dS m^{-1}, low available nitrogen and phosphorus with high available potassium. The initial experimental soil samples were analysed for pH in 1 : 2.5 soil water suspension as per procedure given by Piper (1966). Available Nitrogen by 'Alkaline permanganate method', available phosphorus by 'Olsen extractant P method' (1954) were assessed. Available potassium was extracted by 1 N NH_4OAc solution and it was determined by flame photometer (Jackson 1967). The treatments comprised of T_1-untreated control, T_2-recommended dose of fertilizer (25 : 50), T_3-dose of fertilizer as per the soil test, T_4-targeted yield (15qt/ha for groundnut and 30 qt/ha for wheat), T_5-targeted yield (20 qt/ha for groundnut and 40 qt/ha for wheat), T_6–T_4 + FYM 5 t/ha + BF (PSB), T_7–T_5 + FYM 5 t/ha + BF (PSB) for groundnut-wheat cropping system in randomized block design with for replications. The fertilizer prescription equations developed by STCR scheme M.P.K.V., Rahuri is as under.

For groundnut	*For wheat*
FN = (4.16T) – (0.37SN)	FN = (7.54T) – (0.74SN)
FP_2O_5 = (4.96T) – (4.36SP)	FP_2O_5 = (1.90T) – (2.88SP)
FK_2O = (3.14T) – (0.16SK)	FK_2O = (2.49T) – (0.22SK)

Where, FN, FP_2O_5, FK_2O are the fertilizer doses and SN, SP, SK are the soil test values in kg/ha. '*T*' is the yield target in qt/ha. Based on soil test values of NPK and different yield target (groundnut and wheat) aimed at, fertilizer dose were applied for various treatments. The FYM 5 t/ha was applied basically, NPK full dose for groundnut and 50 per cent N and full dose of P, K for wheat. After 30 days sowing 50 per cent N was top dressed to wheat. The regular management practice and need base plant protection measures were carried out. Pod and kernel yield of groundnut and grain yield of wheat crop was recorded and the economics was worked out on the basis of existing prices of output gained and input used under different treatments. The statistical analysis was carried out by procedure suggested by Panse and Sukhatme (1961).

Results and Discussion

The data from the Table 18.1 revealed that the dry pod and kernel yield of groundnut (23.75 and 16.77 q/ha), grain yield of wheat (42.35 q/ha) and groundnut equivalent yield of the system (61.99 q/ha) was significantly superior over other treatments due to application of inorganic fertilizer (9 kg N, 46 kg P_2O_5 and 8 kg K_2O /ha for groundnut and 151 kg N, 42 kg P_2O_5 and 24 kg K_2O/ha for wheat) with targeted yield equation and FYM 5 t/ha + biofertilizer PSB (T_7). Dudhatra *et al.* (2002) have also reported the significant, effects of integrated input management on the yields of groundnut and wheat wherein the significantly highest pod yield (764 kg/ha) in groundnut as well as in succeeding wheat crop the highest grain yield (3751 kg/ha) which are on the line of results of present investigation. However, Singh *et al.* (1999) also reported the beneficial effect of integrated input management in groundnut and succeeding wheat crop, which attributing to the addition of crop residue as organic matter. The targeted yield of 15 q/ha in groundnut and 30 q/ha in wheat was obtained due to application of inorganic fertilizer (0 kg N, 22 kg P_2O_5 and 0 kg K_2O/ha for groundnut and 76 kg N, 22.5 kg P_2O_5 and 0 kg K_2O/ha for wheat) with targeted yield equation, FYM 5 t/ha and biofertilizer PSB (T_6). Santhi *et al.* (2004) have also reported the similar results in sunflower, the fertilizer prescription equations were achieved the yield target in IPNS. The economic analysis of present studies revealed that the net returns Rs. 63034/ha and benefit cost ratio (2.77) was highest in treatment (T_7) and proved economically best amongst rest of the treatments. Despande and Jadhav (1996) have also reported that groundnut-wheat cropping sequence as the most remunerative system and gave the highest B : C ratio and thus supported the results of present investigation.

Table 18.1: Yield and Economics of the Groundnut-Wheat Cropping Sequence as Influenced by Different Treatments

Treatment	*Groundnut Yield (kg/ha)*		*Wheat Yield (kg/ha)*	*Groundnut Equivalent Yield (kg/ha)*	*Net Returns (Rs/ha)*	*B : C Ratio*
	Dry Pod	*Kernel*				
T_1	1332	958	1901	3074	19217	1.64
T_2	1767	1270	3908	5272	45108	2.16
T_3	1983	1408	3969	5645	53313	2.42
T_4	1479	1073	3204	4372	35047	2.01
T_5	1740	1260	4102	5584	52299	2.38
T_6	2059	1469	3697	5387	51477	2.50
T_7	2375	1677	4235	6199	63034	2.77
Mean	1833	1302	3574	5076	44214	2.21
S.E.	5.28	5.75	1.39	126.8		
C.D. at 5%	15.695	17.11	4.14	370.1		

Conclusion

Application of organic manures as per the STCR equation along with FYM 5t/ha and biofertilizer gave the highest pod yield of groundnut (23.75 qt/ha) and grain yield (42.35 qt/ha) in wheat and the total groundnut equivalent yield (61.99 qt/ha) of the system. Thus the yield of groundnut and wheat was higher to the tune of 18.75 per cent and 5.87 per cent respectively over the targeted yields.

References

Despande, S.B. and Jadhav, A.S., 1996. Productivity of irrigated crop sequences in western Maharashtra.

Indian Journal of Agronomy, 41(1): 22–26.

Dudhatra, M.G., Vaghani, M.N., Kachot, N.A., and Asodaria, K.B., 2002. Integrated input management in groundnut (*Arachis hypogaea*)–wheat (*Triticum aestivum*) cropping system. *Indian Journal of Agronomy*, 47(4): 482–486.

Jackson, M.L., 1967. *Soil Chemical Analysis*, Prentice Hall of India Pvt. Ltd., Delhi, pp. 111–132.

Olsen, S.R., Cole, C.V., Watanabe, F.S. and Dean, L., 1954. *Estimation of Available Phosphorus in Soils by Extraction with Sodium Bicarbonate*. U.S.D.A., Circ. 9 : 39, U.S. Govt. Printing Office, Washington D.C.

Pal, A.R., 1993. A paper presented at Summer Institute on Advances in Agronomy held at Indian Gandhi Krishi Vishwavidyalaya, Raipur, p. 3–21.

Panse, V.G. and Sukhatme, P.V., 1961. *Statistical Methods for Agricultural Workers*, ICAR, New Delhi.

Piper, C.S., 1996. *Soil and Plant Analysis*. Int. Sci., Publisher, Univ, Adelaide, Australia, Asian Ed., pp. 7–46.

Santhi, R., Natesan, R. Andi, K. and Selvakumari, G., 2004. Soil test based fertilizer recommendation for sunflower, *Helianthus annuus* L. in inceptisols of Tamil Nadu. *J. Oilseeds Res.*, 21(1): 78–81.

Santhi, R., Selva Kumari, G., 1999. Yield targeting and integrated plant nutrition system for soil fertility maintenance in rice based cropping sequence. *Madras Agric. J.*, 86(1–3): 138–139.

Singh, R.A., 1999. Integrated input management in groundnut (*Arachis hypogaea*)–wheat (*Triticum aestivum*) cropping system under drought prone areas of semiarid regions. *Indian Journal of Agronomy*, 44(4): 673–676.

Singh, R.P., Das, S.K., Bhaskara Rao, U.M. and Narayana Reddy, M., 1990. *Towards Sustainable Dryland Agricultural Practices*. Central Research Institute for Dryland Agriculture, Hyderabad, pp. 106.

Chapter 19

Extent of Adoption and Economics of Integrated Pest Management on Farmer's Field

M. Chandra Sekhar Reddy[1]*, K.V. Siva Reddy*[1] *and V. Chenga Reddy*[3]

[1]Research Associate, [2]Senior Scientist (Breeding) and CCPI
Technology Mission on Cotton (TMC) Mini Mission (MM 5.1) Project,
Regional Agricultural Research Station, Lam, Guntur – 34

Adoption of farm innovation is a slow process. For instance even in developed countries like USA, over 14 years were required for hybrid seed of cotton to reach 99 per cent adoption in IOWA, USA. Similar time lags were discovered in many other studies on innovations. The interest of promoters of innovations in finding out ways and means of speeding up adoption or reducing time lag is but natural. It is particular in developing countries where the lags can be expected to be even greater but the changes must be brought out quicker. Studies on adoption of innovations and constraints in the process of adoption have got special significance for Guntur region of Andhra Pradesh state.

The phenomena of adoption of improved technology is multidimensional in its basic frame work. It occurs under the complex interplay of economic, social and cultural forces operation on individuals and the group among whom innovations are diffused. As such evaluation studies are needed to find out the level of adoption, constraints for non adoption and partial adoption of innovations. The present work is oriented in such a frame work. The main objectives are:

1. To measure the deviations between the available technology and actual level of its adoption in cotton.
2. To find out the constraints in adoption of recommended technology of cotton.
3. To know the cost of cultivation on farmer field under Integrated Pest Management and Non Integrated Pest Management conditions.

Fifty farmers were selected to implement the IPM technology in the Anathavarappadu of Vatticherukuru mandal in Guntur district of A.P. IPM components (Pheromone traps, Marigold seedlings, castor seeds, NPV, NSKE) were supplied to selected farmers under the project MM 5.1 of Technology Mission on Cotton functioning at Regional Agricultural Research Station, Lam, Guntur during 2004–05 season.

Results

Deviations between the available improved technology and actual level of its adoption on farmer's field were presented in the Table 19.1.

Table 19.1: Deviations Between the Available Improved Technology and Actual Level of its Adoption in Cotton Cultivation

Sl. No.	*Items*	*Recommendation*	*No. of Adopted Farmers*	*Deviation from Recommendation (%)*
1.	Soil type	Heavy/medium	38	24.00
2.	Sowing time	At proper time	50	0.00
3.	Improved cultivars	Variety/hybrids	50	0.00
4.	Spacing	120 × 60 (hybrids)	28	44.00
		90 × 60 cm (varieties)	31	38.00
5.	Proper fertilizer dose (NPK kg/ha)	120 : 60 : 60 (hybrids) 90 : 45 : 45 (varieties)	50 50	100.00 100.00
6.		Time of fertilizer application		
	Basal	At time of sowing	20	60.00
	Top	45 DAS	27	45.00
7.	Bio pesticide used	NSKE/NPV	1	98.00
8.	Pheramone traps		30	40.00
9.	Trap crops	Castor/marigold seedlings	9	82.00
10.	Advice from extension staff in use of inputs		22	56.00
		Total no of farmers (N) = 50		

It can be seen from the Table 19.1 that all farmers were cultivated improved cultivars at proper sowing time. The deviations were more in the adoption of balanced fertilizer dose and adoption of IPM components (Bio pesticides, pheramone traps and trap crops).

Constraints

Analysis of each component indicated that there was reduction in the yields due top non-adoption of package of practices at proper way in the proper time. From the study the following constraints were identified:

1. Non availability of sufficient quantities of Bio pesticides.
2. IPM adoption in the smaller areas was not effective.
3. Non availability of bio agents.
4. With very much difficulty farmers were convinced to spray biopesticides.

Economics of IPM

Farmers need not necessarily to adopt a technology *i.e.,* feasible and alternative if it is not in concurrence with that objective of profit maximization. The profitability is determined by the cost involved in the application of new technology, crop productivity and output price. There was no difference in cotton price received by IPM and Non IPM farmers. Thus given the output price, productivity and cost of technology are the main determinants of profitability. In the following paragraphs we examine the impact of IPM on costs.

Cost of Cultivation

The average cost of cultivation of IPM farms was estimated to be Rs. 25814/ha which was 18.72 per cent less than the Non IPM farmers. The expenditure on plant protection inputs was 38.67 per cent less on IPM farmers than Non IPM farmers.

Value of Output and Net Returns

Even though, there is no significant difference in yields between IPM and Non IPM farmers, the per hectare net returns on IPM farms were 25.82 per cent higher than Non IPM farmers.

The project farmers realized higher cost benefit ratio of 1 : 1.79 due to less investment on plant protection as compared to the cost benefit ratio of 1 : 1.44 obtained by the Non IPM farmers. The details of the economic aspects of the IPM and Non IPM Farmers are given in the Table 19.2.

Table 19.2: Economic Returns of IPM and Non IPM Farmers

Sl.No.	*Particulars*	*IPM Farmers*	*Non IPM Farmers*	*Per cent Change Over Non IPM*
1.	No. of insecticide applications	6.5	9	
2.	Cost of plant protection (Rs/ha)	8885.00	12321	– 38.67
3.	Yield (q/ha)	28.15	28.00	+ 0.53
4.	Gross returns (Rs/ha)	50670.00	50400	+ 0.53
5.	Cost of cultivation (Rs/ha)	25814.00	30646.00	– 18.72
6.	Net returns (Rs/ha)	24856.00	19754.00	+ 25.82
7.	CB Ratio	1 : 1.79	1 : 1.44	

Price of cotton kapas @ Rs. 1800/- per quintal

Policy Implication

1. More emphasis should be given on large farm demonstrations. IPM should be implemented on community approach.
2. Need to educate farmers on need based plant protection strategy with proper plant protection doses and proper use of inputs
3. Empowerment of extension workers and farmers in the use of tools and methods of IPM, since the IPM is a knowledge intensive method.

References

Deaonkar, A.M., 1996. A study on the extent of adoption of improved technology and economics of cotton cultivation on farmer's field.

Grover, R.K., K.N. Rai and K.K. Sharma, 1993. Study on adoption pattern and constraints analysis in paddy. *Indian Journal of Agril. Research,* 26: 96–10.

Chapter 20

Growth of Vegetable Using Vermicompost

G. Nithya and T. Thangaraj
P.G and Research Department of Zoology, Kongunadu Arts and Science College, Coimbatore – 641 029, Tamil Nadu, India

ABSTRACT

Germination and growth of vegetables palakeerai and araikeerai were observed using red soil and red soil amended with vermicompost. Germination was slightly delayed in the vermicompost media than in red soil whereas the growth was more in vermicompost than the red soil.

Keywords: *Araikeerai, Palaikeerai, Pheretima posthuma, Vermicompost.*

Introduction

Organic farming is becoming more and more popular since it prevents soil erosion, protect water quality, save energy, use natural soil amendments and pest controlling agents, minimize farm workers exposure to chemicals, help small farms, promote biodiversity and produce more tasty food with good keeping quality (Zende *et al.*, 2000). The use of vermicompost in improving the soil condition, increasing the crop yield and crop quality is not fairly accepted as a general practice in India. It also minimizing the use of chemical fertilizers and increase the crop yield.

Materials and Methods

Compost is prepared from vegetable and animal refuses, collected from farm in towns or in the villages.

Source of Animal

Earthworms (*Pheretima posthuma*) were collected from the local garden areas in Perinthalmanna, Malappuram (Dist.), Kerala, and also from the college campus and the maintained in a proper moisture conditions.

Collection of Seeds

Seeds of palakeerai and araikeerai are of commercial source and from Tamil Nadu agricultural university and used in the present study.

Preparation of Vermicompost from Cowdung

Vermicompost was prepared as detailed by Jeevan Rao (2005).

8 kg of cowdung was dried and made into powder. This powder was taken in four plastic buckets, 20 kg each and a vermibed was prepared by sprinkling water. A loose bed was maintained to enable easy movement of earthworms. In each bucket half kg of earthworms (80 numbers) were introduced. The tanks were kept in shade and covered with wire mesh to prevent any ant attack and small holes were put at the sides of the bucket for aeration. Water was sprinkled everyday in the vermibed to maintain an optimal conditions for the earthworms to grow and multiply. This setup was maintained for 60 days to harvest vermicompost.

Cultivating the Plants Using Vermicompost

The seeds were sown in the wooden containers containing compost and soil in the ratio 2 : 1.

Comparison of the growth of plants in terms of biomass production was noticed.

Earthworm compost along with the soil (in the ratio 1 : 2) was used for germination of seeds and kept as an experimental. Red soil from the garden was taken and used as control. Growth of plants was noticed using red soil containing vermicompost in the ratio 2 : 1 for 6 to 9 days and the turnover of growth in terms of biomass. Germination property was also noticed. Three sets of plants were grown and each set consisted of 10 plants.

Results and Discussion

Biomass Production in Araikeerai and Palakeerai

Biomass conversion was observed in Araikerrai, and Palakeerai grown in red soil and red soil amended with Vermicompost in the ratio of 2 : 1 for 6 to 9 days and the growth in terms of biomass production was calculated in the experimental as well as in control.

The growth of Araikeerai in the medium consisting of Vermicompost showed 30.33 gm wet weight of the plant when compared to 3.33 mg of wet weight in the control containing red soil alone. The results showed 10.1 times higher yield.

The growth of palakeerai in the experiment showed 27.06 gm wet weight of the plant when compared to 2.1 gm of wet weight in the control containing red soil alone. The results showed 13 times higher yield when compared to the soil containing without vermicompost.

Germination of Araikeerai and Palakeerai

The germination property of the seeds of Araikeerai and Palakeerai were assessed using red soil (Control) and amended with vermicompost in the ratio of 2 : 1.

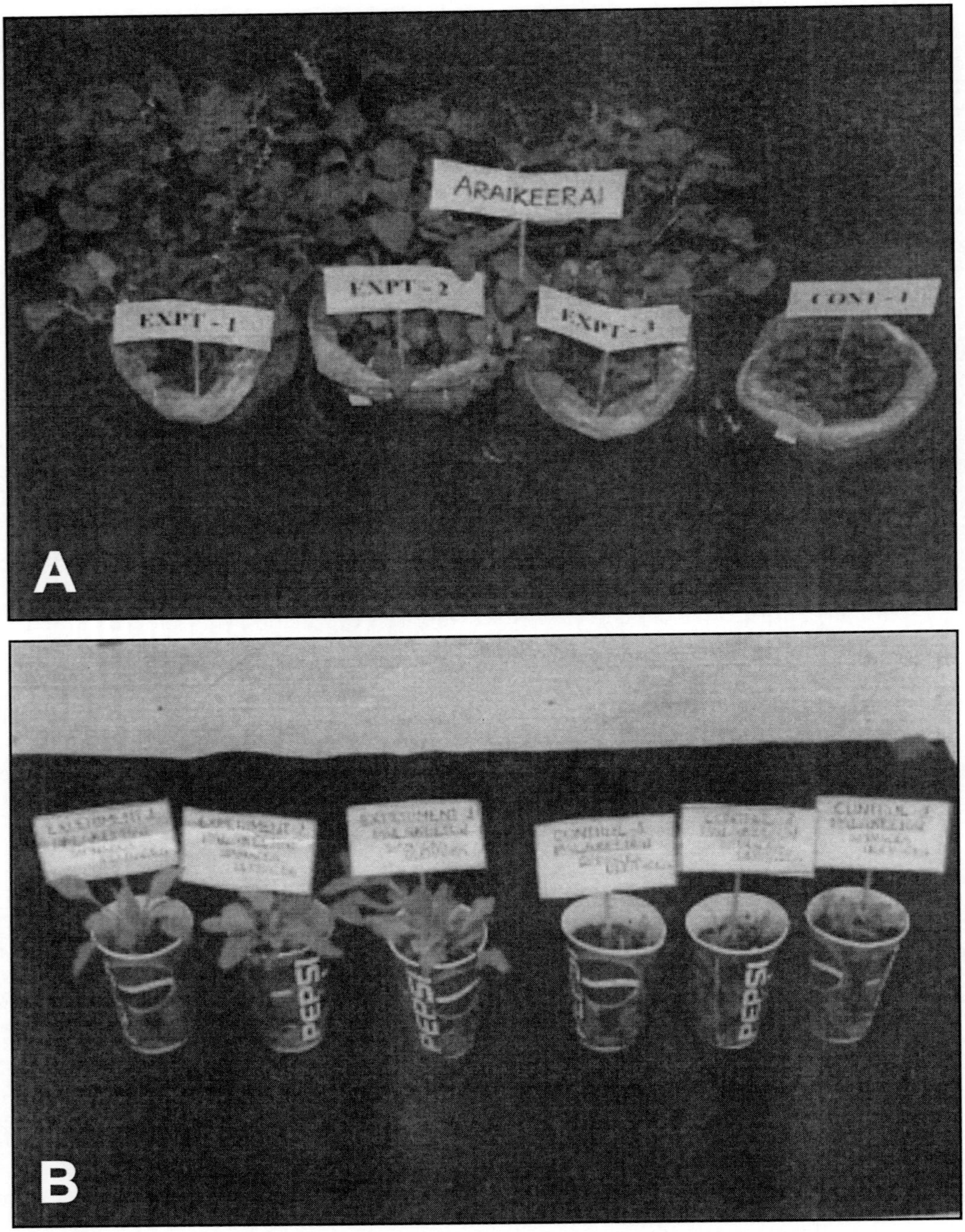

Plate 20.1: Growth of Plants in the Media Amended with Vermicompost and Control

A: Araikeerai (A_1–A_3: Vermicompost amended with red soil; A_4: Red Soil)

B: Palakeerai (B_1–B_3: Vermicompost amended with red soil; B_4–B_6: Red soil)

The germination rate in the araikeerai showed the, mean days of 7.33 in the vermicompost soil whereas, the number of days were taken in red soil was 4–6 days. The percentage of germination in Araikeerai was found to be 76 per cent when sown in vermicompost whereas, in red soil the percentage of germination was found to 46 per cent only. Suggesting that the germination in control was lesser when compared to the experiment. It is noteworthy to mention that the number of days taken for germination of seeds is little prolonged in the media containing vermicompost than in red soil alone in which earlier germination was noticed.

The germination period in the Palakeerai showed the mean days of 9.3 in the vermicompost soil whereas, the days taken for germination in control was 7.3 days. The percentage of germination in Palakeerai was found to be 70 per cent consisting of vermicompost. Whereas, in red soil the percentage of germination was 50 per cent only. The time taken for the germination of seeds is earlier in red soil than in experiment. In general, the percentage of biomass yield and germination of seeds was found higher containing vermicompost media than in red soil [Plate 20.1(a, b)].

Though the germination of seeds in vermicompost was higher, the time taken for germination in the media containing compost was found higher than in the red soil, the reason for delayed germination in amended with vermicompost is not clear at present. Since vermicompost is a rich source of macro and micro nutrients, vitamins, enzymes, antibiotics, growth hormones and immobilized microflora. Vermicomposting is an appropriate media for better growth of the plants.

References

Jeevan Rao, K., 2005. Compost of organic waste. *Kissan World*, 1: 29–30.

Zende, G.K., Desai, V.G., Josh, S.N., Kulkarni, J.H. and Ruikar Jagadis Patwardhan, S.K., 2000. Influence of season on the performance of an Adsali and its Ratoon crop of sugarcane receiving vermicompost and chemical fertilizers in different combinations.

http://www.krishworld.com/html/vermicomp/html.

Chapter 21

Computer Ergonomics: Relooking at Machines vs. Environment

K. Chandra Sekharaiah[1]*, *Md. Abdul Muqsit Khan*[2]**, *Upakaram Gopal*[3]***

[1]*Associate Professor, Department of C.Sc. and Engineering, College of Engineering, Jawaharlal Nehru Technology University, Hyderabad – 500 072, India, E-mail: shakes123@sify.com*

[2]*Lecturer, Department of Computer and Info, Maulana Azad National Urdu University, Gachibowl, Hyderabad – 32 India, E-mail: abdul_muqsit_khan@yahoo.co.in*

[3]*Research Associate, Pullaiah Al Research Foundation, Nellore – 524 004, Andhra Pradesh, India, E-mail: gopal_uv@rediff.com*

ABSTRACT

With the advent of information technology has emanated the concept of knowledge workers which includes all the stakeholders who have a fine, air-conditioned working environment with computers much different from the traditional sweating working environments in mechanical or electronics machines industries. Ironically, the so called knowledge workers in a/c environments are becoming victims of non-ergonomic component designs and conditions. Much more so, in massive programming environments, in large-scale data processing, data-entry applications, promoting human-computer interface is of crucial concern to computer users' working environment. In this paper, ergonomic conditions based on a good quality human-computer interface, human-application interface and design are explored for. Investigation

---

* The first author is a Fellow of Indian Electronics and Telecommunications Engineers (F.I.E.TE) and a Fellow of Society for Environmental Science (F.S.E.Sc.) and a Senior Member of C.S.I., M.I.E., and a Life Member of Association of Medical Physicists of India (M.A.M.P.I.) Indian Science Congress Association (M.I.S.C.A.) and Indian Society for Technical Education (M.I.S.T.E). He is also member of European Association for Software Science and Technology (EASST). He is also member of Indian Society of Ergonomic (I.S.E).

** This author is a life member of ISTE and CSI.

*** This author is a life member of NLP Association of India.

of ergonomic design features for machine components is considered an important factor in making up the state-of-the-art hardware interface. Taxonomy of ergonomics is presented including accounts on the human-robot interface and mobile human-computer interaction. The paper provides a single platform for cross-fertilization of works peculiar to biomechanics, multimodal interactions and virtual books. The reports of our studies on the deleterious effects of non-ergonomic computing environments and desirable ergonomic conditions are presented in the paper. Though this paper presents neither an ergonomic design nor ergonomic solution to the existing problems in computing environments, it will go a long way in facilitating the computer users to take care of the ergonomic aspects of computing environments with special emphasis from the perspective of software ergonomics.

Keywords: *Virtual book, Multimodal Interactions (MMI), Cognitive ergonomics, Software ergonomics, Human Computer Interaction (HCI), Human Robot Interface (HRI).*

Section 1: Introduction

Ergonomics is the study of the environment, conditions and efficiency of workers. It is to study the conditions in which people work most effectively with machines. Ergonomic designs are felt very valuable times owing to increasing machine-working environments for optimum overall system performance and human well being. The work contributes to the improvement of the ergonomic quality of interactive software. It is about increasing available knowledge about users activities and cognitive characteristics as well as about the usability of software systems. It also presents how to improve ergonomic design and evaluation methods (Active Report INRIA, 2004).

The rest of the article is organized as follows. In the next section, we present taxonomy of ergonomics. In section 3 importance of ergonomics and biomechanics of gesturing are discussed. Section 4 presents the ergonomics of multimodal user interaction with 3D design environments. In section 5 we discuss design aspects of a virtual book. Section 6 presents the deleterious effects of non-ergonomic computing environments. Section 7 presents desirable ergonomic conditions for avoiding certain non-ergonomic health hazards. Section 10 gives conclusions.

Section 2: Taxonomy of Ergonomics

Ergonomists contribute to the design and evaluation of tasks, jobs, products, environments, and systems in order to make them compatible with the needs, abilities, and limitations of people. A taxonomy of ergonomics is given in Figure 21.1 (http://www.iea.cc/ergonomics).

Physical Ergonomics

Physical ergonomics is concerned with human anatomical, anthropometric, physiological, and biomechanical characteristics as they relate to physical activity. It includes working postures, materials handling, work related musculoskeletal disorders, repetitive movements, workplace layout, safety and health. *Cognitive ergonomics* is concerned with mental processes such as perception, memory, reasoning, and motor response, as they affect human-machine involving mental workload, decision-making, skilled performance, HCI and human reliability (Active Report INRIA, 2004).

Organizational Ergonomics

Organizational ergonomics is concerned with the optimization of sociotechnical systems including their organizational structures, policies, and processes. It includes communication, crew

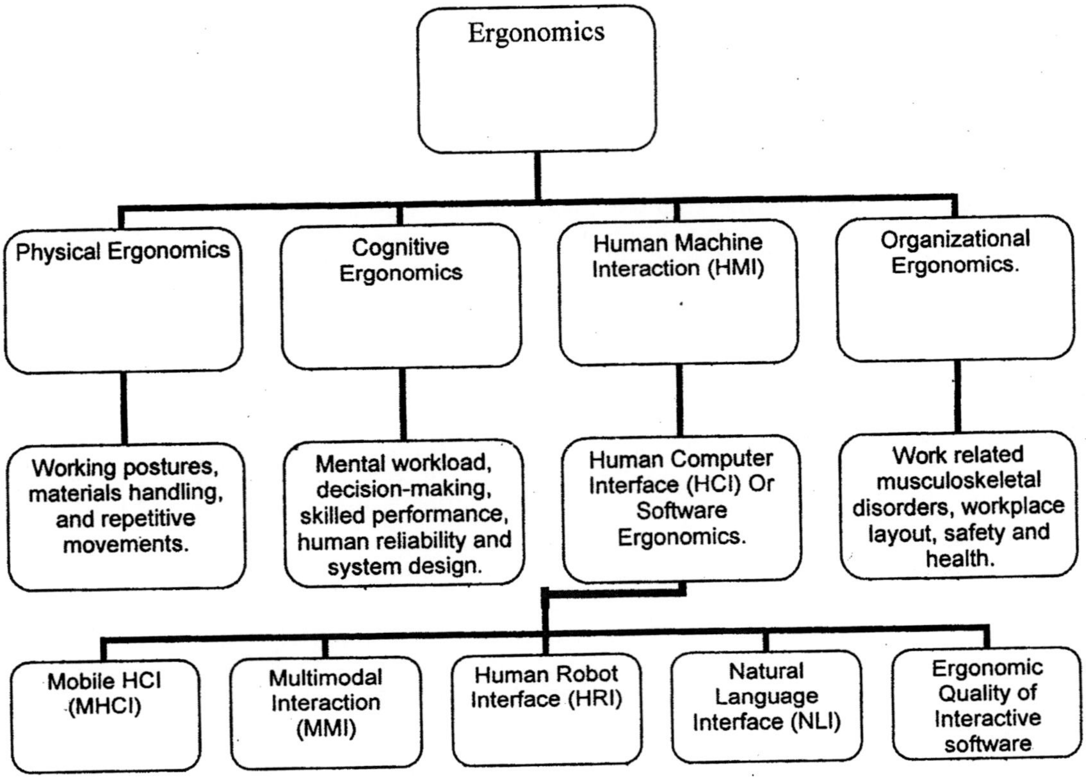

Figure 21.1: Taxonomy of Ergonomics

resource management, and work design, design of working times, teamwork, participatory design, community ergonomics, cooperative work, new work paradigms, virtual organizations, telework, and quality management.

Software/HCI Ergonomics

Software/HCI Ergonomics inherits from the main characteristics of ergonomics. It is a science that contributes to the knowledge necessary to software design, and to computer-based environments, with the overall perspective of human security and well-being, and for facilitating users' tasks, limiting learning time, reducing errors and related cost. Software ergonomics focuses on the improvement of HCIs mainly in terms of cognition. Of late novel interaction techniques (*e.g.*, multimodality) and novel environments (*e.g.*, virtual reality) are considered under physical ergonomics.

Human-Computer Interaction

Human-computer interaction constitutes the set of hardware, software, human and environmental elements that influence the efficiency of systems and products, both from a technological and a human point of view. HCI is a specialization of Human-Machine Interaction. It has three complementary aspects: the human aspect, the computer aspect, and the interaction aspect. The human and interaction aspects belong to ergonomics; the computer and interaction aspects belong to the HCI part of computer science. *Mobile HCI* is of special elevance. Cognitive. ergonomic should be considered In the various phases involved in the development life cycle of HCI. Such services as audio and speech interaction, website design for mobile devices, intelligent environment, multimodal interaction, application related interface techniques (automotive, mobile, TV), user centered design tools and methods for mobile systems, designing systems for different age groups and users with special needs are to be explored. HRI and NLI are specializations of HCI too.

Ergonomic Quality of Interactive Software

Ergonomic quality of interactive software (Scapin & Bastien, 1997) covers all software aspects which have an influence on the users' task completion and facilitates ease of use, *i.e.*, the extent to which the users can easily reach their interaction goals such as presentation and dialogue aspects of the interaction it also covers utility, *i.e.*, the extent to which the users can reach their task goals functional aspects of the interaction such as functions, objects, data. From a software engineering perspective (*e.g.*, architecture models), *Ergonomic Quality* covers not only the classical presentation, dialogue control and application interface aspects, but also some application kernel aspects such as those that have an influence on the users reaching their goals.

Section 3: Ergonomics and Biomechanics of Gesturing

Biomechanics is the field of study, which makes use of the laws of physics and engineering concepts to describe motion of body segments, and the forces, which act upon them during the activity. The gestures in Figure 21.2 are found stressing to perform. Ergonomics and biomechanics of gesturing indicate physically stressing gesture is avoided. In this section, the biomechanics in the hand (Lin *et al.*, 2001; Lee & Tosiyasu, 1995) is described. Figures 21.3 and 21.4 relate the terms. Joint movements are explained.

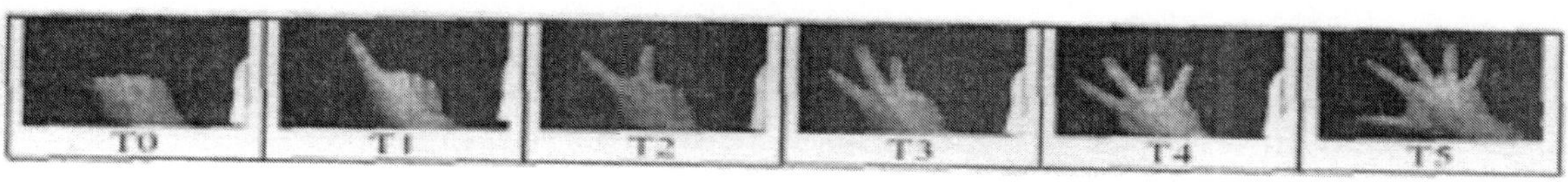

Figure 21.2: Technology Based Gestures (Nielsen *et al.*, 2003)

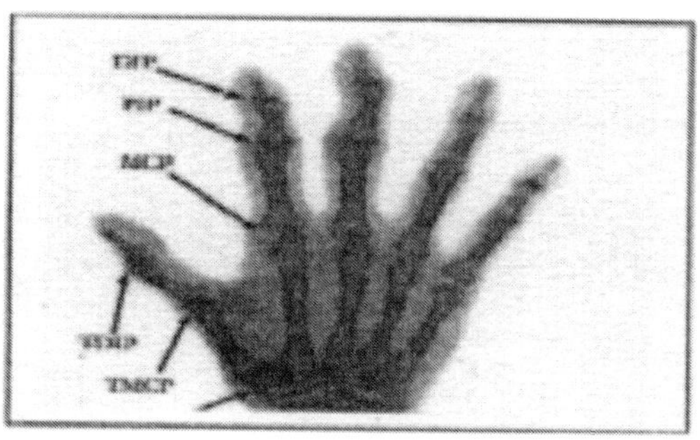

Figure 21.3: X-ray of Right Hand Joint Names and Wrist Movement (Nielsen *et al.*, 2003)

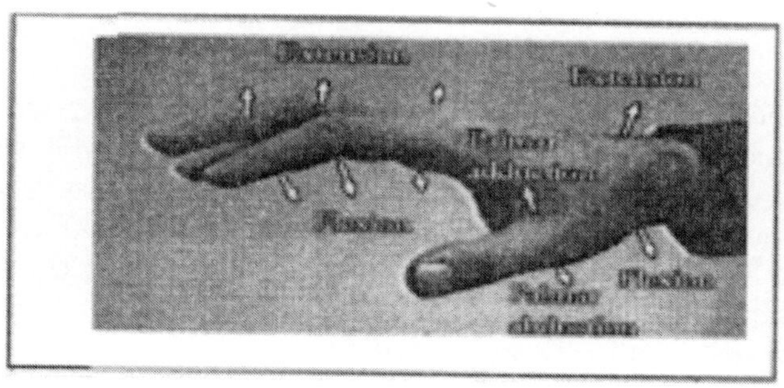

Figure 21.4: Hand from the Side. Finger and Wrist motion (Nielsen *et al.*, 2003)

The gesture postures have static joint constraints (Eaton, 1997; Nielsen *et al.*, 2003) and dynamic constraints. The dynamic constraints are inter-finger and intra-finger constraints. Inter-finger constraints are the dependencies between the postures of the neighboring fingers. Intra-finger constraints are the dependencies between the joints in the particular finger. Furthermore, finger postures affect pressure in the carpal tunnel (Keri *et al.*, 1998). Ergonomics principles in (Grant; Shaw & Hedge; Hedge & Muss, 1999):

1. Avoid outer positions.
2. Relax muscles.
3. Relaxed neutral position is in the middle between outer positions.
4. Avoid repetition.
5. Avoid staying in static position.
7. Avoid internal and external force on joints that may stop body fluids.

Given these aspects, a gesture interface should by nature be ergonomically superior to physical handheld devices, which introduces external force.

The gestures in Figure 21.2 are stressing because they do not follow the inter-finger constraints of the hand. While the limitations of the constraints are different for each person, the user must use more or less force to position the fingers for the system to recognize the correct posture. Furthermore, the recognition is vulnerable to how stretched the fingers are. This means that the users must stretch them to outer positions. The ergonomics show that it is important to make the recognition algorithms tolerant to de-stressing movements, which allows the user to avoid staying fixed in *e.g.* a static "residue" or "pointing" gesture. Tolerance for deviations in gestures is desirable when implementing gesture interfaces, also because of varying hand shapes and posture performance (Nielsen *et al.*, 2003).

Section 4: Ergonomic Design for Multimodal Interactive Systems

Communication through each of the five senses is known as a mode. Computer may process information through modes as well Keyboard, Microphone, Camera etc. Multimodal interfaces combine two or more different modes of communicating. MMI means that the user and system can utilize various types of modalities in order to present information such as natural language (spoken or written), graphics, images, videos, and tables. It is, some times, more efficient than unimodal communication.

MMI systems use multimodal input/output in working environments. Unlike classical interaction techniques they are not based on mouse, keyboard etc. MMI systems capture 3D objects manipulation and virtual environments for design activity. MMI takes support of 3D design environments, speech (vocal commands synergistically combined to gase head direction and hand gastrula deictic mimetic movements), using a communication metaphor. Output techniques for MMI takes support of speech interface, gesture interface, graphics interface or combination of these. Pointing hand gestures has same expressive power as gaze w.r.t object selection in large display such as electronic blackboards, caves etc or in 3D environment.

The work on multimodality addresses such issues as:

1. How to design oral messages that help visual search in cluttered displays?
2. How to design multimodal command languages that use information on gaze movements to disambiguate oral commands, especially those including deictic phrases?

3. Are voice + graphics help messages more effective than standard text + graphics ones? Does this form of output multimodality actually improve the effectiveness and efficiency of online help?

Oral support to visual search (Carbonell & Kieffer) involves oral information on the location of a visual target in a complex, cluttered, display to improve the efficiency of its identification (accuracy and selection times). Targets are either familiar (visual presentation of the isolated target prior to scene display) or unfamiliar (oral characterization of the target only, prior to scene display), monomodal (visual, oral) or multimodal (visual, oral) (Inria, 2004).

Section 5: Ergonomics in Virtual Books

Virtual books make on-line reading more comfortable. A monitor on a desktop to read electronic documents forces readers to sit in a fixed position. Virtual book devices free readers to be able to make small adjustments to their posture. This capability allows readers to relax and to focus on documents, rather than on the delivery technology. The virtual book should be so much handy that it can be easily carried like a textbook. Dynabook (Kay & Goldberg, 1977) and Lectrice are examples of virtual books.

As regards physical characteristics of virtual books, the following guidelines help. A virtual book is approximately the same size as a pad of paper and is clearly a device for viewing things because it is dominated by the display. The ability to display a complete page at close to 1 : 1 size forces the screen to be close to 8.5″ 11″ in practice. Most documents have margins that can be stripped to a screen. The case of the device provides the margin. The ideal weight is less than 1.5 kg. As the weight is increased above 2 kg, it becomes less and less acceptable. The thickness of the device (for a given weight) matters less when it is being used than when it is in transit. To remain comfortable when the virtual book is held in the lap, it is important that the power dissipated by the device be low enough not to annoy the reader. As regards user interface characteristics of virtual books, main features to allow continuous reading are:

1. The low cognitive effort required to drive the interface.
2. The lack of distracting user interface elements on the screen.
3. Responsiveness.

Low cognitive effort involves more than just being easy to use. After a short learning period, common actions have to be performed unconsciously.

Section 6: Deleterious Effects of Non-ergonomic Computing Environment

Computer Eye Strain: The Dilbert Syndrome

Computer usage causes headaches, tired eyes, blurred or double vision, eye strain, increased sensitivity to light, and difficult focusing and refocusing between the monitor and other objects. However, there is no evidence of permanent damage from prolonged use of VDTs, numerous tests indicate no relationship between computer use and cataracts, retinal damage, or permanent nearsightedness. However the strain on eye muscles can, causes discomfort (http://www.vh.org/adult/patient/familymedicine).

Practically, all computer-related eye problems are preventable and correctable. They are mainly owing to lack of appropriate lighting. Even excessively bright overhead lighting frequently causes glare. Use a glare shield, available at most computer stores, to reduce uncomfortable reflection. As regards the position of computer monitor, the optimal screen position is at about 10 to 20 degree angle below eye level. Placing the top of the screen below eye level also prevents head and neck aches caused by tilting head back to look up at the monitor.

Sitting close to the monitor interferes with the eye's ability to focus on the screen. The screen should be about 20"–26" from your eyes. Periodic rest breaks are vital to the health and comfort of eyes. Every 15 minutes, spend about 20 secs looking around the room and refocusing eyes on a distant point. Some stretching exercises help to relieve tension in back, shoulders, and neck. Generally, we blink about 22 times each minute. While at the computer, we only blink about 7 times a minute. Infrequent blinking produces fewer tears and also causes eye moisture to evaporate. Without tears, vision becomes blurrd and eyes feel miserably uncomfortable.

Chronic Fatigue Syndrome: What is it?

Chronic fatigue syndrome involves debilitating symptoms. No definitive test exists for chronic fatigue syndrome unlike for diabetes etc. It is not a disease. Its many causative factors vary from person to person. It causes six months or more of frequent, profound exhaustion that is not relieved by rest. The patients also have at least four of the following symptoms: impairment of short-term memory or concentration, sore throat, tender lymph nodes, muscle pain, multi-joint pain without swelling or redness, headaches, unrefreshing sleep, or post-exertional malaise lasting more than 24 hours. The fatigue and associated symptoms interfere with quality of life more than many debilitating diseases such as congestive heart failure. Chronic fatigue is often misdiagnosed because it resembles other disorders such as fibromyalgia, Lyme disease, mononucleosis, multiple sclerosis, lupus, and candida albicans. In most cases, chronic fatigue is diagnosed when other medical illnesses causing fatigue are ruled out (http://www.vh.org/adult/patient/familymedicine/prose).

Computer Vision Syndrome

Computer Vision Syndrome affects 75 per cent of computer users especially those over the age of 40 (http://www.vh.org/adult/patient/opthalmology). Decrease of blink rate by 66 per cent causes:

1. Eye fatigue
2. Dry eyes
3. Burning eyes
4. Light sensitivity
5. Blurred vision
6. Headaches
7. Pain in the shoulders, neck or back.

Solutions

1. Blink more frequently.
2. Vision break for every 15 minutes.
3. Use eye lubricants

Glare and Reflection

Reflections of the monitor cause eye strain. This causes squinting or developing a posture that is fatiguing in order to "get around the glare"

Solutions

1. Reposition the monitor such that windows are not directly to the front or behind the monitor.
2. Overhead lights can cause glare: change light bulbs to a full spectrum light or a lower wattage light. Regular fluorescent lights emit blue light, which makes it difficult for the eyes to focus, changing the colour of the light can make the eyes more relaxed.
3. Use a glare filter.

Monitor

The redraw or refresh rate of many monitors is 60 Hz. This speed causes a flicker that makes the screen appear to roll. The flicker stimulates the eye to accommodate or to refocus, tiring the visual system.

Solutions

1. Reset the refresh rate of the monitor to 70–85 Hz.
2. Poor positioning of your monitor can make it difficult to focus on the screen. Take care of monitor size, repetitive work, font size and colour, brightness/contrast.
3. Flat screens have fewer glares. Keep the screen free of dust and smudges.

Ergonomics

Poorly set up workstations cause pain due to repetitive injury. In people over 40, who have a decrease in focusing ability and rely on bifocals.

Solutions

Adjust the height of monitor. Monitor positioning should be 20″–26″ away. Computer glass help to give ergonomically correct position, and decrease repetitive motion injuries. A Glare coating on lenses helps (http://www.vh.org/adult/patient/opthalmology).

Section 7: Ergonomic Conditions for Avoiding Eyestrain

The following tips are suggested in (http://www.ctsplace.com/http://www.ergonext.com/) for avoiding eyestrain:

1. Look away from the monitor for 30 seconds, every 15 or 20 minutes. Look at or scan things at least 20 feet away to allow the eyes to focus in rest position.
2. Position the monitor 20″–26″ from eyes.
3. Set the monitor height so that the top edge is even with your view when looking straight ahead. The optimal screen position is 10 to 20 degrees below eye level.
4. Set the monitor screen resolution, internet browser text size, and the zoom and font default on other applications so that text is easy to read (600   800 for the screen resolution is standard). Set the monitor refresh rate to 75 hz or above.
5. Blink often else use an eye moistener.
6. Put the palm of your hands over eyes for a minute, once every half hour. This warms the muscles around the eyes, relaxing them.
7. Minimize glare. Make sure the background light level around the monitor is about the same as the screen light level. Minimize direct sunlight or bright lights in front of the monitor or directly behind it. Attach a glare shield if a bright background or reflection can't be minimized.

8. A bright screen causes eyestrain. Adjust the contrast and brightness to levels you use when reading a book comfortably.

Section 8: Conclusions

Computer ergonomics is a fast emerging arena, which involves study of handling many problems of stress injury due to bad ergonomic designs of machines, bad usage of the devices and bad environmental conditions. Basically, computer users can promote paperless office. Usage of paper can cause environmental problems unless proper care is taken. An old paper degenerates and invites mites and other insects. Nevertheless, long-term, continuous computer usage could lead to stress injury problems. In this context, this paper has discussed solutions and suggestions A taxonomy of ergonomics is presented.

References

____2004. Active report INRIA.

Carbonell, N. and S. Kieffer, Do oral messages help visual search? In: *Natural, Intelligent and Effective Interaction in Multimodal Dialogue Systems.*

Charles Eaton, M.D., 1997. *Electronic Textbook on Hand Surgery.* http://www.eatonhand.com/.

Grant, Chris. *Ten Things you should know about Hand and Wrist Pain.* Ann Arbor, Michigan.

Hedge, A. and T. Muss. 1999. *Comparative Study of Two Computer Mouse Designs.* Cornell University.

http://www.ctsplace.com/http://www.ergonext.com/

http://www.iea.cc/ergonomics.

http://www.vh.org/adult/patient/familymedicine/, "Computer Eye Strain: The Dilbert Syndrome".

http://www.vh.org/adult/patient/familymedicine/prose/, "Chronic Fatigue Syndrome: What is it?"

http://www.vh.org/adult/patient/ophthalmology/, "Computer Vision Syndrome".

Kay, Alan and Adele Goldberg, 1977. "Personal dynamic media. *IEEE Computer.*

Keir, Bach, Rempel, 1998. *Effects of Finger Posture on Carpal Tunnel Pressure During Wrist Motion.* Division of Occupational Medicine, U.C.S.F., 1998.

Lee, Jintae & Tosiyasu, K., 1995. *Model-Based Analysis of Hand Posture.* University of Aizu, IEEE.

Lin, J., Ying Wu and T.S. Huang, 2001. Modeling the constraints of human hand motion. In: *Proc. 5th Annual Federated Laboratory Symposium,* (ARL 2001), Maryland.

Nielsen, Michael, Moritz Storring, Thomas B. Moeslund and Granum, 2003. A procedure for developing intuitive and ergonomic gesture interface for HCI: Gesture-Based Communication in Human-Computer Interaction: 5th International Gesture Workshop, GW 2003, Geneva, Italy, April 15–17, 2003, Selected Revised Papers, LNCS 2915/2004, pages 409–420.

Nielsen, Michael, Thomas Moeslund, Moritz Störring and Erik Granum, 2003. Procedure for Developing Intuitive and Ergonomic Gesture Interfaces for Man-Machine Interaction, Technical Report CVMT 03-01 ISSN 1601–3646, http//www.cvmt.dk/mnielsen.

Scapin, D.L. and J.M.C. Bastein, 1997. Ergonomic criteria for evaluating the ergonomic quality of interactive systems. *Behaviour and Information Technology,* 16: 220–231.

Shaw, G. and A. Hedge. *The Effect of Keyboard and Mouse Placement on Shoulder Muscle Activity and Wrist Posture.* CU Ergo, Cornell University.

Chapter 22

Studies on Feasibility of Different Packaging Materials for Turmeric

V.T. Yamgar and M.S. Shirke
Agricultural Research Station, K. Digraj, District Sangli, M.S., India

ABSTRACT

Storage studies on turmeric were conducted during 1996-1999. Pooled data over three years revealed that the losses observed in storage of turmeric due to use of different packing material were statistically significant. The highest losses (19.38 per cent) were observed in the treatment, use of gunny bags where as the minimum losses (9.19 per cent) are there when the iron bins were used. Therefore for storing the processed marketable turmeric for about eight months or more, iron bins were recommended. Other materials like empty plastic bags of fertilizers, bamboo bin, plastic bags, gunny bags alone and with plastic bags inside it were not found beneficial.

Keywords: *Turmeric, Iron bin, Gunny bag, Bamboo bin, Storage.*

Introduction

Spices like all other agricultural commodities, invariably contain high moisture (55–85 per cent) at the time of harvest which must be brought down to 8–12 per cent. During their post harvest processing they are subjected to different types unit operations like drying, cleaning, grading and packaging until they are ready for the consumer or for market. Such post harvest processing should ensure proper conservation of the basic qualities of spices for which they are valued. In most cases like turmeric they are transported to an appropriate centre for marketing or processing. Storage of course vary with the individual species concerned (Pruthi, 1970, 1980, 1993(b), Purseglove 1981).

Spices like turmeric are hygroscopic in nature and being highly sensitive to moisture, their absorption of moisture may result in discoloration, rancidity, mould growth and insect attack. Furthermore turmeric contains volatile aromatic principles and loss of these and the absorption of foreign colours as a result of insufficient packaging may pose serious problems, especially in ground spices.

Pruthi (1985) has classed the turmeric, which contains curcumin pigment as spice that needs protection against light otherwise fade during storage. Turmeric also needs protection against the ingress of moisture. Besides the essential oil components naturally present are prone to oxidation by atmospheric oxygen, particularly at high storage temperature, resulting in the development of off flavours. The ingress of moisture also encourages mould attack and insect infestation (Pruthi, 1994). They have further described the paper products, polythene flexible films, aluminium foils, tin hesian and timber suitable for packaging of spices. The selection of packing material intrinsically will depend on the nature and other considerations. Hence, experiment was conducted to assess the feasibility of packing materials of turmeric.

Table 22.1: Mean % Losses in Weight of Turmeric as Influenced by Different Packaging Materials

Sl.No.	*Treatments*	*Losses in Weight (Mean)*			*Pooled Mean*
		1996–97	*1997-98*	*1998-99*	
1.	Empty plastic bags of fertilizer	6.75 (15.04)	3.54 (10.88)	13.01 (21.14)	7.30 (15.68)
2.	Bamboo bin	8.50 (16.94)	5.71 (13.81)	11.70 (19.99)	8.50 (16.91)
3.	Plastic bags of 90 gauge	15.75 (13.85)	2.58 (9.28)	4.85 (12.69)	4.30 (11.94)
4.	Plastic bags of 100 gauge	7.25 (15.59)	2.17 (8.56)	3.16 (10.26)	4.00 (11.47)
5.	Plastic bags of 150 gauge	3.25 (12.36)	2.00 (8.23)	3.03 (10.03)	3.20 (10.20)
6.	Iron bin	1.75 (7.59)	0.87 (5.44)	6.30 (14.54)	2.60 (9.19)
7.	Gunny bags lined with plastic bags of 90 guage	3.25 (10.16)	1.92 (8.05)	4.61 (12.47)	3.20 (10.23)
8.	Gunny bags	10.50 (19.12)	4.87 (12.79)	19.59 (26.21)	11.00 (19.37)
	SE±	1.24 (0.05)	0.82 (0.02)	0.61	1.54 (0.08)
	CD	3.76 (0.43)	2.51 (0.19)	1++84	4.68 (0.67)

Materials and Methods

The experiment was conducted at Agricultural Research Station, K. Digraj during 1996–1999. The experiment conducted as per randomized block design with three replication and eight treatments. Before storage the turmeric was processed through boiling, adequate sun drying and polished properly. The turmeric is stored on concrete floor in well ventilated room under ambient conditions. The packaging is undertaken by observing the initial weight of prove beneficial year after year. Tin is described as advantageous for storing the costly spices having delicate thermolabile aroma like saffron (Pruthi, 1994). They have found the polythene lined jute bags to give better protection than that of gunny bags for packing whole spices and supported the results of present study as the comparatively minimum losses were found in gunny bags lined with polythene bags (10.23 per cent).

The results thus concluded and recommended as for minimizing the losses in weight of fingersets of turmeric, the iron bins are useful than other packaging material tested.

References

Pruthi, J.S., 1970. Packaging of spices and condiments. *Jour. Packaging (India)*, 3(1): 110: 3(2): 7.

Pruthi, J.S., 1980(a). *Spices and Condiment: Chemistry, Microbiology and Technology*. Academic Press Ind., New York (USA).

Pruthi, J.S., 1993(a). Major spices of India: Crop management and Post-Harvest Technology. *Indian Council of Agril. Research*, New Delhi.

Pruthi, J.S., 1994. Innovations in post harvest technology of species. In: *Advances in Horticulture* Vol. 10: Plantation and Spice Crops Part 2, (Eds) K.L. Chadha and P. Rethinam. Malhotra Publishing House, New Delhi, India, pp. 1269–1282.

Purseglove, J.W., 1981. *Spices*, Vols. I and II, Longman, London.

Chapter 23

Effect of Organic Manure, Inorganic and Biofertilizers on Yield of Turmeric (Curcuma longa) Cv. Salem.

V.T. Yamgar, M.S. Shirke, B.M. Kamble and P.S. Patil
Agricultural Research Station, K. Digraj, District Sangli, M.S., India

ABSTRACT

A field experiment was conducted during 1999–2002 to assess the performance of organic, inorganic and biofertilizers on the yield of turmeric and its monetory returns. The pooled data revealed that the difference in yield of green turmeric was statistically significant. Among the various fertilizer treatments combinations, the RDF (200: 100 : 100 kg NPK/ha) + 25 tonnes of FYM and 5 kg biofertilizer/ha (T_2) was produced highest yield (386.80 qt/ha) of green turmeric followed by 75 per cent RDF + 35 tonnes of FYM and 5 kg biofertilizer/ha (T_3) which was at par with T_2. The net monetory returns were also found maximum (Rs. 89,420/ ha) due to the treatment T_2 as it has resulted in highest yield.

Keywords: *Turmeric, RDF, FYM, Biofertilizer, Salem, Inorganic.*

Introduction

The indiscriminate use of chemical fertilizers has caused serious damage to the soil rendering them often times, saline and less suitable for cultivation. The heavy cost of production or procurement of nitrogenous and phosphatic fertilizers is also another factor to be reckoned with for a developing countries like ours. Hence, it has become imperative to turn to more ecofriendly methods of nutrient management of horticultural crops. Organic and biofertilizers offer themselves as alternative to chemical fertilizers and are being increasingly used in horticultural crop production today.

To maintain the soil fertility on long term basis integrated nutrient management system (INMS) especially with biofertilizers is crucial in underground crops like potato, turmeric (Kate *et al.*, 2005). Biofertilizers offer an economically attractive and ecologically sound means of reducing external inputs and improving the quality and quantity of internal resources. Biofertilizers are inputs containing micro organisms capable of mobilizing nutritive elements from unusable form to usable, through biological processes. They increase the sustainability of soil and make it more productive. The biofertilizers having significance to horticultural crops are biological nitrogen fixer, phosphate solubilizers (Asokan *et al.*, 2000).

Turmeric responds to heavy dressings of organic matter and many experimental evidences are available on the beneficial effects of organic matter either alone or incombination with inorganic fertilizers on the growth and productivity of turmeric (Rethinam *et al.*, 1994). Usually large quantities of organic manures in the form of FYM are applied in different turmeric growing states. Recommendations differ according to variety, soil type, initial soil fertility and the climatic conditions.

Turmeric, being an exhaustive crop, requires heavy manuring for higher yields. Its manurial and fertilizer requirements vary according to climatic conditions. Many reports revealed that organic manuring for turmeric is essential (Nair, 1982; Inamdar and Diskalkar, 1987; Saha, 1988), detailed studies on role of organic manures on growth and productivity of turmeric is lacking. The information available is very scanty on the effects of biofertilizers on growth, and yield of sweet potato and other tuber and root crops (Jadhav *et al.*, 1998). Hence, the present investigation was undertaken to study the effect of organic, inorganic and biofertilizers on yield and monetory returns from turmeric.

Materials and Methods

The experiment was carried out at Agricultural Research Station K. Digraj, District Sangli, Mahatma Phule Krishi Vidyapeeth, Rahuri (M.S.) during the year 1999–2002 on medium black soils. The initial soil pH was 8.2, its electric conductivity (EC) was 0.38 ds/m, organic carbon 0.52 per cent while available NPK were 218, 8.42 and 524 kg/ha respectively. The plot size was maintained as gross 6 3.75 m^2 and net 4.80 3 m^2. The trial was laid out in Randomized Block Design with six replications using four levels of fertilizers as treatments (Table 23.1). The cultivation was undertaken as recommended agronomical practices. The commonly cultivated variety, 'Salem' was taken under study. The required quantity of NPK fertilizers was calculated and applied in the form of urea (46 per cent), single super phosphate (16 per cent) and muriate of potash (60 per cent K_2O), respectively. The phosphate solubilizing bacterial and Azotobactor cultures were used as biofertilizer.

The observations on growth attributes like plant height and average number of tillers were undertaken (Table 23.1). The data on yield and monetory returns as per prevailing market price in local Sangli market are statistically analysed and presented in Tables 23.1 and 23.2, respectively.

Results and Discussion

The results obtained from the present investigation on green yield of turmeric as influenced by application of various levels of organic manure, inorganic and biofertilizers are discussed hereunder.

The data from Table 23.1 revealed that during the experimental year 2001–2002 the treatments of fertilizer levels affect significantly the plant height, average number of tillers and per cent recovery. The per cent recovery (21.5), plant height (110.25 cm) and average number of tillers (2.30) observed as highest in the treatment of RDF (200 : 100 : 100 kg NPK/ha) + 25 tonnes of FYM + 5 kg biofertilizer/ha

(T_2) as compared to others. However during this third year of experiment yield of green turmeric was found produced highest in treatment of 75 per cent RDF + 35 tonnes of FYM and 5 kg biofertilizer/ha (T_3). The yield difference among the treatments was found statistically significant.

Table 23.1: Growth and Yield Attributes of Turmeric as Influenced by Organic Manure, Inorganic and Biofertilizers

Sl.No.	Treatments	Yield		Recovery (%)	Plant Height (cm)	Av. No. of Tillers
		kg/plot	qt/ha			
1.	RDF + B.F.	41.49	288.13	20.00	95.30	1.96
2.	RDF + 25 ton FYM + B.F.	42.13	292.55	21.5	110.25	2.30
3.	75% RDF + 35 ton FYM + B.F.	44.50	309.03	19.20	100.20	2.10
4.	50% RDF + 45 ton FYM + B.F.	43.72	303.63	20.10	96.74	2.20
	SE±	0.697	4.84			
	CD at 5%	2.092	14.59			
	C.V. %	3.98	3.98			

Table 23.2: Mean Green Turmeric Yield as Influenced by Different Organic Manure, Inorganic and Biofertilizers (Pooled)

Sl.No.	Treatments	Green Yield kg/plot			Pooled Mean	Yield (qt/ha)
		1999–2000	2000–2001	2001–2002		
1.	RDF + B.F.	41.60	65.48	41.49	49.53	343.95
2.	RDF + 25 ton FYM + B.F.	50.91	74.04	42.13	55.70	386.80
3.	75% RDF + 35 ton FYM+B.F.	44.68	73.41	44.50	54.20	376.38
4.	50% RDF + 45 ton FYM+B.F.	40.88	70.78	43.72	51.79	359.65
	SE ±	6.41	1.44	0.697	1.629	11.31
	CD at 5%	19.32	4.33	2.092	5.638	33.93

Table 23.3: Monetary Returns as Influenced by Different Organic Manure, Inorganic and Biofertilizers

Sl.No.	Treatments	Gross Returns (Rs./ha.)	Cost of Production (Rs./ha.)	Net Returns (Rs./ha.)	B : C Ratio
1.	RDF + B.F.	137580	58967	78613	1 : 1.33
2.	RDF + 25 ton FYM + B.F.	154720	65300	89420	1 : 1.36
3.	75 per cent RDF + 35 ton FYM + B.F.	150540	67521	83319	1 : 1.22
4.	50 per cent RDF + 45 ton FYM + B.F.	143860	73076	70784	1 : 0.96

Selling price of green turmeric: Rs. 2000/qt.

However, the data after pooled analysis over three years (Table 23.2) revealed that the difference in yield of green turmeric due to various treatments is again found statistically significant. The highest yield (386.80 qt/ha) was recorded in the treatment RDF + 25 tonnes of FYM + 5 kg biofertilizer/ha (T_2).

The next best treatment due to which the yields recorded was (376.38 qt/ha) is 75 per cent of RDF + 35 tonnes of FYM + 5 kg biofertilizer/ha (T_3) and was at par with T_2. These results on plant growth (Table 23.1) and (Table 23.2) indicating the vital role of higher dose of NPK with FYM and biofertilizer in build up of better plant growth related to miximization of uptakes of nutrients and their synergistic effects on 'P' and 'K' in availability of 'N' to the plants. Manjunathgoud, *et al.* (2002) have also recorded the similar trend of results as proportionate increase in growth parameters like plant height, number of tillers/clump, number of leaves etc. with increased level of NPK fertilizers upto certain level (150: 125: 250 kg NPK/ha) beyond which they found no significant difference in growth of turmeric and supported the result of present study. Sheshagiri (1993) and Venkatesha (1994) have also recorded maximum growth of turmeric when fertilized with the 150 : 125 : 250 kg NPK/ha and thus again supported the present results except in case of potassium as in present investigation initial available 'P' is 524 kg/ha indicating differential potash level in soils of different climatic conditions. Venkatesha *et al.* (1998) reported the linear increase in cured yield of turmeric and B : C ratio from 0 : 0.0 to 150 : 125: 250 kg NPK/ha, beyond this level declining trend was observed. Plants provided with this NPK dose recorded highest cured yield and B : C ratio as compared to other levels of NPK as observed in the present study. They have also reported that the turmeric has found to respond to both organic and inorganic manures.

Umate *et al.* (1984) in Maharashtra reported the significant increase in yields of fresh rhizomes (31 t/ha) with increased levels of nitrogen upto 120 kg/ha. However, Muthuvel *et al.* (1989) did not observed any significant effect due to varying levels of potassium on Co-1 Turmeric at Bhavanisagar (TN) and suggested only 60 kg K_2O/ha as adequate. Later, Thamburaj (1991) reported that application of potassium at 90 kg/ha in four splits recorded higher yields of rhizomes 30 t/ha and supported the results of yield 38.68 tonn/ha in present study with application of 100 kg K/ ha. The increased yield in this study could attributed to increased dose of 'NPK' due to addition of organic matter 'FYM' and biofertilizer having vital role in improving soil properties and soil fertility due to availability of micronutrients by biological processes as described by (Asokan *et al.*, 2002).

The data on maximum monetory returns from the green turmeric produced by applying various fertilizer doses indicated that, net returns of Rs. 89,420/ha were obtained due to fertilizer treatment of RDF + 25 tonnes of FYM + 5 kg biofertilizer/ha (T_2) was highest. The Cost : Benefit ratio was also found highest (1 : 1.36) in the same treatment. Hence, the treatment is recommended for production of maximum turmeric yield in the western Maharashtra conditions. By considering the uptake of NPK, yield and B : C ratio application of 150 : 125 : 200 NPK kg/ha is suggested for maximum production as optimum irrespective of cultivars (Venkatesha 1998).

Jadhav *et al.* (1998) reported that the 75 kg Nitrogen /ha consugated with Azotobactor (AZT) + Azospirilum (AZP) recorded more monitory benefits and net additional returns over control than application of RDF of Nitrogen alone in sweet potato. They have further reported that the biofertilizer inoculants and increasing levels of nitrogen enhanced the sprouting, vine length, total number of tubers / plant in potato and these results are in accordance with those of present study where in biofertilizer may attributed for increased growth and yield of turmeric.

References

Asokan, R., Sukhada Mohandas and Lalitha Anand, 2000. Biofertilizer and biopesticides for horticultural crops. *Indian Horticulture*, pp. 44–47.

Inamdar, K.S. and Diskalkar, P.D., 1987. Turmeric, its cultivation and marketing in Tasgaon Taluka

(Sangli District). *J. Maharashtra Agric. Univ.*, 12: 241–242.

Jadhav, A.C., Memane, S.A. and Konde, B.K., 1998(a). Economics of the use of Bacterial inoculants in sweet potato. *J. Maharashtra Agric. Univ.*, 23(1): 33–35.

Jadhav, A.C., Memane, S.A. and Konde, B.K., 1998(b). Evaluation of biofertilizers in respect to growth and yield attributing parameters in sweet potato.

Kate, D.M., Solanke, A.V., Tiwari, T.K. and Nemade, S.M., 2005. Growth and yield of potato cultivars as affected by Integrated nutrient management system. *J. Maharashtra Agric. Univ.*, 30(2): 236–237.

Manjunathgoud, B., Venkatesha, J. and Bhangavantagoura, K.H., 2002. Studies on plant density and levels of NPK on growth, yield and quality of turmeric Cv. Bangalore local. *Mysore J. Agric. Sci.*, 36: 31–35.

Muthuvel, P., Rajakanna, B.K., Selvaraj, V., Kulandaivelu, R. and Chamy, A. 1989. Irrigation nitrogen and potash requirements of turmeric south. *Indian Hort.*, 37: 61–63.

Nair, P.C.S., 1982. Agronomy of ginger and turmeric. In: *Ginger and Turmeric*, (Eds.) M.K. Nair, T. Premkumar, P.N. Ravindra and Y.R. Sharma, Central Plantation Crops Research Institute, Kasaragod, pp. 63–68.

Rethinam, P., Srinivasan, K. and Sushama, P.K., 1994. Nutrition of Turmeric In: *Advances in Horticulture Vol. 9: Plantation and Spice Crops. Part 1*, (Eds.) K.L. Chadha and P. Rethinam. Malhotra Publishing House, New Delhi, India.

Sheshagiri, K.S., 1993. Studies on nutrition of turmeric in hill region. *M.Sc. (Agri.) Thesis*, Univ. of Agril. Sci., Bangalore.

Venkatesha, J., 1994. Studies on the evaluation of promising cultivars and nutritional requirements of turmeric *Ph.D. Thesis*, Univ. of Agril. Sci., Bangalore.

Venkatesha, J., Khan, M.M. and Chandrappa, H., 1998. Studies on uptake of NPK Nutrients by Turmeric cultivars. *J. Maharashtra Agric. Univ.*, 23(1): 12–14.

Chapter 24

Utilisation of Delonix regia (Bojer ex Hook) Raf, as a Fuelwood Resource and its Impact on Health Among the Rural Folk Around Mysore City, India

N.R. Rajendra Prasad[1], *A.G. Devi Prasad*[1] *and S.P. Hosmani*[2]

[1]*P.G. Department of Environmental Sciences, University of Mysore, Mysore – 570 006, Karnataka, India*

[2]*P.G. Department of Botany, University of Mysore, Mysore – 570 006, Karnataka, India*

ABSTRACT

Energy is a fundamental tool for social and economic development to attain improved quality of life. About 33 per cent of a developing country's energy needs are fulfilled by biomass. India, a country of villages, the rural folk mainly depend on biomass for cooking and heating. Investigation carried out about the fuelwood species utilized in and around villages of Mysore in Karnataka State have shown that the woody branches and pods of *Delonix regia* are being widely exploited. The paper analyses about the calorific values of these parts in comparison with traditional fuels used and also highlights about the characteristics and effects of combustion of *Delonix regia* wood and dried pods including human health hazards.

Keywords: *Energy, Delonix regia, Calorific value, Biomass, Health hazards.*

Introduction

Energy is the thread sustaining and integrating all life and its use determines the level of abuse to organisms as well as the life supporting systems. It is essential to economic and social development and improved quality of life in India. Over the past few centuries there has been rapid transition from the traditional biomass fuels (wood, animal dung, crop residues such as rice husks, etc) to fossil fuels such as kerosene, LPG or to electricity. About half of the human population has been able to afford the fossil fuels, which burn cleanly and efficiently compared to biomass fuels. The remaining half of the population almost all in developing countries today rely on biomass fuels for cooking or home heating in colder places. In fact, household fuel needs, continue to make up more than half of total energy demands in more than 100 countries (Merson *et al.*, 2001). It is a surprising fact that biomass provide up to 15 per cent of the total world energy equivalent. For about half of the world population, living mainly in the third world, biomass in the form of fuelwood is overwhelmly the most important traditional energy source (Rajendra Singh and Dinesh Kumar Mishra, 1996).

From the past several years, the firewood has been replaced by much cleaner energy sources, such as conventional fuels–kerosene, LPG, CNG, etc., and now renewable energy sources such as solar, wind, biogas, etc. Amidst the changing energy sources, the firewood stands the dominating energy source used for domestic purposes in the rural area of India. Around 180 million tons of biomass fuel is used for cooking with inefficient and smoky cook stoves or chulas (Rajvanshi, 2003).

As on March, 2001, the annual average per capita consumption of fuelwood in forested rural areas of Karnataka was 406 kg. The projected population dependent on forests in Karnataka by 2006 is 11.4 million with an annual consumption of 4.6 million tons of fuelwood. Rural households derive biomass fuel from natural forests, plantations, woodlots and trees around the homestead (Agarwal, 1998).

Biomass is the main source of the energy for many developing countries, providing more than 90 per cent of the energy supply in some developing countries (Kayagusuz and Ahmet Sari, 2003). Biomass fuel stands the dominating energy source used for domestic purposes in the rural and poor households especially in the developing countries (WHO, 1997 and WRI, 1998). In Karnataka, non-commercial energy sources like firewood, agricultural residues, charcoal and cowdung account for 53.2 per cent (Ramachandra *et al.*, 2004).

It is observed that the rural people residing around the urban fringes of Mysore mainly depend on the local trees as firewood for cooking and heating water. Due to increased demand for fuelwood, people have started exploiting the avenue trees to meet their requirement. The most commonly exploited species in these areas is *Delonix regia*, a legume tree. The present article analyses the products of this species utilized as fuelwood and its impact on the health of consumers.

Materials and Methods

About 20 villages were surveyed around the Mysore City to know about the species of firewood that are being frequently used by the villagers. About 200 households in each village were interviewed to know about the commonly used firewood. Observations were made on the plant parts collected especially of *Delonix regia*, a most exploited species as firewood resource.

The calorific value of dried fruits and wood of *Delonix regia* were determined using Bomb Calorimeter following the standard methods of British Standards Institution (BS 1016 : Part 5, 1967).

People involved in burning of the wood and dried fruits in each village were screened for their health hazards. Data pertinent to the consumption of different types of fuel sources and consumer's opinion (expressed in percentage) regarding the characteristics and effects of combustion in the villages around Mysore taluk were obtained.

Results and Discussion

The survey carried around the rural areas of Mysore city revealed that the woody branches and dried pods of *Delonix regia* are extensively exploited as a source of firewood especially during summer. The calorific values of woody branches and dried pods estimated were found to be 13,557 KJ/kg and 12,041 KJ/kg respectively. A comparative picture of average calorific values of different fuel sources used in these villages around Mysore are shown in Table 24.1. The characteristics and effects of burning or combustion of woody branches and pods of *Delonix regia* are shown in Table 24.2.

Table 24.1: Average Calorific Values of Different Fuels Consumed in Villages Around Mysore

Fuel	*Consumption in the Year 2002*	*Average Calorific value (KJ/kg)*
LPG	21.22 million kgs	48000
Kerosene	14.88 million litres	43450
Firewood	0.34 million tones	15000
Biogas (for 3 and 4 m^3 capacity plant)	2.7×10^4 cubic meter	27 MJ/m^3
Delonix regia		
Stem	*NA	13557
Fruits		12041

* NA: Data not available.

Table 24.2: Characteristics and Effects of *Delonix regia* Parts Used as Fuelwood

Sl.No.	*Characteristics and Effects of Combustion*	*Consumer's Opinion (Expressed in %)*	
		Dry Fruits	*Dry Stem*
1.	Ignition fast	90	75
2.	Burns well	100	85
3.	Heat production less	65	75
4.	Produce smoke	85	80
5.	Produced ash	95	60
6.	Produces smell (pungent smell)	70	35
7.	Health hazards (couch, eye and nose irritation)	100	95

The studies have shown that the wood is yellow-brown, weak, brittle and soft with a specific gravity of about 0.3. Although the species is not a good timber source, the wood is widely used as firewood. The pericarp of the fruit is also used as fuel source. The investigations carried out have shown that the energy values of woody stem and fruits of *Delonix regia* were least as compared to that of LPG or kerosene or good firewood. Their energy values are 2–3 folds lesser than biogas, LPG and kerosene as already detailed in Table 24.1. This may be one of the probable reasons for more utilization and exploitation of this species. This also signifies that the rural folk are ignorant of energy values of fuel resources which in turn has made them to burn more quantity of woody branches and pods till their requirements are met with.

The investigations were also made with respect to the characteristics and effects of combustion of the woody parts and pods. The survey made on the health hazards among the users revealed that more than 95 per cent of the people associated with burning or cooking with these parts are affected by cough, eye and nose irritation. The cooking or water heating places in these houses are not properly ventilated. Domestic cooking has become an important source of indoor air pollution and is associated with significant mortality (Samet *et al.*, 1987). It was observed that the pods ignite faster and burn quickly compared to woody branches. But the calorific or energy values are found to be higher in woody stem than pods. Burning of pods produce more smoke and ash without the formation of any charcoal. The wet woody stem and pods have produced mild pungent smell while burning. The smoke emitted during this have blackened the cooking vessels and walls of the room to a greater extent. It is observed that women in these villages suffer more from the smoke related health problems. Also it was found that the woodfuel consumption per capita for large households is greater than for small households. The surveys estimated average per capita domestic energy consumption to be 0.38 and 0.40 Tonne of Coal Replacement (TCR) in the rural and metropolitan areas, respectively (Ramachandra *et al.*, 2000).

The rural folk exploit this species mainly during summer which has also resulted in the prevalence of respiratory disorders to a greater extent in summer. There is no utilization of this species or fuel source during rainy season.

The people use kerosene and LPG more during rainy and winter seasons and try to strike a balance with their economy by exploiting *Delonix regia* as fuelwood source during summer as it is available in their vicinity for no cost. Moreover, most of the very low income households in "these villages depend on this species for a long period of time irrespective of the season. This has led to the decline in population of this species. Cooking energy has the major share in the total household energy consumption (Vijay Laxmi, *et al.*, 2003). It has been rightly suggested by Yohji Uchiyama (2002) that stable energy supply is absolutely essential for leading a high quality of life. A well-balanced utilization of energy resources must be achieved through economic growth, a stable energy supply and environmental preservation.

The forest around the villages of Mysore cannot supply enough for heating and cooking. Hence, there is a pressure on the availability of firewood in these villages. Therefore, owing to its demand, it has been suggested to grow this species in the wastelands available in their areas as energy plantations and also to encourage this to grow as an avenue tree. Also, the people are being suggested to use eco-friendly cookstoves of better quality for efficient burning of fuelwood.

References

Agarwal, A., 1998. Myths about forest depletion. *Down to Earth*, 7(1): 29–42.

Bogcich, S.V., 1985. *Wood as Fuel-Energy for Developing Countries*. Praegar Publishers, USA.

Kayagusuz, K. and Ahmet Sari, 2003. Renewable energy potential and utilization in Turkey. *Energy Conservation and Management*, 44: 459–478.

Rai, S.N and S.K. Chakraborthi, 2001. Demand and supply of fuelwood and timber in India. *The Indian Forester*, 127(3): 263–279.

Ramachandra, T.V., G. Kamakshi and B.V. Shruthi, 2004. Bioresource status in Karnataka. *Renewable and Sustainable Energy Reviews*, 8: 1–47.

Ramachandra, T.V., D.K. Subramanian, N.V. Joshi, S.V. Gunaga and R.B. Harikantra, 2000. Domestic energy consumption patterns in Uttara Kannada District, Karnataka State, India. *Energy Conversion and Management*, 41: 775– 831.

Samet, J.M., M.C. Marbury and J.D. Spengler, 1987. Health effects and source of indoor air pollution. (Part I). *Am. Rev. Respir. Dis.*, 136: 1486–1508.

Singh, Rajendra and Dinesh Kumar Mishra, 1996. *Development and Environmental Changes in India.* APH Pub. Corporation, New Delhi, pp. 293–301.

The Wealth of India, 1952. *Raw Materials*, Vol. III: D–E, CSIR, New Delhi, pp. 30.

Vijay Laxmi, Jyoti Parikh, Shyam Karmakar and Pramod Dabrase, 2003. Household energy, women's hardship and health impacts in rural Rajasthan, India: need for sustainable energy solutions. *Energy for Sustainable Development*, 7(1): 50–68.

WHO, 1997. Health and Environment in Sustainable Development. WHO/EHG/97, Geneva: World Health Organisation.

World Resources Institute, 1998. United Nations Environment Programme, United Nations Development Programme, World Bank. World Resources 1998–1999: A Guide Environment. Oxford University Press, New York.

Yohji Uchiyama, 2002. Present efforts of saving energy and future energy demand/supply in Japan. *Energy Conservation and Management*, 43: 1123–1131.

Chapter 25

Biomedical Waste Generation and Impact on Environment and Health in Small Towns

J.A. John Paul[1] *and Thilagavathy Daniel*[*2]

[1]*Department of Biotechnology, J.J. College of Arts and Science, Pudukkottai – 622404, Tamil Nadu, India*

[2]*Department of Biology, Gandhigram Rural Institute–Deemed University, Gandhigram – 624 302, Tamil Nadu, India*

ABSTRACT

The wastes produced by the hospitals, clinics and medical laboratories are called biomedical wastes. There are two big hospitals (more than 250 beds), 46 small hospitals (5-30 beds), 75 clinics and 15 medical laboratories in Dindigul Town which is a small town with less than 2 lakh population. The biomedical wastes survey carried out over a period of 12 months *i.e.*, January to December in 2002 indicated an output of 20.16 per cent infectious wastes and 79.84 per cent non-infectious wastes. The infectious wastes such as tissues (uterus, umblical chord, muscles and blood), metals, sharps and needles, cotton and gauze, sanitary napkins and blood bags alone constitute 29.575 tonnes/year.

Keywords: *Biomedical wastes, Infectious wastes, Health hazards, Disposal methods.*

Introduction

Sometimes infectious waste from hospitals is mixed with the general municipal solid waste and it creates health problems to the sanitary workers, rag pickers and the public. Proper management of biomedical waste can minimize pollution. Improper disposal leads to environmental hazards also.

Corresponding Author.

Burning of mixed rubbish and incineration of hospital wastes emit super-toxic chemicals such as dioxins and polychlorinated biphenyls (Anon, 2001). Hence efforts must be made for proper scientific management of biomedical wastes in cities and towns. Preliminary survey about the sources of biomedical wastes, total quantity of biomedical wastes produced, composition of biomedical wastes and present disposal method are the important aspects to be considered for planning effective biomedical waste management.

Materials and Methods

Biomedical wastes produced from clinics, private and government hospitals of the fifteen sanitary divisions of Dindigul Town, Tamil Nadu were collected in two types of plastic bags, one for infectious wastes and the other for non-infectious wastes. The wastes were separated into different components and weighed. The sample was collected twice in a month for a period of one year *i.e.,* January to December, 2002. The mean values of the various components of the infectious and non-infectious wastes such as tissue waste, metals, sharps and needles, cotton and gauze, sanitary napkins, blood bags, bottles and bottle covers, rubber, carton, food waste and plastics and miscellaneous were estimated.

Results and Discussion

Different sources of biomedical wastes generated in Dindigul Town is given in Table 25.1. Two big hospitals (more than 250 beds), 46 small hospitals (5–30 beds), 75 clinics and 15 medical laboratories produce biomedical wastes in Dindigul Town. Only one big hospital, which is located in sanitary division two, has an incinerator. The composition of the biomedical wastes originating from the hospitals and the clinics of the fifteen sanitary divisions of Dindigul Town is given in Table 25.2. Of the total municipal solid waste 0.87 per cent is biomedical waste. The composition of the biomedical wastes of Dindigul Town showed 79.84 per cent of non-infectious wastes containing glass, rubber, food wastes, carton material, plastics and miscellaneous wastes and the remaining 20.16 per cent contained the infectious wastes such as tissue wastes, metal wastes, cotton and gauze, sanitary napkins and blood bags. Of this 3.73 per cent was highly infectious and it contained tissue wastes like uterus, umbilical chord, muscles and blood. The WHO (2002) also has observed the presence of about 20 per cent of hazardous materials and 80 per cent of benign materials in the biomedical wastes. The annual production of the infectious wastes are 29.575 tones. Different components of the infectious wastes of Dindigul Town in different months (January–December, 2002) are shown in Figure 25.1. Among the total wastes, metals, sharps and needles were observed in higher quantum during the months of September and October *i.e.,* 9.89 and 9.92 per cent respectively and these were lesser during the months of April and May *i.e.,* 6.83 and 6.75 per cent respectively. Different components of the non-infectious wastes of Dindigul Town observed in different months (January–December, 2002) are shown in Figure 25.2. Among the total wastes, food wastes dominated the other non-infectious wastes in all the months. It was observed maximum in the months of April and May *i.e.,* 56.68 and 57.76 per cent respectively and it was low in the months of September and October *i.e.,* 51.64 and 51.32 per cent respectively and that was because April and May months are the holiday season and draw more visitors to hospital. Deepa *et al.* (2002) had observed an increase in the organic wastes during non-working days than the working days in Mysore City, India.

The flow chart indicating the recommended disposal methods for various components of the biomedical waste of Dindigul Town is given in Figure 25.3. The biomedical wastes generated in the hospitals should be segregated at source into infectious and non-infectious wastes in different coloured containers as per the bio-safety guidelines. Infectious wastes should be disinfected, incinerated and buried. The non-infectious wastes can be transported to sub-depots and further segregated into non-degradables and degradables. Recycling, preparation of refuse derived fuel, land filling and incineration are the disposal methods suggested for the non-degradables. Conventional composting, vermicomposting and composting using microbial cultures and preparation of mixed manure for plant growth are the suggestions given for the disposal of degradables.

Table 25.1: Different Sources of Biomedical Wastes in Dindigul Town

Sanitary Divisions of Dindigul Town	*Big Hospitals (> 250 beds)*	*Small Hospitals (5–30 beds)*	*Clinics*	*Medical Laboratories*
1	–	12	5	1
2	1*	6	7	3
3	–	–	4	–
4	–	8	5	2
5	–	4	10	1
6	–	–	4	1
7	–	4	8	–
8	1	3	5	–
9	–	0	4	1
10	–	2	5	1
11	–	–	1	1
12	–	2	2	2
13	–	1	6	1
14	–	1	2	–
15	–	2	2	1
BML	–	1	1	–
Total	2	46	75	15

BML: In the border of municipal limit.

*: Incinerator is used.

Table 25.2: Composition of the Biomedical Wastes Produced in Dindigul Town

Sl.No.	*Category of Wastes*	*Weight (kg)*	*Percentage (%)*
	Infectious		
1.	Tissue wastes	15.19±0.24	3.78
2.	Metals, sharps and needles	33.56±0.65	8.35
3.	Cotton and gauze	14.59±0.20	3.63
4.	Sanitary napkins	8.48±0.17	2.11
5.	Blood bags	9.21±0.19	2.29
	Non-infectious		
6.	Bottles and bottle covers	43.53±0.83	10.83
7.	Rubber	21.06±0.38	5.24
8.	Carton	24.24±0.49	6.03
9.	Food wastes	220.02±4.02	54.74
10.	Plastics and miscellaneous	12.06±0.15	3.00

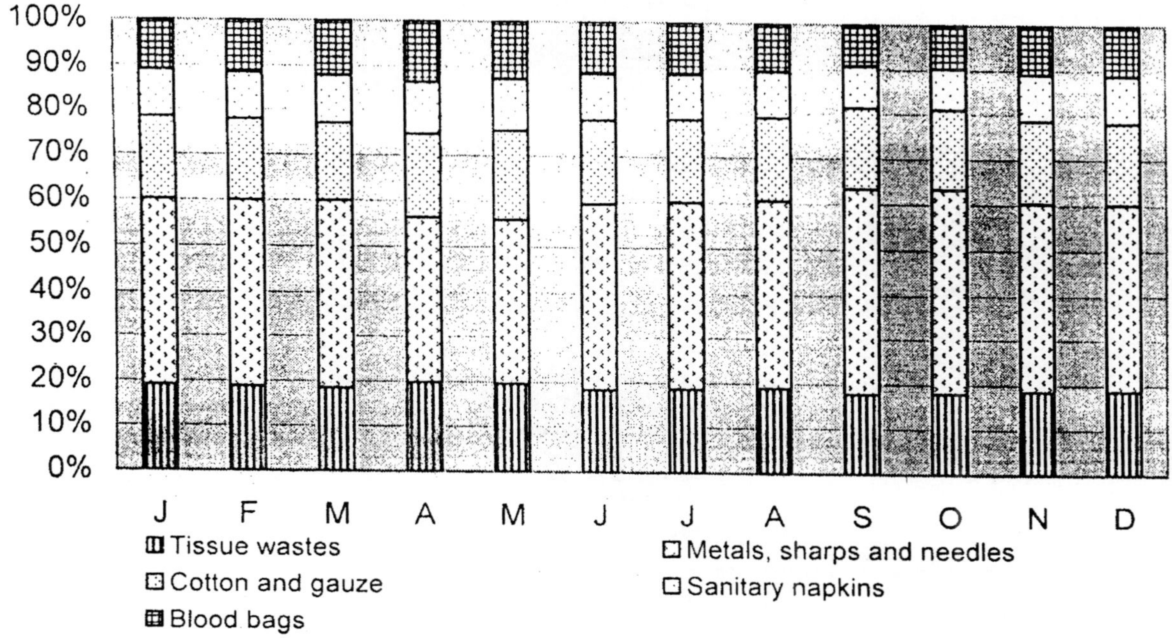

Figure 25.1: Different Components of the Infectious Wastes of Dindigul Town in Different Months (January–December, 2002)

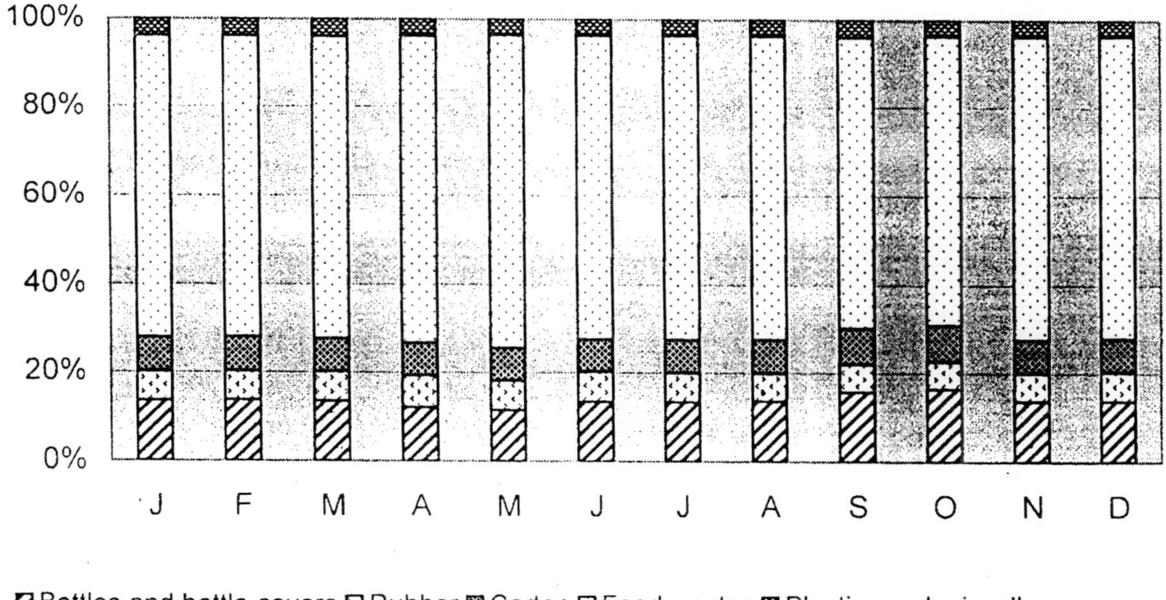

Figure 25.2: Different Components of the Non-infectious Wastes of Dindigul Town in Different Months (January–December, 2002)

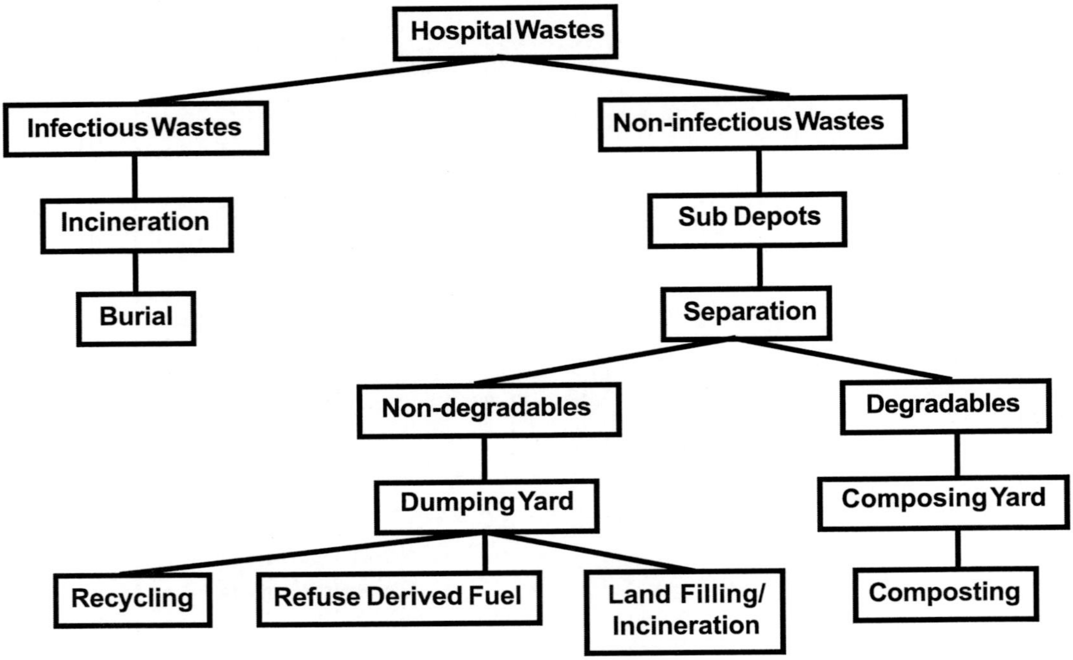

Figure 25.3: Flow Chart Indicating the Recommended Disposal Methods of Various, Components of the Biomedical Waste for Dindigul Town

Except one big hospital all the others have formed an association and send the everyday wastes to the dumping yard which is located in a corner of the Dindigul Town but well within the municipal limit. The wastes are neither disinfected nor incinerated, but are half burnt and left in the open air which form an inviting meal for the dogs. When they are carried by the dogs to the poorer sections of the town, the people are exposed to diseases and the environment is also dirtied. Strict regulations are needed to prevent the indiscriminate dumping of the biomedical wastes.

References

Anon, 2001. Waste burning hurts sexual development. *Science India Eco News*, 4(4): 20.

Deepa, A., Hemalatha, S., Sindhu, N. and Shetally, R., 2002. Solid waste management: A case study. In: *Proceedings of National Seminar on Solid Waste Management: Current Status and Strategies for Future*, (Eds.) R.K. Somashekar and M.A.R. Iyengar. Bangalore, India, pp. 14–18.

WHO, 2002. *Wastes from Health Care Activities*. World Health Organization Information, Fact Sheet No. 253.

Chapter 26

Impact of Composted Agro-industrial Waste on Biometric and Yield Parameters of Dolichos lablab (L.)

Y.K. Sridevi and A. Vijayalakshmi
Department of Botany, Avinashilingam Deemed University, Coimbatore – 641 043

ABSTRACT

A study was conducted to estimate rates of impact of the conjunctive use of organic manures like pressmud and coirpith on *Dolichos* lablab (L). A pot culture experiment was conducted in red soil with Bean seeds. The investigation included 5 treatments in various doses of pressmud, coirpith and *Rhizobium*. The yield parameters such as root length, shoot length, root volume, number of leaves, plant fresh weight, plant dry weight, no of fruits plant^{-1}, weight of fruits plant^{-1}, length of fruit was significantly increased by all the treatments and the combined application of composted pressmud, coirpith and *Rhizobium* was found to be most effective in increasing these yield parameters.

Keywords: *Composted, Pressmud, Coirpith, Rhizobium, Pleurotus florida, Urea, Bean seeds.*

Introduction

The philosophy of sustainable agriculture is to maximize crop production through scientific methods of farming, growing crop varieties and hybrids of high-yield and high-quality potential, using optimum inputs of manures, fertilizers, biofertilizers and agricultural chemicals without exploiting the natural resources of soil, water and environment. Waste if properly utilized is rich sources of plant nutrients. Biofertilizers are the products containing living cells of different microorganisms, which have an ability to mobilize nutritionally important elements from non-usable to usable forms through biological processes. Biofertilizers are apparently environmental and farmer-friendly renewable sources.

Pressmud contains mixture of sugarcane fibre, sucrose, cogulated colloides, cane wax, aluminoid with inorganic salts and soil particles. Pressmud cake (PMC) contains about 1.0 to 3.1 per cent nitrogen, 0.6 to 3.6 per cent phosphorus and 0.3 to 1.8 per cent potassium on oven dry basis. Coirdust is having attractive moisture retention property (500–600 per cent), higher potassium and micronutrient contents which could be exploited for agricultural use. This organic waste can be converted into valuable organic manure.

The protein occupies an important and unique place in human diet and legumes are major food sources of protein. Bean (*Dolichos lablab*) is an important multipurpose leguminous crop being rich in protein and other nutrients. It is used for various purpose *viz.*, pulse, immature seeds, green vegetables and the rest of the plant are used as animal fodder. In view of this, the present work was carried out to asses the resource potential of composted agro industrial waste on *Dolichos lablab.*

Materials and Methods

The agro industrial wastes, Coirpith was collected from Pollachi, Coimbatore District and Pressmud from Shakthi Sugar factory, Sathyamangalam, Coimbatore District. An area of 5 metre length and 3 metre width was selected. One hundred kilogram of the agro wastes was spread uniformly. One bottle of spawn (*Pleurotus florida*–300 gms) was spread in the agro wastes. This spawn was covered with one kg of urea. This process of sandwitching the *Pleurotus florida* and urea alternatively with the agro wastes was repeated until the heap reaches a height of one meter. Water was sprinkled on the heap to maintain the moisture content of the wastes. At the end of 50 days the wastes turn to black mass of compost.

The pots were filled with 7Kg of red sandy loam soils. Coirpith and Pressmud (alone and in combination with *Rhizobium*) were applied to the respective pots and mixed thoroughly. Viable seeds were selected and about 5 seeds were sown in each pot.

Treatment Details

There were five treatments as detailed below, which were evaluated against the control:

T_1: Absolute Control

T_2: Composted Coirpith 5 tha^{-1}

T_3: Composted Pressmud 5 tha^{-1}

T_4: Composted Pressmud (5 tha^{-1}) + Composted Coirpith (5 tha^{-1})

T_5: Composted Pressmud (5 tha^{-1}) + Composted Coirpith (5 tha^{-1}) + Rhizobium

On the 30th and 45th day after sowing (Vegetative Stage) and 60th day after showing (yield stage) the plants were uprooted from the pot and the vegetative and yield parameters were noted.

Results and Discussion

Effect of Composted Pressmud, Composted Coirpith and Biofertilizer (*Rhizobium*) on Vegetative Characters of *Dolichos lablab* (L.) (Tables 26.1 and 26.2)

Shoot Length and Root Length

On the 30th and 45th day of showing all the treatments increased the shoot length significantly. However the combination of composted pressmud (5 tha^{-1}) + composted coirpith (5 tha^{-1}) + Rhizobium was very effective in increasing the shoot length when compared to other treatments. The root length was significantly raised by T_5 (38.6 cm) when compared to control T_1 (25 cm). On the 45th day a favourable increase in the root length was noted with the treatments like T_4 and T_5 (T_4–34 cm and T_5–44.6cm) over the control (T_1–25 cm). This result is in agreement with the results of Jeyabal *et al.* (1998).

Table 26.1: Vegetative Parameters of *Dolichos lablab* (L.) Influenced by Pressmud, Coirpith and Biofertilizer (Rhizobium) 30 DAS

Treatment	*Shoot Length (cm)*	*Root Length (cm)*	*Root Volume (cu.mm)*	*No. of Leaves*	*Plant Fresh Weight (g)*	*Plant Dry Weight (g)*
T_1	25.00	23.33	0.83	6.33	4.70	2.09
T_2	29.66	28.33	1.00	6.00	6.19	1.30
T_3	28.00	31.00	1.16	6.33	7.44	2.11
T_4	34.66	35.33	1.40	7.33	8.85	2.83
T_5	38.66	41.33	1.56	7.66	9.69	2.96
S.Ed.	2.0331	2.0111	0.0516	0.4595	0.9039	0.4826
CD (5%)	4.5300	4.4810	0.1151	1.0238	2.0140	1.0752
CD (1%)	6.4436	6.3740	0.1637	1.4563	2.8648	1.5294
	*	**	**	*	**	*

*: Significant; **: Highly Significant.

Table 26.2: Vegetative Parameters of *Dolichos lablab* (L.) Influenced by Pressmud, Coirpith and Biofertilizer (Rhizobium) 45 DAS

Treatment	*Shoot Length (cm)*	*Root Length (cm)*	*Root Volume (cu.mm)*	*No. of Leaves*	*Plant Fresh Weight (g)*	*Plant Dry Weight (g)*
T_1	30.33	25.00	1.03	6.66	6.89	2.23
T_2	35.66	25.33	1.30	7.33	7.92	3.49
T_3	35.33	26.66	1.40	7.66	9.97	3.45
T_4	40.66	34.00	1.53	8.00	10.57	3.53
T_5	43.33	44.66	1.70	8.33	13.10	4.50
S.Ed.	3.3400	2.6833	0.1135	0.4216	0.8784	0.5329
CD (5%)	7.4420	5.9788	0.2350	0.9395	1.9572	1.1873
CD (1%)	10.5859	8.5045	0.3598	1.3364	2.7840	1.6889
	*	**	**	*	**	*

*: Significant; **: Highly Significant.

Root Volume and No. of Leaves

On the 30th day root volume was significantly influenced by the treatment of composted pressmud + composted coirpith + Rhizobium (T_5–1.56 cu.mm). On the 45th day the treatment such as (T_4–1.53 cu.mm and T_5–1.70 cu.mm) had a greater influence than control in increasing the root volume of Bean. Similar results were reported by Padmaja (2004). The number of leaves influenced by the treatment were slightly over control on the 30th day however a significant increase ware observed with treatments T_5 (8.33) on the 45th day.

Plant Fresh Weight and Dry Weight

On the 30th day the plant fresh weight were significantly increased in all the treatments (T_3, T_4 and T_5). On the 45th day plant fresh weight were increased significantly in T_3 (9.97 gm), T_4 (10.57 gm) and T_5 (13.10 gm) when compared with T_1 (6.89 gm) and T_2 (7.92 gm). All the treatments increased the plant dry weight of bean over control. On the 45th day, the treatment T_5 composted pressmud (5 tha^{-1}) + composted coirpith (5 t ha^{-1}) + Rhizobium was increased (4.50 gm) when compared with the control T_1 (2.23 gm).

Effect of Composted Pressmud, Composted Coirpith and Biofertilizer (*Rhizobium*) on Yield Characters of *Dolichos lablab* (L.) (Table 26.3)

Shoot Length and Root Length

At the harvest stage the combination of Composted pressmud (5 tha^{-1}) + Composted coirpith (5 tha^{-1}) + Rhizobium (T_5–52.33 cm) had a very significant effect on the root length. All the other treatments are on par with each other and superior than Control (T_1–36.66 cm). A similar finding was reported by Yang and Chang (1976) in sugarcane. The treatments were on par with each other and with control in influencing the root length of bean. At the harvest stage the treatments [Composted pressmud (5 tha^{-1}) + Composted coirpith (5 tha^{-1}) + Rhizobium augmented the root length significantly when compared to control.

Table 26.3: Yield Parameters of *Dolichos lablab* (L.) Influenced by Pressmud, Coirpith and Biofertilizer (Rhizobium) 60 DAS

Treatment	*Shoot Length (cm)*	*Root Length (cm)*	*Plant Fresh Weight (g)*	*Plant Dry Weight (g)*	*No. of Fruits Plant^{-1}*	*Wt. of Pods Plant^{-1} (g)*	*Length of Pods (cm)*
T_1	36.66	31.00	9.66	5.51	2.00	3.66	8.00
T_2	39.66	34.33	10.38	5.80	2.66	6.28	8.33
T_3	42.66	44.00	17.65	8.10	4.33	8.40	8.50
T_4	44.33	51.00	18.45	9.91	6.00	9.17	8.83
T_5	52.33	58.00	20.82	11.87	8.00	11.43	9.33
S.Ed.	2.3570	2.3476	0.8183	0.6082	0.5963	0.6215	0.3651
CD (5%)	5.2518	5.2307	1.8233	1.3552	1.3286	1.3848	0.8136
CD (1%)	7.4704	7.4405	2.5936	1.9276	1.8899	1.9698	1.1573
	**	**	**	**	**	**	*

*: Significant; **: Highly Significant.

Plant Fresh Weight and Dry Weight

At the harvest stage of bean the Composted pressmud + composted coirpith + Rhizobium significantly favoured the plant fresh weight. The other treatments showed slight increase in this parameter when compared with control. As in the present study the increased growth due to compost application was reported by Kore *et al.* (1992) in pulses.

The plant dry weight was also enhanced by all the treatments when compared with control and highest value was noted in the treatment T_4 (T_4–9.91 gm and T_5–11.87 gm). These results were in close conformity with the findings of Prabhakaran (1998) in red gram.

Number of Fruits Plant^{-1}

At the harvest stage the number of fruits/plant was influenced by the treatment T_4 and T_5 (T_4–6.0 and T_5–8.00) whereas the other treatments had a slight increase in this parameter over the control. According to Prabhakaran and Subramanian (2001) in soyabean, Sathish Kumar *et al.* (2003) in mungbean the composted pressmud and coirpith individually and in combination with Rhizobium effectively improved the number of fruits plant^{-1}.

Weight of Pods Plant^{-1} and Length of Pods

A slight increase was observed with all the treatments when compared with control (T_1–3.66 gm, T_2–6.28 gm, T_3–8.40 gm, T_4–9.17 gm, T_5–11.43 gm). At the harvest stage the length of pods *I* plant was influenced by the treatment T_4 and T_5 (T_4–8.83 cm and T_5–9.33 cm) where as the other treatments has a slight increase in this parameter over the control (T_1–8 cm). These results were in close conformity with the findings of Satish Kumar *et al.* (2003).

References

Jeyabal, A., Palaniappan, S.P. and Chelliah, S., 1998. Evaluation of pressmud based biocompost in maize. *Madras J.*, 85(2): 148–149.

Kore, S.S., Deshmukh, G.V. and Jauhav, 1992. Green House evaluation of Rhizobium strains of moong bean. *Biofertilizer Technology Transfer*, pp. 141–147.

Muthukumar, V.B., Gunasekar, N., Omkumar, K. and L. Gurusamy, 2001. To control Parthenium in sugarcane. *Kisan World*, 28(10): 47.

Padmaja, C.K., 2004. Effects of coirwaste along with NPK on the growth and yield of chick pea (*Cicer arietinum*). *J. Ecotoxicol. Environment*, 14(4): 315–318.

Prabhakaran, J., 1998. Effect of pressmud and sheep manure on the growth and yield of red gram in alfisol. *Madras Agric. J.*, 85(6): 303–304.

Prabakaran, N.K. and Subramanian, P., 2001. Performance of soyabean under integrated management of irrigation, composted coirpith and nutrients. *Madras Agric. J.*, 88(7–9): 538–540.

Sathiskumar, R.C. Singh and Kadian, V.S., 2003. Performance of mungbean as influenced by seed inoculation with rhizobium and levels of organic and inorganic sources of nutrients. *Indian J. Pulses Res.*, 16(1): 67–78.

Sneha, S., Raut, C.N., Deotale, R.D., Hatmode, C.N. and H.V. Waghmare, 2003. Response of seed dressing with Biofertilizers and nutrients on morpho-physiological parameters and yield of soyabean. *J. Soils and Crops*, 13(2): 309–313.

Yang-Shia and Wang Yue-chang, 1976. The effect of organic matter on physiochemical properties of soild and cane yield in red soil. *Rep. Tiwan. Sug. Res, Inat.*, 77(76): 1–3.

Chapter 27

Impact of Plant Growth Promoting Rhizobacteria on Plant Growth and Root Rot Incidence in *Vigna radiata* (L.) Wilczek.

N.S. Kulkarni[1], *M.G. Bodhankar*[2], *R.B. Somani*[3], *J.V. Jaiswal*[1]
[1]*Department of Microbiology, R.A. College, Washim*
[2]*Department of Microbiology, Bharti Vidyapeeth's M.B.S.K. College, Kadegaon, Sangli*
[3]*Department of Plant Pathology, Dr. P.D.K.V., Akola*

ABSTRACT

A field experiment was conducted to study the effect of plant growth promoting rhizobacteria (PGPR) on plant growth and disease incidences in green gram. The results in the field experiments have been revealed that *P. putida* followed by *P. fluroscence* and *S. marcescens* have not only exerted the beneficial effects on plant development, but reduced significantly the root rot disease in *Vigna radiata*.

Keywords: *PGPR, Rhizoctonia infection, Plant growth, Disease incidence.*

Introduction

Microbial disease associated with seed and plant reduces the crop yield (Shrinivasan, 1994). Among these diseases *R. botaticola* infection in green gram, cotton and tomato crops were frequently noticed. Chemical formulations were ttied, but desired results were not obtained besides these, chemicals has set back due to carcinogenicity, mutagenicity and toxic hazards (Innes *et al.*, 1960). Replacement of chemical formulations has been done by the utilization of medicinal plant extracts for plant disease management (Doutt *et al.,* 1971). Since, the mass cultivation of medicinal plants depends on the myriad of agro climatic conditions, the availability of such plants may be the major constraint. The importance of PGPR for control of plant pathogens in green house system showed considerable promise (Paulitz *et al.*, 2001). Hence, studies on the effect of PGPR in the management of Rhizoctonia root rot in green gram was undertaken.

Materials and Methods

The field experiment on green gram was conducted in micro plots of size 1 1 m² at a local farm. *Pseudomonas fluorescens, Pseudomonas putida,* and *Serratia marrcessence* were isolated from soybean rhizosphere, and their identity was tentatively confirmed adopting conventional methods of biochemical and cultural characters (Bergey, 1999; Portyrata, 1992). These isolates were used for seed treatments. Completely Randomized design was used with four treatments *viz.* T_1–uninoculated control, T_2–*P. putida,* T_3–*P. fluorescence,* T_4–*Serratia marcescens.* The bacterial cultures of 10^9 cell/ml each were used for seed bacterization separately. For the induction of infection in plant, three weeks old culture of *Rhizoctonia* multiplied on Rawa (Semolina) sand medium was added to the soil at the rate 10 gm per running feet of row at the time of sowing. Simultaneously, sick soil was prepared by mixing the culture grown on Rawa sand medium at 10 per cent w/w and incubated for week. This soil was also added to the micro plot at the rate of 1 kg per plot prior to sowing and thoroughly mixed with the plot soil. The calculated quantity of FYM at the rate of 10 tones/ha was applied manually to the respective treatment plots. Germination percentage, root-shoot length and SVI was recorded at 30th DAS (Hatwalne, 1993). Disease incidence was recorded from each plot at 15 days interval up to harvest as suggested by Gillizan (1983).

Results and Discussion

Growth Parameters (Table 27.1)

In case of germination all the treatments proved to be statistically significant over uninoculated control. The significantly higher germination (88 per cent) was recorded in the treatment with *Pseudomonas putida,* followed by *Pseudomonas fluorescens* (76 per cent) and *Serratia marcescens* (68 per cent). Similarly, in case of root-shoot length, the significant improvement was recorded in the treatment with *P. Putida* followed by *P. fluorescens.* However, effect of *S. marcescens* was at par with un inoculated control in the improvement of plant growth. The beneficial effect of *P. putida* and *P. fluorescens* may be due to the secretion of auxins, cytokinins, which have exerted the plant growth enhancement. The role of phyto hormone synthesis by PGPR in rhizosphere is the direct mechanism of plant growth enhancement (Glick *et al.*, 1995). However, in the treatment with *S. marcescens* the insignificant findings may be due to its inability to multiply in the spermosphere in response to seed exudates so as to achieve significant count for plant growth promotion. Several environmental implicit in the colonization process of plant growth promoting rhizobacteria. These results are in accordance with the reports of Kloepper *et al.* (2002).

Disease Incidence (Table 27.2)

The plant disease controlling potential of PGPR in *Vigna radiata* was studied in micro plots of *Rhizoctonia* sick soil. The results clearly indicated that application of PGPR had protected the plants from *Rhizoctonia* infection. The treatment with *P. putida* proved highly significant in reducing plant disease incidence followed by *P. fluorescence* over uninoculated control. The control of *Rhizoctonia* infection in *V. radiata* may be due to the antagonastic activity of PGPR or due to their successful competition with pathogen for nutrient. It was also observed that as the age of crop increases the

Table 27.1: Effect of PGPR Seed Bactrization on Biometric Observations in *Vigna radiata* (L.) Wilczek

Treatment	Mean			SVI
	Germination %	Root Length (cm)	Shoot Length (cm)	
T_1 Control	64	4.2	7.7	1238.8
T_2 *P. putida*	88	6.5	11.2	1557.7
T_3 *P. fluorescence*	76	5.1	10.8	1208.0
T_4 *S. marcencense*	68	4.7	8.7	911.2
S.E mean±	1.02	0.26	0.67	
C.D. (P = 0.05)	4.23	0.88	2.32	

Table 27.2: Effect of PGPR Seed Bacterization on Mean Per cent Disease Incidence

Treatment	Per cent Disease Incidence			
	15 DAS	30 DAS	45 DAS	Harvest
T_1 Control	5.6	6.9	11.2	11.9
T_2 *P. putida*	3.5	4.7	7.7	8.9
T_3 *P. fluorescence*	4.6	5.0	8.7	9.2
T_4 *S. marcencense*	4.7	6.6	10.8	11.2
S.E mean±	0.48	0.21	0.67	0.71
C.D. (P = 0.05)	1.67	2.72	2.32	2.45

disease incidence also increases in all the treatments including control. The results on the present studies are in concurrence with the experimental findings of Tuzun *et al.* (1993). However, their work, was associated with protection against angular leaf spot disease of cucumber. As the present studies were carried out in *Rhizoctonia* sick soil, it might show variation in natural (non-sick) soils. However, the results in the present investigation indicate the beneficial effects of PGPR in controlling plant disease and may offer an environmentally sustainable approach to increase crop production and health.

References

Bergey, 1989. *Manual of Systematic Bacteriology*, Vol. 4. Williams and Wilkins, London.

Doutt, R.L. and Smith, R.F., 1971. The pest syndrome diagnosis and suggested prophyton. In: *Biological Control*, (Ed.) C.B. Hutt, Plenum Press, New York and London, pp. 3–15.

Gilligan, C.A., 1983. Modeling of soil borne pathogens. *Ann. Rev. Phytopathol.*, 21: 45–64.

Glick, B.R., 1995. The enhancement of plant growth by free-living bacteria. *Can. J. Microbiol.*, 41: 109–117.

Hathwalne, 1993. Effect of Biological nitrogen fixer and YAM on the yield and dry matter production of sorghum SPH (468). *Ph.D. Thesis*, Dr. P.D.K. V., Akola (M.S.) pp. 29–31.

Innes, J.R., B.M. Ulland, M.G. Vollerio, L. Petrucelle, L. Ashbeing, L. Hart, E.R. Pallota, M. Mitchell and J. Peters, 1960. Bioassay of pesticides and Industrial chemicals for tumirogenicity in nice preliminary note. *J. Nat. Cancer Insti.*, 42: 1101–1114.

Kloepper, J.W. and Jetiyanon, J., 2002. Mixture of plant growth promoting rhizobacteria for induction of systemic resistance against multiple plant disease. *Biol. Control*, 24: 28–29.

Paulitz, T.C. and Belanger, R., 2001. Biological control in greenhouse systems. *Ann. Rev. Phytopathol.*, 39: 103–133.

Portyrata, D.A. and Krichevsky, 1992. Micro-Is: A microbiological database management and analysis system. *Binary*, 4: 31–36.

Shrinivasan, K.V., 1994. *Cotten Disease*. Indian Society for Cotten Improvement, Bombay, pp. 311.

Tuzun, L., Kloepper, J.W., Liu, S. and Wei, G., 1993. In: *Pest Management Biologically Based Technologies*, (Eds.) R.D. Lumsden, J.L. Vaughu. American Chemical Society Books, Washington DC, pp. 156–165.

Chapter 28

Water and Wastewater Management: A Membrane Bioreactor Technology: A Review

K.L. Prakash and R.K. Somashekar

Department of Environmental Sciences, Bangalore University, Bangalore – 56

ABSTRACT

MBRs provide a reliable, high-quality, reusable effluent. Its particle-free effluent allows more effective post-disinfection, as required before reuse. MBRs provide excellent pretreatment when reverse osmosis (RO) is needed to generate very high-quality reclaimed water. MBRs may also remove fouling fractions of organic matter more effectively than micro filtration prior to RO.

Keywords: *Wastewater Treatment, Membrane Bioreactor, Pollutants.*

Introduction

In the evolution of life on earth, the membrane was an essential component in that it allowed the formation of cells, and later followed by the compartmentalization of processes in cells. As humans have progressively learned how to produce more complicated and efficient synthetic membranes, so also have developed the ability to compartmentalise various processes. In this way, for example membranes can be used to filter cells from wastewater. If the filtered cells playa role in breaking down additional waste flowing through such loaded membranes, a membrane bioreactor may be said to have been created. A membrane bioreactor consists of some biological constituent or constituents in association with a membrane. A membrane is a surface that has the ability to selectively let some components through it while blocking some others (Crites and Tchobanoglous, 1998).

Since the early 1960s the membrane technology encountered a broad pattern of applications in the field of water, wastewater treatment and other systems of solid-liquid and solid-gas separation (Cornel and Krause, 2003 and Gunder, 2001). The maximum application was the Reverse-osmosis (RO) high-pressure process for seawater desalination and the high-flux however, low pressure systems for saline water desalination. The Reverse-osmosis technology allows removing nearly all dissolved solids contained in the liquid. Membrane treatment of groundwater, surface waters and wastewater is based on the removal of particles, bacteria and cysts, and, consequently turbidity reduction.

Microfiltration (MF) and Ultra filtration (UF) can also maintain these tasks. The advantage of MF or UF for organic removal with the selective salt removal by Reverse-osmosis membranes turn integrated systems into a very promising and challenging development approach.

Recently membrane technology is developed for low quality water and wastewater treatment in both industrial and agricultural applications (Stephenson, *et al.*, 2000). Many groups have conducted studies on performance of pilot and full-scale wastewater; develop and encourage international cooperation towards optimal membrane performance and performance evaluation; support operational research studies and management modelling of membrane systems; identify research needs and develop international technology transfer processes related to polishing low quality water for unlimited uses (Femandez *et al.*, 2000 and Daigger, *et al.*, 2001).

In order to maintain selective solute removal, anew type of membranes was developed finally. The intermittent size Nano Filtration (NF) membranes enable to maintain an improved selective dissolved solids removal. The NF membranes retain hazardous materials and allow other favorable constituents to flow with the permeate. An attractive area of application for NF membranes application is groundwater and surface water treatment that contain pollutants such as soluble organic compounds and heavy metals.

Objectives

The objectives of the Membrane Bioreactor technique are to advance knowledge and experience in the area of development and implementation of membrane technology for low quality water and wastewater treatment. It is associated with the development of membranes and adjustment to the low quality to be treated and subject to the required qualities.

Salient Features of MBR

The use of different membrane processes in combination with biological and biotechnological treatment processes has been developed intensively over the last decade. The use of different membrane clusters can optimize removal of organic constituents such as BOD5, COD, TOC, and TSS and reuse criteria for reclaimed effluent for irrigation. This new concept concerning various membrane combinations in compact vessels generates the efficient Membrane Bioreactor. The MBR is the new advanced trend in wastewater treatment technology. Several treatment systems based on the use of the MBR technology are already in operation. The simple procedure involves:

1. Combine the activated sludge found in high throughput sewerage treatment plants with membrane filtration (see image below), to facilitate removal of biodegradable organics, suspended solids and inorganic nutrients (such as nitrogen and phosphorus).
2. MBR retains particulate and slow-growing organisms (thereby treating more slowly biodegraded organics) and removes a very high percentage of pathogens (thereby reducing chemical disinfection requirements).
3. They also require less space than traditional activated sludge systems because less hydraulic

residence time (HRT) is needed to achieve a given solids retention time (SRT).

4. In addition, MBRs are more automated, making them ideal for decentralized treatment because they are simpler to operate.

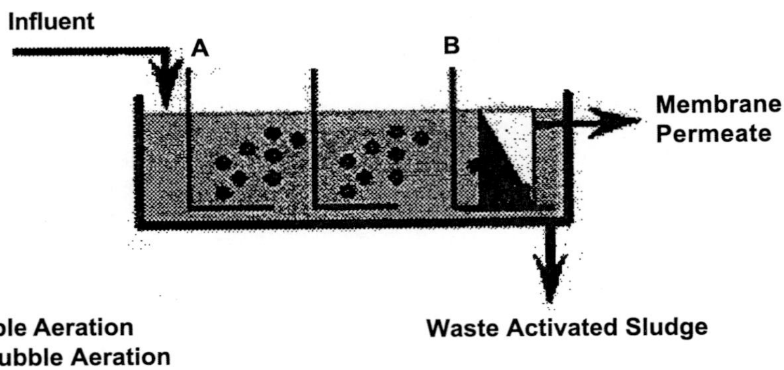

Description of MBR Technology

An MBR is a combination of the activated sludge process, a wastewater treatment process characterized by a suspended growth of biomass, with a micro- or ultra-filtration membrane system that rejects particles. The membrane system replaces the traditional gravity sedimentation unit (clarifier) in the activated sludge process. The turbidity and suspended solids concentration of the effluent is far lower than in conventional treatment. All biomass is retained and becomes returned activated sludge. Biological growth leaves the system as waste activated sludge.

MBR Technology is Based on the Following Principles:

The engineering principles underlying MBRs are familiar enough to ensure reliability. Because MBRs combines two familiar technologies:

1. Activated sludge and
2. Membrane filtration

Significant engineering expertise can be applied to MBR design and operation. Several studies already have applied activated-sludge-related biology to MBRs, although current investigations suggest potentially important differences in growth, population diversity, cell activity, and competition. One noticeable difference is that MBR membranes have to be cleaned periodically to minimize biological and chemical fouling, and MBR manufacturers have already developed some cleaning methods.

MBRs have been used to treat a wide range of municipal and industrial wastewaters, and currently are installed at more than 1000 sites in Asia, Europe, and North America, according to a database assembled by the Water Environment Research Foundation. Most units currently treat a few hundred m^3/d (the largest treats less than 40,000 m^3/d). But plans are underway to build MBRs that will treat 30,000 to 150,000 m^3/d, and the technology could be used to treat 300,000 to 800,000 m^3/d.

Enough reliable equipment and technological support are commercially available to meet existing and developing demand. Membrane-manufacturing capacity is expanding, consequently unit costs are declining. The long-term trend is a "virtuous cycle" in which declining costs spur more demand, enabling further cost reduction.

Management Aspects

The key tasks in wastewater treatment are associated with generating high quality waters under cost effective methods. Particularly, construction and maintenance of treatment facilities plays a central role in the planning and operating process of wastewater treatment system for disposal and/or reuses (Tchobanoglous, 2003 and Van der Roest *et al.*, 2002). Optimal condition can be obtained by adopting the holistic approach. The integrative approach is based on trying to encompass all relevant aspects of the membrane units under consideration. The various aspects of membrane treatment plants can be viewed at the following levels:

1. The local level of the isolated treatment and the Point of Use (POU) of the reclaimed effluent;
2. The regional level which refers to the entire picture of the water resources, namely treatment options that include also the membrane alternative, and;
3. Other phases which include environmental issue, health risks, regulative and traditional aspects. The membrane element is only one link in a multi-component treatment system.

Intangible benefits and disadvantages of membrane bioreactor are those features that can hardly be quantified by financial terms, nevertheless are relevant in the decision-making procedure. Although they are difficult (or impossible) to quantify, the intangible benefits and/or disadvantages of a technology that can outweigh the results of the tangible cost evaluation, thus affecting the go/no-go decision of potential users. For example, potential users may find it beneficial to implement the new technology (concerning increased plant safety and reliability, reduced future liability, promoting public perception, etc.) even it is more expensive than the baseline technology.

There are two basic approaches concerning water management:

1. Water is a public good–consequently the prices are subject to national considerations.
2. Water prices are subject to the market forces. The impact of the market forces played major role in the UK where the farmers halted the irrigation of cereal crops and preferred to irrigate crops (mainly potatoes) which provide high shadow-price yields. Current use of effluent for low-grade agricultural products is often justified mainly on social grounds, however, when there will be a clear substitute in imported products, the outlook for these investments does not look bright.

The Challenge for Sustainable Sanitation Services in Developing Countries

This is emerging as a major problem so far as developing countries are considered. It is noticed that:

1. Half of the world's people (about 3 billion) live on less than US$1 per day.
2. About 800 million people lack access to health care.
3. About 10 million children under 5 years died in 1999, mostly from preventable diseases.
4. In 2002, an estimated 1.1 billion people lacked access to a safe water supply and 2.4 billion to improved sanitation.
5. Africa has 38 per cent of its population unserved by safe water and 40 per cent by sanitation, the figures for Asia are 19 per cent and 52 per cent, and 15 per cent and 22 per cent for Latin America and Caribbean.

6. Over the next 25 years, the urban populations of Africa and Asia will almost double; the urban populations of Latin American and the Caribbean will increase by nearly 50 per cent.
7. Delegates to the 2002 Johannesburg Summit agreed to cut in half the proportion of people without basic sanitation; this means providing sanitation to 2 billion more people.
8. The provision of full water and wastewater connections and primary wastewater treatment to the urban population would entail an annual cost of US$ 17 billion for water and US$ 32 billion for sanitation. To serve 2 billion more people by 2015 will require connections for more than 350,000 individuals each year.
9. The recent Third World Water Forum highlighted the fact that there are a further 3 billion people who only use pit toilets, flush toilets, or sewers without any treatment before discharge to the environment (World Water Forum, Rich Nations Get Wealth by Polluting Poor Nations, 17th March, Kyoto, 2003).

References

Cornel, P. and Krause, S., 2003. State of the art on MBRs in Europe. In: *Proceedings Applications and Perspectives of MBR in Wastewater Treatment and Reuse*, 28–29 April 2003, Cremona, Italy.

Crites, R.W. and Tchobanoglous, G., 1998. *Small and Decentralized Wastewater Management Systems.* McGraw-Hill Book Company, New York, NY.

Daigger, G.T., Crawford, G., Fernandez, A., Lozier, J.C. and Fleischer, E., 2001. WERF Project: Feasibility of Membrane Technology for Biological Wastewater Treatment B Identification of Issues and MBR Technology Assessment Tool. In: *Proceedings of the Water Environment Federation 74th Annual Conference and Exposition*, Atlanta, GA, CD-ROM.

Fernandez, A., Lozier, J.C. and Daigger, G.T., 2000. Investigating Membrane Bioreactor Operation for Domestic Wastewater Treatment: A Case Study. In: *Proceedings of the Water Environment Federation 73rd Annual Conference and Exposition on Water Quality and Wastewater Treatment*, Anaheim, CA, CD-ROM.

Gunder, B., 2001. *The Membrane-Coupled Activated Sludge Process in Municipal Wastewater Treatment*, Technomic Publishing Co, Lancaster, PA.

Stephenson, T., Judd, S., Jefferson, B., and Brindle, K., 2000. *Membrane Bioreactors for Wastewater Treatment.* IWA Publishing, London.

Tchobanoglous, G., 2003. The strategic importance of decentralized wastewater management in the twenty-first century. In: *Proceedings IDA Conference on Water Reuse and Desalination*, Singapore (CD-Rom).

Van der Roest, H.F., Lawrence, D.P. and van Bentem, A.G.N., 2002. *Membrane Bioreactors for Municipal Wastewater Treatment*, IWA Publishing, London.

Chapter 29

Recovery of Membrane Bound Enzymes in Soleus and EDL Muscles of Palm-Squirrel after Experimentally Induced Sciatic Neuropathy

K. Pratap Reddy and K. Praveen Kumar*

Neurobiology Lab, Department of Zoology, University College of Science, Osmania University, Hyderabad – 500 007, India

ABSTRACT

This study reports the time course changes of weight and activities of membrane bound enzymes, acetyl cholinesterase (AChE), $Na^{+}K^{+}$ ATPase, Mg^{++} ATPase and Ca^{++} ATPase in tonic soleus and phasic Extensor digitorum longus (EDL) muscles of palm squirrel at 1,4,8,15,30,60 and 120 days after experimentally induced peripheral (sciatic) mono neuropathy. The muscles showed distinct and characteristic progressive atrophy from day 1 to 60, the rate of decrease is being more in soleus (63 per cent) than EDL (44 per cent), followed by recovery of weights to control levels at day 120. The AChE activity of soleus increased progressively up to day 60 (156 per cent) with reversal to control on day 120, in contrast to decrease of AChE in EDL up to day 60 (71 per cent) with recovery to control on day 120. The Na^{+} ATPase activity shows a contrasting pattern in both muscles to that of AChE. The Ca^{++} ATPase was not altered up to day 60 in soleus and EDL with the decreased activity in EDL on day 120. Mg^{++} ATPase was not significantly altered in soleus while in EDL it gradually decreased (35 per cent). The results show that the time dependent and muscle specific differential reversal of atrophy, transport ATPases and AChE activities in soleus and EDL over 120 days after mononeuropathy.

Keywords*: Mononeuropathy, ATPases, AChE, EDL, Soleus, Sciatic nerve.*

Corresponding Author: E-mail: pratapkreddy@rediffmail.com.

Introduction

Neuropathy represents a pathological state of the nervous system with the potential to interfere with normal end organ function and metabolism. The neuropathy is initiated by primary lesion of nerve due to trauma and neuroma or dysfunction of the nervous system or chronic diseases like diabetes, post therapeutic (viral) neuralgia or rheumatoid inflammation (Bridges *et al.*, 2001). An animal model of peripheral mono neuropathy in the rat has been described which has pain sensation reminiscent of those seen in man (Attal *et al.*, 1990; Bennet and Xie, 1983; Jazat *et al.*, 1998; Xie and Xiao, 1990). The end organs like muscles exhibit corresponding responses in neuropathies, such alterations in nerve physiology may have impact on muscle metabolism and manifest as quantitative enzymatic changes (Korr, 1981). The different type of muscles exhibit characteristically varied enzymatic responses during disease or experimental denervation and reinnervation (Gutamam, 1976). There are no discernable literature on the responses of fast phasic and slow-tonic type of muscles after neuropathic pain in experimentally produced mono neuropathy with sciatic nerve ligature, which mimics the human neuropathies. Therefore, this study is designed to report the time course alterations in atrophy, membrane transport ATPases, and acetyl cholinesterase activities in slow tonic soleus and fast phasic EDL of Indian palm squirrel with experimentally induced sciatic mononeuropathy over 120 days.

Materials and Methods

Adult male Indian palm-squirrels, *Funambulus palmaram* were used in all experiments. The animals were housed in groups of 6 in standard laboratory cages. The animals were anesthetized with sodium pentobarbital (40mg/Kg) and mono neuropathy was induced as described by Bennet and Xie (1983). The common sciatic nerve was exposed at the level of middle-thigh region and about 1cm nerve was freed of adhering tissue and 4 ligatures (4.0 chromic gut) were tied loosely around it with about 1.5 mm spacing. An identical dissection was performed in each animal on the contra lateral side except that the sciatic nerve was not ligated. The animals were returned to cages and maintained in aseptic conditions for selective time intervals up to 120 days.

The animals were sacrificed at 1, 4, 8, 15, 30, 60 and 120 days and the soleus as well as EDL were dissected out for experimental analysis. The ligated limb was not in proper use up to 30 days, after which it had shown sign of usage and its usage was almost normal at 120 days. The rate of atrophy of muscles (dry-weight) was obtained by allowing the muscles in a dessecator containing hygroscopic P_2O_5 at 60°C for 24h. The acetyl cholinesterase activity was assayed according to Ellmann's method (1961). The muscles were homogenized in phosphate buffer containing EDTA, Triton X100 and centrifuged at 16,000 rpm to obtain crude membrane fraction which was used for enzyme assay. The Na^+K^+ ATPase and Mg^{++} ATPase activities were estimated as described by Kaplay (1979). The muscles were homogenized in sucrose buffer containing Histidine and centrifuged at 16,000 rpm to obtain crude membrane preparation. The Na^+K^+ ATPase and M^{++} ATPase activities were assayed by using ouabain, which inhibits the formers activity. The Ca^{++} ATPase activity was assayed as described by Samaha and Yunis (1973). The phosphorous content in ATPases activity was estimated as described by Tasuaky and Shorr (1953) and protein content in enzyme preparations was estimated as described by Lowry method (1951).

Results and Discussion

As shown in Figure 29.1, the weights of soleus and EDL were progressively decreased up to day 60 but the rate of the decrease was more in former. The weights of both muscles were however recovered to that of control muscle on day 120. The AChE activity (Figure 29.2) in soleus was not significantly

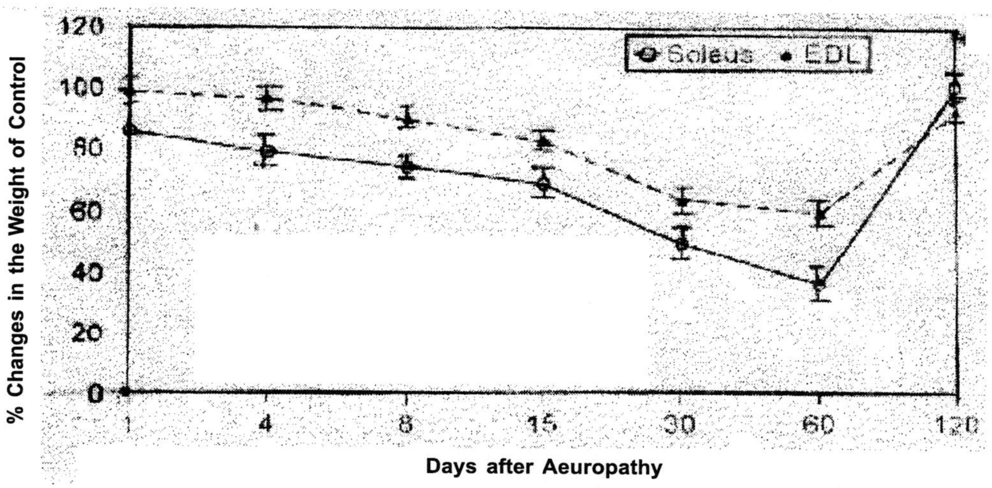

Figure 29.1: The Change in Dry Weights of Soleus and EDL Muscles after Experimentally Induced Sciatic Mononeuropathy

The dry weights of muscle calculated as mg of muscle/animal. The numbers of the ordinate represent per cent of control value and on abscissa represent time in days after neuropathy. Each point represent mean ± S.D of 6 animals. The dry weights of soleus and EDL are 12.01±2.0 and 19.5±2.05 mg of muscle/animal respectively.

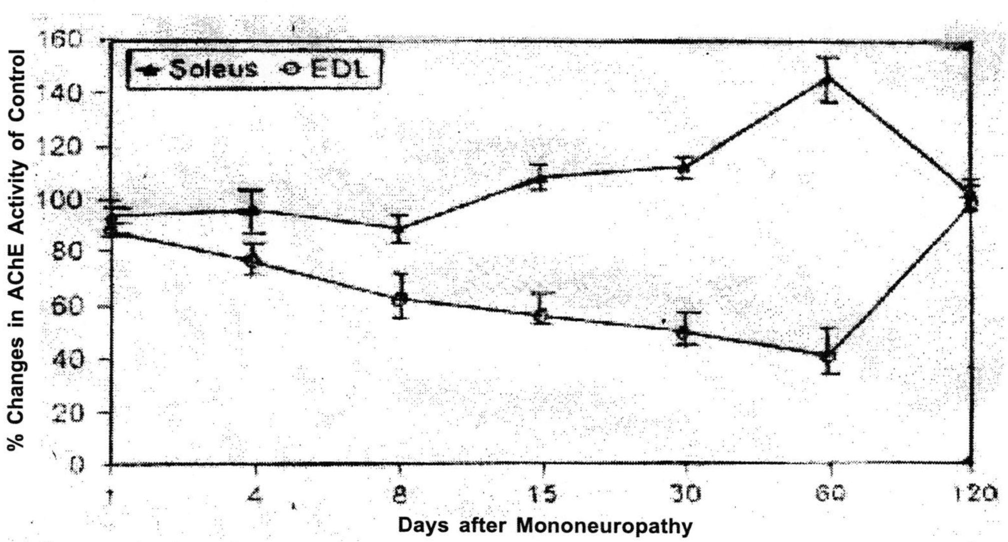

Figure 29.2: The AChE Activity of Soleus and EDL Muscle of Mice During Experimentally Induced Sciatic Mononeuropathy

The Acetylcholinesterase activity is calculated as nano moles of AChE hydrolyzed mg of muscle/ min. Numbers of the ordinate represent per cent of control value and on abscissa represent time in days after neuropathy. Each point represent mean±S.D of 6 animals. The AChE activity in control soleus and EDL is 0.4642±0.032 and 0.7972±0.024 n moles of AChE/mg of muscle/min respectively.

altered from control up to day 8 followed by gradual increase upto day 60 with reversal to control value on day 120. The EDL exhibited progressive decrease of AChE activity upto day 60 with reversal to control value on day 120. Na^+K^+ ATPase (Figure 29.3) activity was increased on day 1 in soleus, with reversal to control value upto 15 with progressive decrease up to day 60 followed by reversal to control on day 120. In contrast, the enzyme activity increased (22 per cent) in EDL on day 60 followed by reversal to control on day 120. Mg^{++} ATPase activity (Figure 29.4) showed insignificant decrease from control up to day 60 with reversal to control value. The EDL showed elevated level on day 4 followed by progressive decrease. The Ca^{++} ATPase activity (Figure 29.5) was not altered upto day 15 followed by progressive decrease in both muscle up to day 60. However, soleus showed recovery on day 120 in contrast to decreased activity in EDL (35 per cent).

The time dependent and muscle specific alterations and reversal of weight and membrane bound enzyme activities of soleus and EDL of palm squirrel to normal levels over 120 days were not reported earlier are in line with behavioral and neuroanatomical evidence of peripheral mono neuropathy found in rat (Attal *et al.*, 1990; Xie and Xiao, 1990; Jazat *et al.*, 1998; Gautron *et al.*, 1990; Sugimoto *et al.*, 1989). The recovery of weight and AChE in soleus and EDL in long term reinnervation after single crush in rat (Misulis and Dettbarn, 1985) and palm-squirrel (Reddy, 1994) were reported. The observed elevated AChE activity in soleus and sudden decrement in EDL up to 60 days may be due to conversion of slow fibers to fast type of fibers but not vice versa (Guth Brown 1965), conversion of slow muscle to neonatal state which dominated by fast fibers (Rubinstein *et al.*, 1978) or preferential faster reinnervation of slow muscle than fast muscle (Guth *et al.*, 1965) and reversal of activities to control values during further period (120 days) may be related to reestablishment of neural connection with simultaneous elimination of hyper neuralization (Misulis and Dettbarn, 1985). In addition the differential pattern of AChE in soleus and EDL may be due to characteristic neural regulation of muscle specific AChE molecular forms in soleus and EDL (Groswald *et al.*, 1985).

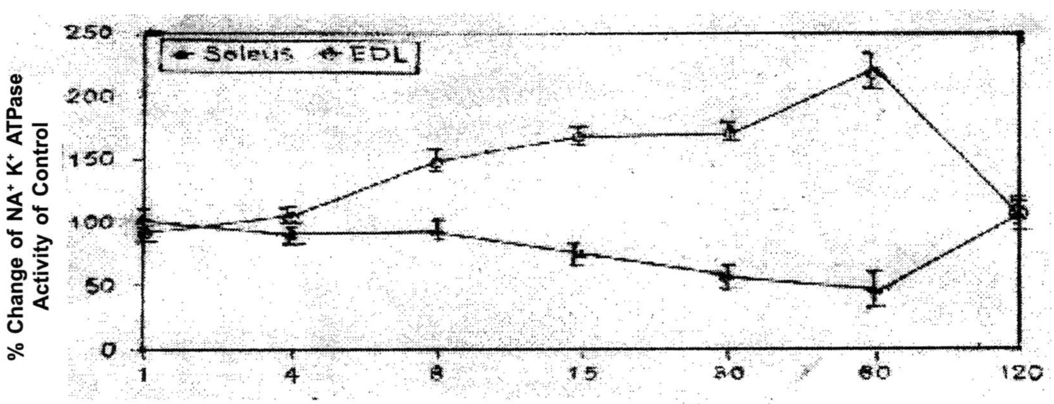

Figure 29.3: The Na^+K^+ ATPase Activity of Soleus and EDL Muscle of Mice During Experimentally Induced Sciatic Mononeuropathy

The Na^+K^+ ATPase activity is calculated as n moles of Pi/mg protein/hr. Numbers of the ordinate represent per cent of control value and on abscissa represent time in days after neuropathy. Each point represent mean±S.D of 6 animals. The Na^+K^+ ATPase activity in control soleus and EDL is 116±4.30 and 92±5.70 n moles of Pi/mg of muscle/min respectively.

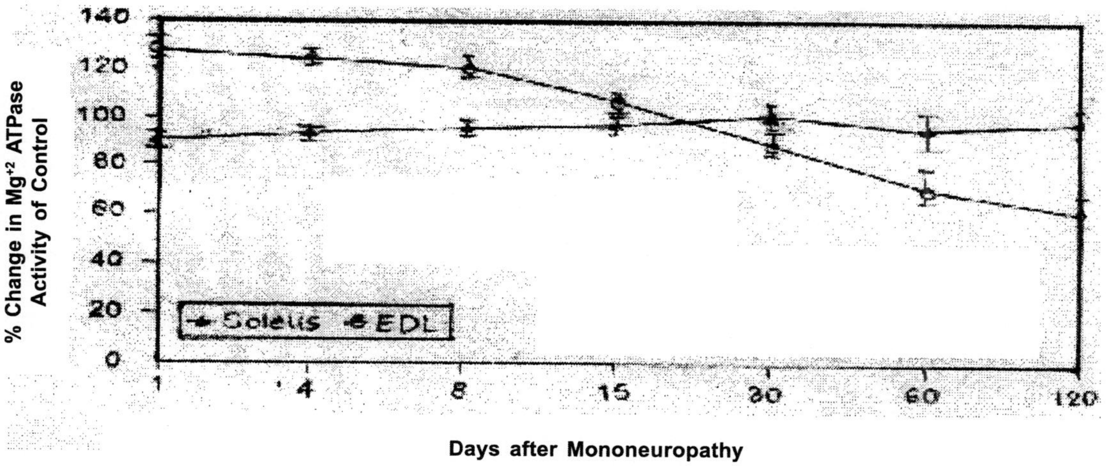

Figure 29.4: The Mg^{2+} ATPase Activity of Soleus and EDL Muscle of Mice During Experimentaly Induced Sciatic Mononeuropathy

The Mg^{2+} ATPase activity is calculated as n moles of Pi/mg protein/hr. Numbers of the ordinate represent per cent of control value and on abscissa represent time in days after neuropathy. Each point represent mean±S.D of 6 animals, The Mg^{2+} ATPase activity in control soleus and EDL is 828±24.0 and 1415±23.0 n moles of Pi/mg of muscle/min respectively.

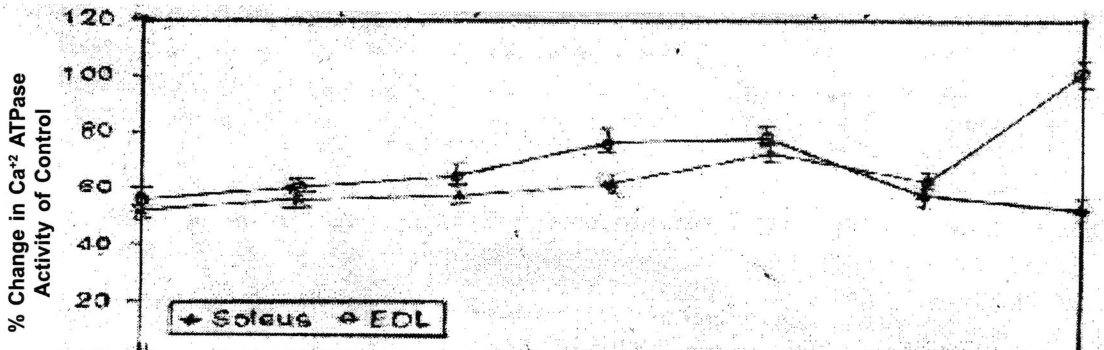

Figure 29.5: The Ca^{2+} ATPase Activity of Soleus and EDL Muscle of Mice During Experimentally Induced Sciatic Mononeuropathy

The Ca^{2+} ATPase activity is calculated as nano moles of Pi/mg protein/hr. Numbers of the ordinate represent per cent of control value and on abscissa represent time in days after neuropathy. Each point represent mean±S.D of 6 animals. The Ca^{2+} ATPase activity in control soleus and EDL is 828±24.0 and 1415±23.0 n moles of Pi/mg of muscle/min respectively.

The disturbance in neural connection found to alter the membrane permeability and its active capacity to pump cations (Broadie and Sampson 1985), The conduction and contractile properties of slow and fast muscles observed to show varied altered pattern after neural disturbance and these are due to alterations in ion specific transport, ATPases (Brandel *et al.*, 1987). The observed activity pattern of Na^+K^+ ATPase, Mg^{++} ATPase and Ca^{++} ATPase in soleus and EDL of palm squirrel, which serves as an important tool for exploring the potential pathogenic role of metabolic factors, confirms the involvement of altered membrane function in time dependent progressive degeneration and j reversal of muscle structure and function during the long-term mononeuropathic pain. The differential responses observed in membrane bound enzymes of soleus and EDL may be due to characteristic membrane, structural and neural properties of the muscles. In this animal mode the loosely constrictive ligatures around sciatic nerve may result in loss or damage of nerve fibers consequently showing the symptoms of pain like hyperalgesia, allodynia, dysasthesia which may appear in about 2 days and last for 2–3 months (Attal *et al.*, 1990; Xie and Xiao, 1990; Gautron *et al.*, 1990). The dramatic loss of myelinated fibres in the lesioned nerve and reduction in their conduction velocities could result in observed responses (Xie and Xiao, 1990; Gautron *et al.*, 1990). Abnormal ectopic discharges in primary myelinated afferent fibres of ligated nerve and consequential occurrence of more dark neurons in spinal cord with trans synaptic degeneration after experimental painful neuropathy may account for some aspects of observed pathological significance (Sugimoto *et al.*, 1980). The tendency of normalization of altered properties could be due to the involvement of neural factors particularly regeneration and remyelination of sciatic nerve in later period after possible demyelination and axonal degeneration as a result of ligation (Gautron *et al.*, 1990). These altered probable mechanisms are in confirmation to observed time course changes reported in this animal model.

Acknowledgement

Authors express thanks to UGC [No. F: 30–191/2004(SR)] for financial assistance.

References

Attal, N., Jazat, F., Kayser, V. and Guilbaud, G., 1990. Further evidence for pain related behaviours in a model of unilateral peripheral mononeuropathy. *Pain*, 41: 235.

Bennett, G.J. and Xie, Y K., 1983. A peripheral mononeuropathy in rat that produces disorders of painsensation like those seen in man. *Pain*, 33: 87.

Brandel, C.J., de Legn, S., Martin, S. and MacLehnan, B.H., 1987. Adult forms of Ca^{++} ATPase of *Sarcoplasmic reticulum* expression in developing skeletal muscle. *J. Biol. Chem.*, 262(8): 3767.

Bridges, D., S.W.N. Thompson and A.S.C. Rice, 2001. Mechanisms of neuropathic pain. *Br. J. Anaesth.*, 87(1): 12–26.

Broadie, C. and Sampson, S.R., 1985. Contribution of electrogenic Na^+K^+ ATPase to resting membrane potential of cultured rat skeletal myotubes. *Brain Res.*, 28: 35.

Ellmann, G.L., Courtney, D.K., Amders Jr. V. and Eestherstone, Rm., 1961. A new and rapid calorimetric determination of acetylcholineesterase activity. *Biachem. Pharmacal.*, 7: 88

Gautron, M., Jazart, F., Ratinahirana, H., Hanco, J.J. and Gilbard, G., 1990. Are there some natomical correlates of hyperalgesia and anodynia exhibited by rats with a unilateral mononeuropathy: A preliminary light microscopic study of the sciatic nerve. *Neurosci. Lett.*, 111: 28.

Groswald, D.E. and Dettbarn, W.D., 1985. Nerve induced changes in AChE molecular forms of soleus and EDL muscles. *Exp. Neural*, 89: 198.

Gutamam, E., 1976. Neurotrophic relation. *Ann. Rev. Physial.*, 38: 177.

Guth, L., and Brown, W.C., 1965. The sequence of changes in cholinesterase activity during reinnervation of muscle. *Exp. Neurol*, 29: 198.

Jazat, F., Attal, N., Gautron, H. and Guilbaud. G., 1998. Behavioural evidence that a peripheral experimental neuropathy in rat induces abnormal pain sensation. *Eur. J. Neurasci. Supp.*, pp. 179.

Kaplay, S.S., 1979. Erythrocyte membrane $Na^{+}K^{+}$ activated ATPase in protein calorie malnutrition. *Am. J. din. Nutri.*, p. 579

Korr, I.M., 1981. The spinal cord as organizer of disease processes, 1 V Axonal transport and neurotrophic function in relation to somatic dysfunction. *Jam Ostepath Assoc.*, 84: 415.

Lowry, O.H., Rosebrough, N.J., Farr, A.L. and Pandel, R.J., 1951. Protein measurement with Folin Phenol reagent. *J. Biol. Chem.*, 193: 265.

Misulis, K.E. and Dettbam, E.D., 1985. Is fast fibre innervation responsible for increased acetylcholinesterase activity in reinnvervating soleus mucle. *Exp. Neural*, 89: 204.

Reddy, K.P., 1994. Transient elevations of AChE in soleus, plantaris and EDL muscles of palm squirrel during prolonged reinnervation. *Indian J. Exp. Biol.*, 32: 364.

Rubinstein, N.A. and Kelly, A.M., 1978. Myogenic and neurogenic contribution to the development of fast and slow twitch muscles in rat. *Develop. Biol.*, 62: 473.

Samba, E.J. and Yunia, F.J., 1973. Quantitative and histochemical demonstration of calcium activated ATPase in skeletal muscle. *Exp. Neural.*, 41: 431.

Sugimoto, T., Bennett, G.J. and Kajander, K.C., 1989. Strychnine enhanced transynaptic degeneration of dorsal horn neurons in rats with and experimental painful peripheral neuropathy. *Neuroscience Lett.*, 98: 139.

Tussaky, E.M. and Shorr, E., 1953. A micro calorimetric method for the determination of inorganic phosphorus. *J. Biol. Chem.*, 220: 675.

Xie, Y.K., and Xiao, W.H., 1990. Electrophysiological evidence for hyperalgesia in the peripheral neuropathy. *Sci. China B.*, 33: 663.

Chapter 30

Folk Medicine Used to Cure Diseases Like Jaundice, Leucorrhoea, Dysmenorrhoea, Malaria and Filaria in Rural Jamtara District of Jharkhand (India)

Ashok Kumar Mandal and Bijan Bihari Dutta
P.G. Department of Botany, P.K.R.M. College, Dhanbad

ABSTRACT

This article includes information on 29 indigenous medicinal plants used to cure Jaundice, Leucorrhoea, Dysmenorrhoea, Malaria and Falaria by the rural people of Jamtara district of Jharkhand (India). As there is very poor modern medical facilities available, tribal people are mostly dependent on witch-doctors who practice on herbal medicines to cure various ailments. The witch doctors treat successfully many diseases using plant based medicines with a touch of chanting of some "mantras". This article is based on through a survey conducted during 2003–2004 for spot collection, identification and ethno-medicinal enumeration of indigenous plants used by the rural people.

Keywords: *Folk-medicine, Witch-doctors, Ethnomedicine, Rural people, Tribal people.*

Introduction

Jamtara district of Jharkhand (India) is one of the important coal belts of Jharkhand. It is also an important stone-belt of the state with many stone mining and stone crushing units. The district borders Birbhum and Burdhman districts of West Bangal. The district is inhabitated by Santhal, Paharia, Ghatwar, Oraon and Birhore like aboriginal tribal population along with others.

Modern health facilities in the rural areas is very poor due to deficient infrastructure and negligence of the authorities. Not only this there are some tribal people who are scared of taking allopathic medicine. Use of allopathic medicines is considered to lee a sin as they are not purified by "mantras" of their witch-doctor or medicine man. Instead of going to the hospitals they prefer to visit the village witch-doctor to solve the health problems.

As there are many coal mines and stone mines in the district, there are many cases of illegal mining practices. Due to this, a large part of viable land has been devastated, destroying a large number of useful plants. As a result, people are now using some new plants as a substitute for the age-old medicinal plants for the treatment of various diseases. There is a need to explore the newly used medicinal plants by the rural people. Usually the witch-doctors do not disclose the know-how of using the plants and their identity to others except near and dear ones. It is very tiresome job to extract this knowledge from them.

Methodology

By personal contact a rapport was established with the chief of a village community and his guidance was sought to establish contact with the medicine man, locally known as witch-doctor. He keeps every thing secret about the medicinal plants and methods of treatment. In such a secretiveness, it was decided to interview a number of elderly people who have some influence on the witch-doctors. Thus a link man was selected for different localities. Birth-attendants, woodcutters and cow-boys were also consulted. Accompanying them the near by forest, hillocks, and field sites were visited. Frequent field surveys were carried out during deferent seasons during 2003 and 2004. Cross check of collected information from different people were made to know the utility of a plant. Data were also collected through questionnaires in their local languages. In addition to the vernacular names and medicinal uses, detailed information, ingredients, dose, mode of uses were also collected. The plants were identified by their local names, photographed and sample specimen were collected for preparation of herbarium.

Taxonomic confirmation of the collected plants were done with the help of Flora of Bihar and Orissa–H.H. Hains and Flora of Bhagalpur–S.K. Varma. Help was also taken from BSI, Shibpur (Howrah) and T.M. Bhagalpur Univ. Bhagalpur with their herbarium for confirmation of identified plants.

During survey some interesting facts came to light which are not mentioned in any ethnobotanical literature. Some medicine man claim that he had been using some new plants successfully in treatment of diseases which are unknown to others and is the secret of his being a famous and powerful witch doctors.

Enumeration

29 species of plants belonging to 16 families have been recognised for their medicinal properties used in the treatment of diseases like Jaundice, Leucorrhoea, Dysmenorrhoea, Malaria, and Filaria.

For Jaundice

***Andrographis paniculata*, Nees**

Local Name: Kalmegh, Kirayat

Family: Acanthaceae

Mode of Uses: The whole plant is useful. Generally leaf-paste is used.

***Boerhaavia diffusa* L.**

Local Name: Punarnava

Family: Nyctaginaceae

Mode of Uses: The root paste.

***Hydrocotyl asiatica* L.**

Local Name: Brahmibuti, Thamkuni

Family: Apiaceae

Mode of Uses: Fresh or shade dried leaf and root paste or powder

***Leucus aspera*-Spreng**

Local Name: Ghalghasi, Halkusi

Family: Lamiaceae

Mode of Uses: Leaf paste with black pipper seeds

***Oroxylon indicum* Vent**

Local Name: Sona gachha

Family: Bignoniaceae

At present the plant is not available in the locality due to human greed.

Mode of Uses: It was said that the bark of the tree was used

***Phyllanthus niruri* L.**

Local Name: Bhuiamla, Jaramla

Family: Euphorbiaceae

Mode of Uses: Root paste is used

***Scoparia dulcis,* L.**

Local Name: Banchini

Family: Scrophulariaceae

Mode of Uses: Flower are used

***Streblus asper* L.**

Local Name: Shaora/Shehur

Family: Moraceae

Mode of Uses: Stembark paste with curd

For Dysmenorrhoea

***Abroma augusta* L.**

Local Name: Ulat kambal

Family: Sterculiaceae

Mode of Uses: Rootbark as well as stembark is used

***Aloe indica* Willd**

Local Name: Ghrit kumari

Family: Liliaceae

Mode of Uses: Leaf pulp with turmeric powder and salt

***Ficus benghalensis* L.**

Local Name: Bat/Bar

Family: Moraceae

Mode of Uses: Root bark with milk of goat

***Ficus religiosa* L.**

Local Name: Peepal

Family: Moraceae

Mode of Uses: Rootbark or Stembark

***Ficus infectoria* Roxb**

Local Name: Pakar

Family: Moraceae

Mode of Uses: Decoction of stembark and latex

***Saraca indica* L.**

Local Name: Ashok

Family: Caesalpiniaceae

Mode of Uses: Decoction of stembark

***Sida cordifolia* L**

Local Name: Berela/Bariar

Family: Malvaceae

Mode of Uses: Root paste with honey or milk

For Leucorrhoea

***Adhatoda visica* Nees**

Local Name: Basak/Adalsa

Family: Acanthaceae

Mode of Uses: 10–20 gm of leaf or flower is boiled with 250 ml of water. When water is reduced to half of its volume, it is given to the patient to consume.

***Blumea odorata* DC**

Local Name: Kokronda/Barokuksima

Family: Asteraceae

Mode of Uses: Leaf juice is used.

***Evolvulus alsinoides* Wall**

Local Name: Shyam kranta

Family: Convolvulaceae

Mode of Uses: The whole plant is useful. Some medicine man suggested that with *Asparagus racemosus* (Satavar) it is more effective

***Phyllanthus niruri* L**

Local Name: Bhuiamla

Family: Euphorbiaceae

Mode of Uses: The whole plant is used. Root alone is equally effective.

***Sida cordifolia* L**

Local Name: Berela/Bariar

Family: Malvaceae

Mode of Uses: Root paste with sugar and milk.

***Saraca indica* L**

Local Name: Ashok

Family: Caesalpiniaceae

Mode of Uses: 15–20 gm decoction of the stembark is used thrice daily.

***Vinca rosea* L**

Local Name: Sadabahar, Nayantara

Family: Apocynaceae

Mode of Uses: Decoction of root is useful

For Malaria

***Achyranthus aspera* L**

Local Name: Chirchiri/Apang/Latjira

Family: Amaranthaceae

Mode of Uses: Leaf pest with black pepper and garlic is used

***Clerodendron infortunatum* Gaertin**

Local Name: Ghetu/Bhant

Family: Verbenaceae

Mode of Uses: Leaf juice is used

Glycosmis pentaphylla/arborea

Local Name: Bannimbu/Ash shaora/Daton

Family: Rutaceae

Mode of Uses: Decoction of leaf and root bark is useful. It was also reported that those who brush their teeth daily with the twig of the plant, develops immunity against malaria.

***Nyctanthes arbor-tristis* L**

Local Name: Harshingar / Sheoli

Family: Oleaceae

Mode of Uses: Decoction of leaf with ginger and honey.

***Strebbus asper* L**

Local Name: Shaora / Shehur

Family: Moraceae

Mode of Uses: Stembark is used

***Vernonia cinerea* Less**

Local Name: Sheal motra / Kuksima / Sahadevi

Family: Asteraceae

Mode of Uses: One gram of leaf with seven number of black pepper are mixed to prepare the medicine. It was suggested by a medicine man that the medicine should be taken on Sunday or Tuesday morning in empty stomach.

For Filaria

***Cassia occidentalis* L**

Local Name: Kasunda / kalkasunda

Family: Caesalpiniaceae

Mode of Uses: Decoction of the root is useful. With black pepper it is more effective.

***Streblus asper* L**

Local Name: Shaora / Shehur

Family: Moraceae

Mode of Uses: Stembark decoction is useful. The paste of the stembark is applied locally on fissured swollen leg of a filaria patient.

References

Varma, S.K. and Srivastava, D.K., 1981. An ethnobotanical studies of Santhal Pargana. *Ind. For.*, 107(1): 30–41.

Jain, S.K., 1935. *Dictionary of Indian Folk Medicinal Plants* I–IV. Information Book Distribution, Dehradun.

Jain, S.K., Banerjee D.K and Pal, D.C., 1973. Medicinal plants among certain Adivasis in India. *Bull. Bot. Survey, India*, 15: 85.

Biswas, K.N., 1924. *Bhartiya Aushdhavali*, Hanneman. Publ. Co., Kolkata.

Kirtikar, K.R. and Basu, B.D., 1984. Indian Medicinal Plants, Volume I–IV (Allahabad) India.

Haines, H.H., 1961. CIE FCH, FLS: The Botany of Bihar and Orissa, BSI, Kolkata.

Witch-Doctors and medicine men of the locality.

Chapter 31

In vitro Study of Herbal Extracts and Oil on Phytopathogens

B.K. Dutta, S. Karmakar and M. Begam

Defence Research Laboratory, Post Bag No. 2, Tezpur – 784 001, Assam, India

ABSTRACT

In vitro study was conducted to find out the antiphytopathogenic property of *Terminalia chebula* dry fruit aqueous and methanolic extracts and eucalyptus oil against five test fungi *viz., Corynespora cassiicola, Stemphylium solani, Rhizoctonia solani, Colletotrichum lindemuthianum* and *Alternaria brassicicola. T. chebula* aqueous extract is more potent than methanolic extract against phytopathogens. Eucalyptus oil revealed more potent in case of *C. cassiicola*. Although two different experimental substances were tried on five test pathogens, *C. cassiicola* was more susceptible and *A. brassicicola* recorded no zone of inhibition *i.e.* more resistant in both cases.

Keywords: *Antiphytopathogenic, Terminalia chebula, Eucalyptus citriodora, Zone of inhibition.*

Introduction

To get sustainable crop production, integrated crop management system is very essential having controlled use of chemicals, bio fertilizers and bio pesticides. In most of the cases ignorance of farmers causing indiscriminate use of chemical products lead adverse effect on human health and environment, subsequently pathogens achieve resistance capability against pesticides. Due to mishandlings, farmers may also face many physical ailments. The residual effect of these chemical products may destroy directly or indirectly many environment friendly microorganisms even the existence of chemicals also reported in mother's breast milk. To overcome these problems, scientists throughout the world emphasize on the exploitation of more and more botanicals for getting a healthy environment.

Occurrence of plant disease happens when the pathogen get suitable climatic conditions, especially more relative humidity encourages incidence of fungal infections. A burning example is Irish famine in 1845 that destroyed their principle crop potato, due to potato blight disease by the influence of perfect climatic conditions and that disaster actually gave a focus on science and the scientist were compelled to think about this outbreak and ultimately gave birth of plant pathology.

Foliar blight symptoms were observed on the leaves of *Brassica rapa* sub sp. *chinensis* in Yunnan, China where primary lesions consisting small, circular necrotic spots along with the secondary lesions with large irregular borders under high moisture conditions and the pathogenecity tests identified the causal organism as *Rhizoctonia solani* (Yang *et al.*, 2004). Azevedo *et al.* (2000) reported that the economically important vegetable cabbage production was impaired in Brazil due to the occurrence of diseases like black rot and Alternaria leaf spot caused by *Xanthomonas campestris* and *Alternaria brassicicola*, respectively. *Stemphylium solani* causes gray leaf spot of tomato, infections usually appears from lower older leaves to the upper, eventually leading to defoliation of the plant. Humid and wet weather are favourable to spread this disease and the fungal spores spread from the surface of the infected tissues by wind and splashing water, even a chance through long dew period or sprinkler irrigation. Anthracnose of french bean caused by *Colletotrichum lindemuthianum* is a serious disease occurs, in every bean growing area of the world. This pathogen produces elongate, angular spots in veins and ventral surface of the leaves spreading into the surrounding tissues and stem including pods. *C.lindemuthianum* caused the anthracnose outbreak of white beans in USA and Canada in 1976 (Gangopadhyay, 1984). *Corynespora cassiicola* causes infection in foliar plant parts and fruits. Symptoms develop on leave as brown spot and increase in size with a sunken dull green halo around lesions. Fruit spot occur on shoulder, initially as white and ultimately enlarge to a zonatic lesion with dark green to black appearance. This fungus can spread from crop debris by wind or rain splashing and the moist climate is most favourable for its transmission. The aim of the present study was to evaluate the antiphytopathogenic property of aqueous and methanolic extracts of *Terminalia chebula* and non-diluted *Eucalyptus citriodora* oil against five phytopathogens *viz.*, *Corynespora cassiicola* (*Cc*), *Stemphylium solani* (*Ss*), *Rhizoctonia solani* (*Rs*), *Colletotrichum lindemuthianum* (*Cl*) and *Alternaria brassicicola* (*Ab*), in laboratory conditions. Antimicrobial study with *Artemisia nilagarica* (Clarke) Pamp. oil against six bacteria and seven pathogenic fungi were studied and recorded larger inhibition zone against two bacteria and one fungi (Thoppil *et al.*, 2002). Dutta *et al.* (2004) reported, in laboratory conditions *Terminalia chebula* dry fruit aqueous extract is more effective than *Piper betel* fresh leaf extract against two phytopathogens. A mixture of white pine (*Pinus strobes*) and redpine (*P. resinosa*) bark mulch reduced the incidence of damping off of ginseng (*Panax quinquefolius* L.) seedling caused by fungal pathogen *Rhizoctonia solani* under field and controlled environmental conditions (Reeleder *et al.*, 2004).

Materials and Methods

Essential oil was extracted from the fresh leaves of *Eucalyptus citriodora* Hook. by using clevangar's apparatus for conducting the study.

Terminalia chebula aqueous extract was prepared by boiling the clean ground dry fruit (small type) with double distilled water at a ratio 1 : 4 (w/v) for 10 minutes. After cooling, filtered through Whatman filter paper No.1 and the filtrate was sterilized for conducting the study.

Methanolic extract was prepared by dipping the ground dry fruits in a reagent bottle just enough to submerge and kept for five days with vigorous shaking. Filtered through Whatman filter paper No. 1 and filtrate was allowed for complete evaporation of methanol. The dry extract was diluted by double distilled water at a ratio 1 : 10 (w/v) and after sterilizing considered for the study. The test pathogens were collected from Assam Agricultural University, Jorhat (Assam).

Antiphytopathogenic activity of eucalyptus oil was studied by 'disc diffusion method' and extracts by 'agar well diffusion method'. Twenty five ml sterilized Potato Dextrose Agar (PDA) was taken aseptically in each 90 mm dia sterilized petriplates in the laminar flow clean air bench, 0.3 ml of 96 hours old broth culture of that pathogens was evently distributed over the surface of the settled medium by the help of glass spreaders. In case of eucalyptus oil, 5 mm diameter Whatman filter paper No 1 sterile discs were thoroughly moistened and placed over the surface of the medium. In case of *T. chebula* extracts, using separate sterile borer, 8 mm dia holes were made in each culture containing petriplates and bottoms of the holes were sealed by sterile melted PDA medium to prevent leaching or spreading. Aqueous and methanolic extracts were poured into the holes upto the neck. Control was also maintained by pouring sterile distilled water into the holes in the same manner. Petriplates providing oil moistened discs and extracts were kept in BOD incubator at a temperature 28 ± 2 C for 96 hours. Data was recorded as a mean of six replications by accounting the zone of inhibition on the surface of the culture containing plates.

Results and Discussion

In Table 31.1, the result showed that, *T. chebula* aqueous extracts exhibited highest zone of inhibition (ZI) 13.70 per cent in *Cc* followed by 11.86, 6.86, 6.11 per cent in *Ss, Rs* and *Cl*, respectively. Methanol extract of the same also showed highest Zl in *Cc* (13.14 per cent) followed by 11.11 in *Ss*, 8.33 in *Cl* and 5.56 per cent in case of *Rs*.

Table 31.1: Efficacy of *T. chebula* Extracts Against Phytopathogens

Test Pathogens	**Zone of Inhibition at 96 hrs Incubation*					
	T. chebula Aqueous Extract			***T. chebula Methanol Extract*		
	[1]Dia (mm)	*[2]Dia (mm)*	*Net (%)*	*[1]Dia (mm)*	*[2]Dia(mm)*	*Net (%)*
Corynespora cassiicola	20.33	12.33	13.70	19.83	11.83	13.14
Stemphylium solani	18.67	10.67	11.86	18.00	10.00	11.11
Rhizoctonia solani	14.17	6.17	6.86	13.00	5.00	5.56
Colletotrichum lindemuthianum	13.50	5.50	6.11	15.50	7.50	8.33
Alternaria brassicicola	0	0	0	0	0	0

*Mean of six replications; **: Dilution 1 : 10 (w/v)

[1]Including 8 mm dia hole, [2]Excluding dia of hole.

The effect of the extracts on *Rs* and *Cl* was vise versa. *T. chebula* aqueous extract was more effective (13.70 per cent) on *Cc* rather than methanolic extract (13.14 per cent). Zl of aqueous extract, in *Ss* and *Rs* were more (11.86, 6.86 per cent) than methanolic extract (11.11, 5.56 per cent). But in case of *Cl* methanolic extract was more effective (8.33 per cent) than aqueous extract (6.11 per cent). *Cc* was more susceptible towards aqueous and methanolic extracts, showing greater Zl (20.33, 19.83 mm) but in case of *Ab* no Zl was recorded revealing more resistance.

In consideration of Table 31.2, eucalyptus oil exhibited the highest Zl in *Cc* (14.44 per cent) followed by 8.89, 8.33, 5 per cent in *Cl, Rs* and *Ss*, respectively. *Cc* was more susceptible to the oil and Zl was 18.00 mm. *Ab* is totally ineffective to the oil and no Zl was recorded.

Table 31.2: Efficacy of Eucalyptus Oil Extracts Against Phytopathogens

Test Pathogens	*Zone of Inhibition at 96 hrs Incubation**		
	[1]Dia (mm)	*[2]Dia(mm)*	*Net (%)*
Corynespora cassiicola	18.00	13.00	14.44
Stemphylium solani	9.50	4.50	5
Rhizoctonia solani	12.50	7.50	8.33
Coletotrichum lindemuthianum	13.00	8.00	8.89
Alternaria brassicicola	0	0	0

*Mean of six replications.

[1]Including 5 mm dia disc, [2]Excluding dia of disc.

In comparison of Table 31.1 and Table 31.2, it was proved that, although experimental substances were totally different *i.e.* aqueous and methanolic extracts of *T. chebula* and eucalyptus oil, highly susceptible (*Cc*) and highly resistant (*Ab*) species were same in all cases. Highest Zl was recorded in *Cc*, 14.44, 13.70 and 13.14 per cent by the influence of eucalyptus oil, *T. chebula* aqueous and methanolic extracts, respectively. The most resistant test pathogen (*Ab*) recorded nil Zl (0 per cent) in all three cases.

Today there is an interest in the use of antifungal and antibacterial botanicals because of the necessity of finding safer herbal alternatives in combination of the need of preventing environmental degradation. So it may be concluded that the experimental findings may be considered for any eco-friendly products.

Acknowledgement

The authors would like to convey their sincere thanks to Dr. S.N. Dube, Director, Defence Research Laboratory, Tezpur (Assam) for his constant support and encouragement throughout the study and thanks are also due to the plant pathology dept of Assam Agriculture University, Jorhat (Assam) for providing the test cultures for conducting the experiment.

References

Azevedo, S.S., Mariano, R. De L.R. and Michereff, S.J., 2000. Survey of the intensity of black rot and Alternaria leaf spot of cabbage in Agreste of pernambco determination of sample size for disease quantification. *Summa Phytopathologica*, 26: 299–306.

Dutta, B.K., Rahman, I. and Das, T.K., 2004. Growth inhibition of phytopathogens by the application of plant extracts. *Env. and Eco.*, 22: 618–620.

Gangopadhyay, S., 1984. *Advances in Vegetable Disease.* Associated Publishing Company, New Delhi.

Reeleder, R.D., Capell, B.B., Roy, R.C., Grohs, R. and Zukey, B., 2004. Suppressive effect of bark mulch on weeds and fungal diseases in ginseng (*Panax quinquefolius* L.). *Allelopathy Journal*, 13: 211–232.

Thoppil, J.E., Denna, M.J., Tajo, A., Sreeranjini, K., Kochuthressia, M.V. and Leeja, L., 2002. Antimicrobial potential of the essential oil of *Artemisia nilagerica* (Clarke) pamp. *Geobios*, 29: 181–182.

Yang, G.H., Chen, X.Q., Chen, H.R., Naito, S., Ogoshi, A. and Zhao, J.F., 2004. First report of foliar blight in *Brassica rapa* sub sp. *chinensis* caused by *Rhizoctonia solani. Pl. Pathol.*, 53: 260.

Chapter 32

Studies on Identifying Suitable Donors for High Iron and Zinc Content in Rice (Oryza Sativa L) Grains

P. Raghuveer Rao[1] and S.R. Voleti[2]*

[1]*Scientist (SS),* [2]*Senior Scientist*

Division of Plant Physiology, Directorate of Rice Research, Rajendranagar, Hyderabad – 500 030, A.P., India.

ABSTRACT

A study was initiated during kharif-2004 with the objective of identifying suitable donors for high iron and zinc content in rice grains. This study was conducted at four locations *viz.*, Coimbatore (CBT), Hyderabad (Hyd), Maruteru (MTU) and Panthnagar (PNR) with forty entries comprising of released and popular HYV of 120-150 days duration group from various states. Standard procedure was followed for analyzing iron and zinc content by acid digestion of pre milled rice grains, followed by estimation over an Atomic Absorption Spectrophotometer (AAS). After analysis, it was found that the whole grains which includes the husk, pericarp, embryo and endosperm and no attempt was made to do organ wise analysis as this is the preliminary study and needs further confirmation, it is also well known fact that after milling much of the micronutrients are lost in the process of obtaining white rice. There was variation in iron and Zinc content among locations as well as among the genotypes. It was observed that Mandyavijaya had high iron (56 ppm) and zinc (28 ppm) at Maruteru, but the same genotype had low iron (8 ppm) at Hyderabad and low zinc (11 ppm) at Panthnagar. This shows that iron and zinc uptake is mostly location specific. It was also observed that genotypes of rice have iron content from 8 to 56 ppm while zinc content varied from 11 to 28 ppm. Among the genotypes Yamini (42 ppm), Suraksha (40 ppm) and Swarnadhan (39 ppm) had high mean value for iron while Kasturi (21 ppm) Shalivahana (20 ppm) and Manasarovar (20 ppm) had high Zinc content across the locations.

Corresponding Author: E-mail: prrao2005@yahoo.co.in.

Introduction

Rice, the staple diet of millions in the developing world, is a poor source of micronutrients. Where rice is the staple, about two billion people suffer from iron and zinc deficiency leading to anemia in case of iron, and zinc deficiency leads to stunting and morbidity in children who become prone to number of childhood infections such as diarrhoea, pneumonia and malaria. Efforts have thus focused on "biofortifying" rice to make it nutritionally better with fortified iron, zinc and vitamin A (golden rice) through various approaches. Genetic engineering rice to increase iron and zinc content has been one course of action. But is genetic engineering needed to develop iron-rich rice? There are already success reported in naturally breeding and selecting rice with high iron and zinc content, which would not carry the risks associated with genetic engineering (Lim Li Ching, 2004). Plant breeders at the Philippines based International Rice Research Institute (IRRI) have identified rice varieties that are naturally high in iron and zinc. IR 68144 is reported to be a good source for high iron, zinc and also vitamin A. The available scientific evidence (Verma *et al.*, 2004) indicates the positive approach to all these nutrition related breeding to produce mineral and vitamin enriched varieties of rice. Keeping the above facts in view a preliminary study was undertaken to assess the availability of iron and zinc in rice grains and in the long run suitable donors may be identified helping the breeders to use them as parents in their local breeding programme.

Materials and Methods

A study was taken up in All India Coordinated Rice Improvement Programme (AICRIP) with nodal office at Directorate of Rice Research, Rajendranagar, Hyderabad (AP) to analyze rice grains for iron and Zinc availability and also find out the variability across locations and genotypes. For the said purpose this study was conducted at four locations in India *viz.*, Coimbatore (CBT), Hyderabad (HYD), Maruteru (MTU) and Panth Nagar (PNR) with a little over forty entries comprising mostly released and popular HYV of 120–150 days duration group from various states with a popular HYV in their locations as local cheek. The field study was conducted under normal irrigated conditions with a fertilizer dose of 100N–45P_2O_5–60 K_2O Kg/ha grown in a RBD plots with four replications. At maturity the grains more harvested and sent to DRR, Rajendranagar where they were analyzed for iron and zinc content in rice grains. The methodology followed for chemical analysis of iron and zinc in rice grains was from Yoshida *et al.* (1976). From the 40 genotypes with four replications made total number of entries as 120. Chemical analysis was done from four different locations and the methodology involved was from each replication 500 mg grain sample was weighed in acid digestion tubes and the acid mixture was sulphuric acid (1 part), Nitric acid (10 parts) and Perchloric acid (4 parts) *i.e.* 1 : 10 : 4 ratio was used to make the acid mixture. 500 mg of grains were dissolved in 10 ml of acid mixture and heated over kjel dal digestion oven, gradually rising the temperature from 50 to 200°C until the sample was digested and only 1 ml of acid mixture was left in the digestion tube. After digestion which lasted 2 hours the test tubes were added with distilled water to make up the volume to 50 ml. There after the sample was filtered into conical flashes using Whatman No.1 filter paper. Thus the analyzed sample was measured for iron and zinc over an Atomic Absorption Spectrometer (AAS) and the values were expressed in parts per million (ppm) or (mg/kg).

Results and Discussion

The results furnished in Table 32.1 and Figures 32.1–32.3 shows the iron and zinc content of the whole grains in rice and the results show us that genotypes of rice have iron content from 8 to 56 ppm while zinc content varied from 11 to 28 ppm. After performing the statistical analysis, the results of

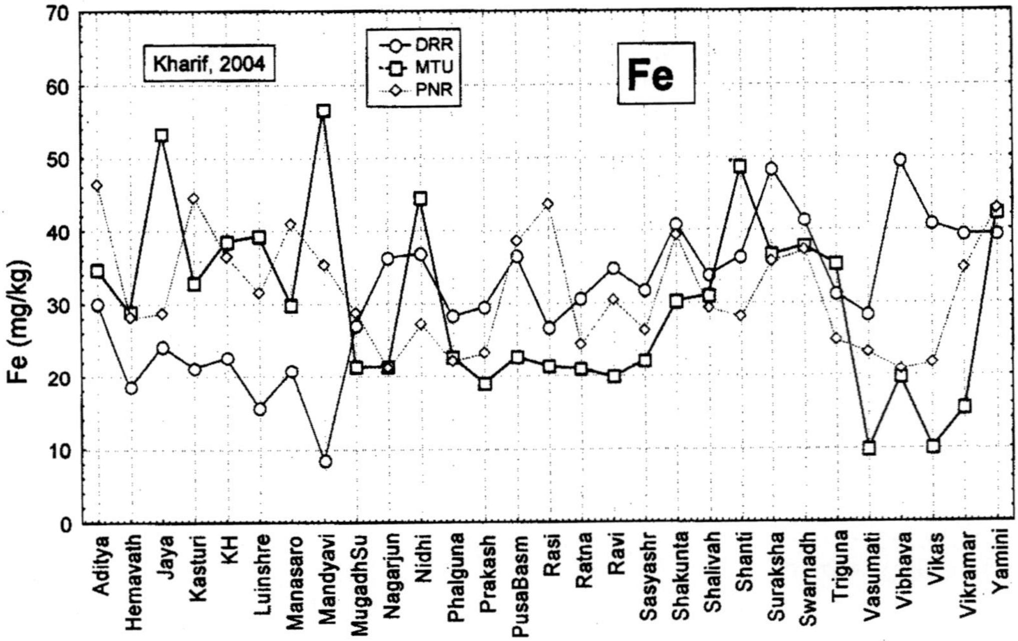

Figure 32.1: Mean Values of 29 Common Entries for Fe at Three Locations

Coimbatore center were not taken into account due to very high absolute values and also high CV (per cent) values. From the 40 entries only 29 entries were found suitable with low SE. Studies at Philippines based International Rice Research Institute (IRRI) have identified IR 68144–3B–2–23 with high iron and zinc concentration. The grain has 21 ppm iron and about double the normal content in rice, and also about 34 ppm of zinc (Lim Li Ching, 2004). The results from our study included whole grains *i.e.* before milling and it was found that Yamini (42 ppm), Suraksha (40 ppm) and Swarnadhan (39 ppm) had high mean values for iron while Kasturi (21 ppm), Shalivahana (20 ppm) and Manasarovar (20 ppm) had high zinc content across the locations. Rice is eaten after milling unlike pulses where much of the micro nutrients are lost, hence efforts should be made to increase iron and zinc after milling (Wolnik *et al.*, 1983 and Augustine *et al.*, 1981). This being a preliminary study the authors makes it a point to mention that further confirmation studies are needed before identifying suitable donors for using them in breeding programme.

Table 32.1: Genotypes having High and Low Mean Values for Iron and Zinc Across Locations

Mean Values Across Locations	*Genotypes for Iron*	*Genotypes for Zinc*
High mean	Yamini 41 (ppm)	Kasturi (21 ppm)
	Sureksha (40 ppm)	Shalivahana (20 ppm)
	Swarnadhan (39 ppm)	Manasarovar (20 ppm)
Low Mean	Vikas (24 ppm)	Vikramarya (14 ppm)
	Prakash (24 ppm)	Triguna (14 ppm)
	Vasumati (21 ppm)	Vibhava (13 ppm)

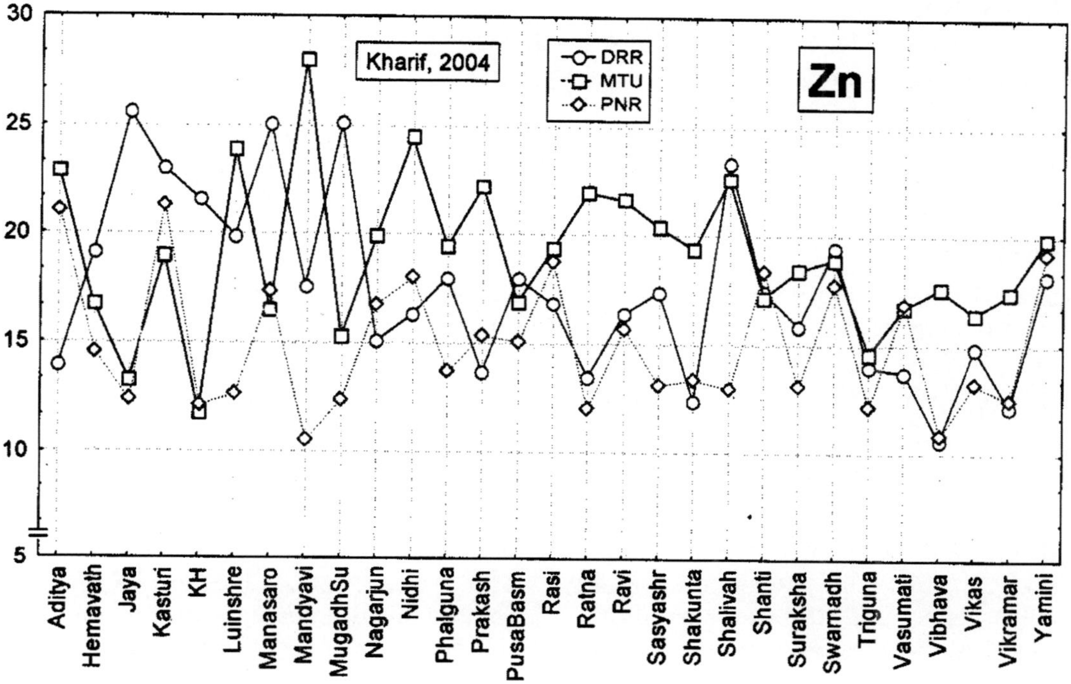

Figure 32.2: Mean Values of 29 Common Entries for Zn at Three Locations

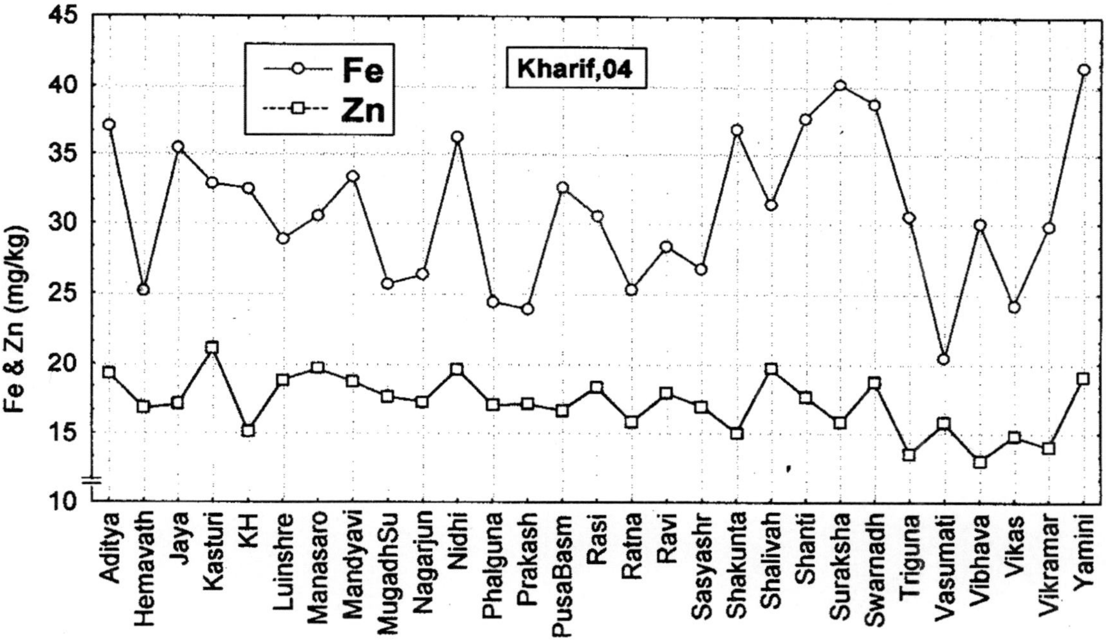

Figure 32.3: Mean Fe and Zn of 29 Varieties at Three Test Locations

Acknowledgements

The first author wishes to thank Dr. S.C. Sankhyadhar, Rice Physiologist at Panthnagar, Dr. S. Sivarama Prasad at Maruteru, Dr. C. Vijayalakshmi at Coimbatore, Dr. C. Keshava Reddy, Head, Plant Physiology, DRR, Hyderabad and Dr. S.V. Subbaiah acting Project Director, DRR, Hyderabad. The authors also wish to thank their technical staff G.V.S.S. Prasad and K. Ramulu at DRR, Rajendranagar, Hyderabad for their efforts in conducting the experiments and helping the first author in conducting chemical analysis.

References

Augustin, J., Beck, C.B., Kalbfleisah, G., Kagel, L.C. and Matthews, R.H., 1981. Variation in the vitamin and mineral content of raw and cooked commercial *Phaseolus vulgaris* classes. *J. Food Sci.*, p. 11701–11706.

Lim Li Ching, 2004. *GM Iron Rice: A Solution*? ISIS Press Release 14/09/04.

Shouichi Yoshida, Donglas A. Forno, James H. Cook and Kwanchai A. Gomez, 1976. *Laboratory Manual for Physiological Studies of Rice*, 3rd Edition, IRRI, Los Banos, Phillippines.

Verma, K.K., N.K. Singh and S.B. Mishra, 2004. Breeding for nutritionally enriched food grain production: A review. *J. Interacad.*, 8(4): 644–674.

Wolrik, K.A., Fricke, F.L., Capar, S.G., Braude, G.L., Meyer, M.W., Satzger, R.D. and Kuennen, R.W., 1983. Elements in major raw agricultural crops in the United States. 2. Other elements in lettuce, Peanuts, Potatoes, Soyabeans, Sweet Corn, and Wheat. *J. Agric Food Chem.*, 31: 1244–1249.

Chapter 33

Assessment of Multiple Antibiotics Resistant Airborne Pathogens in Hospital's Environment

[1]Dilip H. Tambekar, [1]Pranita B. Gulhane, [1]S.R. Gulhane, [1]D.D. Bhokre and [2]Y.S. Banginwar

[1]Department of Microbiology, S.G.B. Amravati University, Amravati – 444 602, India
[2]Institute of Pharmacy, Kaulkhed, Akola

ABSTRACT

The emergence of nosocomial infection and multiple antibiotic resistant bacteria has become a major challenge in the treatment of infectious diseases. Bacterial isolates from hospital air of Amravati were examined to assess multiple antibiotic resistance patterns. Isolated 100 bacterial spp. included *Staphylococcus aureus, Staph. saprophyticus, Staph. epidermidis, Micrococcus luteus, Micrococcus roseus, Pseudomonas aeruginosa, Proteus vulgaris, Escherichia coli, Citrobacter freundii, Enterobacter aerogenes* and *Klebsiella pneumoniae* were examined for antimicrobial sensitivity against 15 different types of antibiotics such as amikacin, augmentin, ceftazidime, ceftriaxone, cephotaxime, ciprofloxacin, chloramphenicol, erythromycin, fusidic acid, gentamycin, lincomycin, netilmicin, ofloxacin, penicillin and vancomycin. The multiple antibiotic resistant (MAR) indexes of the isolates showed the highest (0.0073) with *Enterobacter aerogenes.* The 1500 tests of antibiotics were made against 100 organisms. Out of them 1015 antibiotic tests were found to be resistant, 103 tests were found to be intermediate sensitive and 382 tests were found to be sensitive to various antibiotics tested. The 80 per cent isolates were resistant to augmentin and penicillin, 79 per cent to fusidic acid and lincomycin, 76 per cent to erythromycin, 74 per cent to ciprofloxacin, 71 per cent to vancomycin, 66 per cent to ceftazidime and gentamycin, 62 per cent to chloramphenicol, 60 per cent to netilmicin, 58 per cent to ceftriaxone and ofloxacin, 53 per cent to amikacin and cephotaxime.

Keywords: *Hospital air, Multi-resistant pathogens, Indoor environment, Air-borne pathogens.*

Introduction

Airborne transmission refers to infections, which are contracted from microorganisms contained in droplet nuclei produced by coughing, sneezing or some other form of aerosolation and also apply to dust particles and skin squamae carrying pathogenic microorganisms. The contribution of airborne microorganisms to the spread of infection is likely to be greater than is currently recognized. This is because many airborne microorganisms remain viable and may not detected and some infections arising from contact transmission involve the airborne transportation of microorganisms onto inanimate surfaces (Beggs, 2003). When a person coughs or sneezes many thousands of droplets are expelled at high velocity into the atmosphere (Wells, 1995). According to Beggs, 2002, droplet nuclei are so small that they settle slowly and remain suspended in air for a considerable period of time and distributed widely throughout in indoor hospital buildings.

Hospital air commonly contaminated with various microbes based on the patients of different diseases admitted in the hospital. These pathogens may contaminate the air and are able to survive in the adverse condition and can cause hospital infection and develop resistance to antibiotics. Even in the developed countries despite so many advances in the treatment of infectious diseases "cross infection" in the hospital tends to be high (Nanoty *et al.*, 2003). Bioaerosol particles are usually present in indoor and outdoor air of various sections of hospitals, although their composition and concentration may vary. Human exposure to these airborne microorganisms may resulted in variety of infectious diseases, allergic and irritant responses, respiratory problems and hypersensitivity reaction (Tambekar and Gulhane, 2003).

Hospitals have always acted as a source of infection to patients admitted and about 2–12 per cent incidences of hospital infections have been reported in advanced countries, but it is much higher in crowded hospitals in developing countries (Domale *et al.*, 2005). Mathias *et al.*, 2000 reported that the emergence of multiple antibiotic resistant bacteria has become a major challenge in the treatment of infectious diseases. The important bacterial pathogens associated with hospital infections are *S. aureus, Ps. aeruginosa* that may develop antibiotic resistance. Hospital acquired infection caused by these organism causes serious problem (Thomas, 1988). The hospital environment is an important source of cross infection. The sources are patients suffering from infectious diseases, staff working within hospital environment (Park, 1987). Increased emergence of resistant organisms with hospital and geographical differences has been shown to be due to consumption of antimicrobial agents (Garciadominquez *et al.*, 1994). Brachman, (1970) reported that the airborne route of transmission accounts 10 to 20 per cent of endemic nosocomial infections. Methicillin resistant *Staphylococcus aureus* isolates remain a major clinical and epidemiological problem for the patients of surgical wound infection, burns, pneumonia and orthopedic surgeries (Kurup *et al.*, 2005). According to Basustaoglu *et al.* (1995) antibiotic resistance is an ever-increasing problem for *Pseudomonas aeruginosa,* which is normally associated with increased morbidity and mortality, particularly in hospitalized patients. *Pseudomonas aeruginosa,* a virulent microorganism is susceptible to only a limited number of antibiotic agents (Bennett, 1974) and can survive and replicate within the hospital environment, where it colonizes sinks (Whitby and Rampling, 1972), hospital distilled water systems (Favero *et al.*, 1971), vegetables (Kominos *et al.*, 1972), mattresses (Fujita *et al.*, 1981), hand wash basins, humidifiers, floor mops, plastic washing bowels, soap dishes, nail brushes, syringes, bedrails (Lowbury *et al.*, (1970) and even disinfectants (Burden and Whitby, 1967). Duration of hospital stay and the treatment with newer antibiotics also leads to nosocomial infection by the members of Enterobacteriaeceae like *Escherichia coli,* and species of *Proteus, Klebsiella, Enterobacter, Citrobacter* and *Serratia* (Mathias *et al.*, 2000).

The prominent pathogenic microorganisms found in hospital air are multidrug resistant strains of *Staphylococcus aureus,* which remained a major clinical and epidemiological problem among hospital personnel and patients (Cynthia *et al.,* 1986). Mathias *et al.,* 2000 studied reservoirs of multi-resistant nosocomial pathogens in a Secondary Care Hospital, Ramnagar and measured the indoor and outdoor air contamination of various sections in hospital and recorded most contaminated site was labour room followed by dressing room and operation theatre. Tambekar, (2004) conducted a study at 50 private and general hospitals in Amravati city (India) and reported the maximum bacterial contamination was recorded in general ward (indoor), followed by general ward (outdoor), OPD, private room (indoor), private room (outdoor), operation theatre (indoor), operation theatre (outdoor), labour room (indoor), labour room (outdoor) and least in pathology laboratory and ICU.

Furthermore, changing patterns of susceptibility and the availability of new antimicrobial agents require continuous updating of knowledge concerning treatment of diseases caused by such pathogens. The impact of airborne microorganisms on indoor and outdoor air quality of hospital and impact on human health remains poorly understood. Therefore the present study was conducted in 76 hospitals of Amravati city to assess air contamination by bacterial pathogens and to make aware the people from the multi-resistant airborne pathogens, which can give information on various hospital borne infections for proper treatment.

Materials and Methods

Sample Collection

The aero-biological survey was carried out in indoor and outdoor environment at 76 hospitals in Amravati which includes 7 general, 37 maternity and children, 5 multi-specialty, 9 cardiac, 6 each of orthopaedic and ENT, 1 each of cancer and mental hospitals and 4 clinics.

Microbial Analysis

The total 981 organisms are isolated from 716 air samples by performing sedimentation method. The air samples were analyzed from indoor and outdoor environment of hospitals by exposing the mannitol salt agar, MacConkey agar and Cetrimide agar plates. Qualitative estimation of *Escherichia coli, Cirobacter freundii, Serratia marcescens, Enterobacter aerogenes, Klebsiella pneumonia* and species of *Staphylococcus, Micrococcus, Pseudomonas,* and *Proteus* were identified by applying various cultural, morphological and biochemical tests.

Antibiotic Sensitivity Test

The total hundred clinical isolates were tested for antimicrobial susceptibility tests against the following antibiotics: Ceftriaxone (30 mcg), Ceftazidime (30 mcg), Cephotaxime (30 mcg), Lincomycin (2 mcg), Netilmicin (30 mcg), Ofloxacin (2 mcg), Vancomycin (30 mcg), Amikacin (30 mcg), Penicillin (10units), Gentamycin (10 mcg), Augmentin (30 mcg), Ciprofloxacin (5 mcg), Erythromycin (10 mcg), Fusidic acid (10 mcg), Chloramphenicol (30 mcg). Using the agar diffilsion assay performed susceptibility test and the disks were obtained from Hi-media Laboratories Pvt. Ltd., Mumbai. Antibiotic sensitivity tests were conducted following Davis and Stouts (1971) procedure.

Results and Discussion

The most dominant among the aerial contaminants are *Staph. aureus* followed by *Ps. aeruginosa, M. luteus* and *Staph. epidermidis.* It is generally the case that gram-positive bacteria, such as *S. aureus,* possess a peptidoglycan-rich cell wall, which gives them relative resistance to desiccation. It can also remain viable on aerosolized skin squamae for long periods of time (Sands and Goldmann, 1998).

Wagenwoort, 1993 reported MRSA on ventilation grilles in an orthopedic ward and Cotterill, 1996 identified colonies of MRSA in the exhaust air from an isolation room as the source of an outbreak in an intensive care unit; the MRSA bacteria were re-entering the unit via an open window. *S. aureus* grows on the nasal mucosa; hands then touch the nose and *S. aureus* are transferred to the skin; they colonize the skin and are ultimately disseminated back into the air on skin squamae. Even sometimes Doctors stethoscope or surgeons hair are also the source of *Staph. aureus.* During lengthy operations and orthopedic surgery, it may get disseminated from the hair and hands and can be transmitted into the wound because wound is the susceptible site for infection (Tambekar, 2004).

The dominance of *Ps. aeruginosa* denotes its minimal growth requirements and its survival and replication within the hospital environment. The disperse of *Ps. aeruginosa* from colonized patients and personnel of the hospital might have resulted in further contamination of the environment of the hospital as well as the hands of the medical staff (Mathias *et al.,* 2003). According to Tambekar *et al.* (2005) *Pr. mirabilis, Esch. coli* and other gram-negative bacteria are found in less number in hospital environment. Some species of Enterobacteriaeceae such as *Pr. vulgaris, Morganella morganii, Citrobacter freundii, Serratia marcescens* and *Klebsiella pneumoniae* showed least air contamination, as the main source is the contaminated water droplets.

The various bacterial strains were isolated in hospital environment on a variety of media belong to the species of 11 different types of microorganisms. The per cent distribution of different organisms resulted into 10 per cent each of *Staphylococcus aureus, Staph. epidermidis,* and *Micrococcus luteus,* 4 per cent *Staph. saprophyticus,* 6 per cent *M. roseus,* 38 per cent *Pseudomonas aeruginosa,* 2 per cent *Proteus vulgaris,* 5 per cent *Escherichia coli,* 11 per cent *Enterobacter aerogenes,* 2 per cent each *Citrobacter freundii* and *Klebsiella pneumoniae* (Figure 33.1).

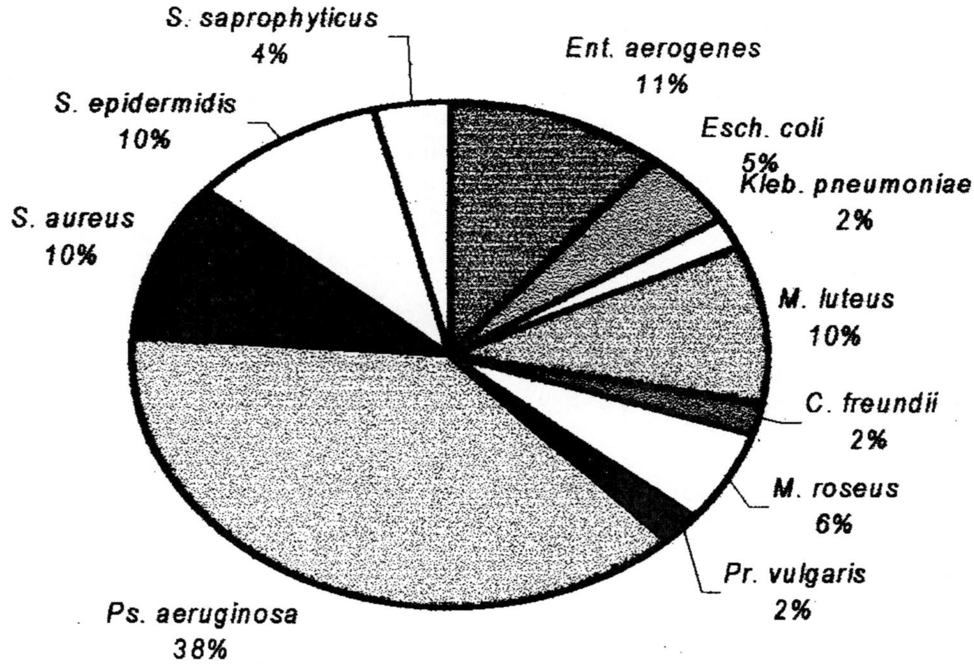

Figure 33.1: Bacterial Pathogens Isolated from Hospital Environment

A total of hundred isolates were tested for antimicrobial susceptibility tests against 15 antibiotics. Out of which near about 80 per cent isolates were resistant to commonly used antibiotics such as penicillin, augmentin, ciprofloxacin, erythromycin, fusidic acid, ceftriaxone and netilmicin. The 25 per cent isolates were resistant to all 15 antibiotics, which include *Ps. aeruginosa, Micrococcus luteus, Staph. aureus.* Nearly 80 per cent *Staph. aureus* were resistant to penicillin. This is due to enzyme beta-lactamase or penicillinase, which destroy the drug.

Out of the isolated strains of bacteria 80 per cent strains were resistant to augmentin and penicillin, 79 per cent to fusidic acid and lincomycin, 76 per cent to erythromycin, 74 per cent to ciprofloxacin, 71 per cent to vancomycin, 66 per cent to ceftazidime and gentamycin, 62 per cent to chloramphenicol, 60 per cent to netilmicin, 58 per cent each to cepholaxime and ofloxacin 53 per cent each to amikacin and cephotaxime (Figure 33.2). The study showed that *Pr. vulgaris* was found highly resistant to tested antibiotics. Next to them were *M. luteus, E. coli, M. roseus, Ps. aeruginosa, Staph. aureus, Ent. aerogenes, S. epidermidis, Kleb. pneumoniae, Staph. saprophyticus* and finally *C. freundii.*

The study showed the highest MAR index (0.0073) with *Enterobacter aerogenes* against augmentin followed by *Micrococcus luteus,* which was 0.0066 against ofloxacin and chloramphenicol. *Staph. aureus* was highly resistant to fusidic acid (MAR 0.006) and least against ofloxacin and Vancomycin. *Staph. epidermidis* was highly resistance (MAR index 0.0053) to penicillin while *Micrococcus roseus* had highest MAR 0.004 against ceftazidime. *Escherichia coli* had MAR index 0.0033 against fusidic acid, penicillin, vancomycin, netilmicin and chloramphenicol. *Staph. saprophyticus* had highest MAR index 0.002 against ceftazidime, erythromycin, fusidic acid, lincomycin and penicillin while *Pseudomonas aeruginosa* had 0.022 MAR index against erythromycin. *Proteus vulgaris, Citrobacter freundii* and *Klebsiella pneumoniae* had highest MAR index 0.0013 against lincomycin. *Proteus vulgaris* was highly resistance against commonly used antibiotics such as lincomycin, fusidic acid, penicillin, augmentin, vancomycin,

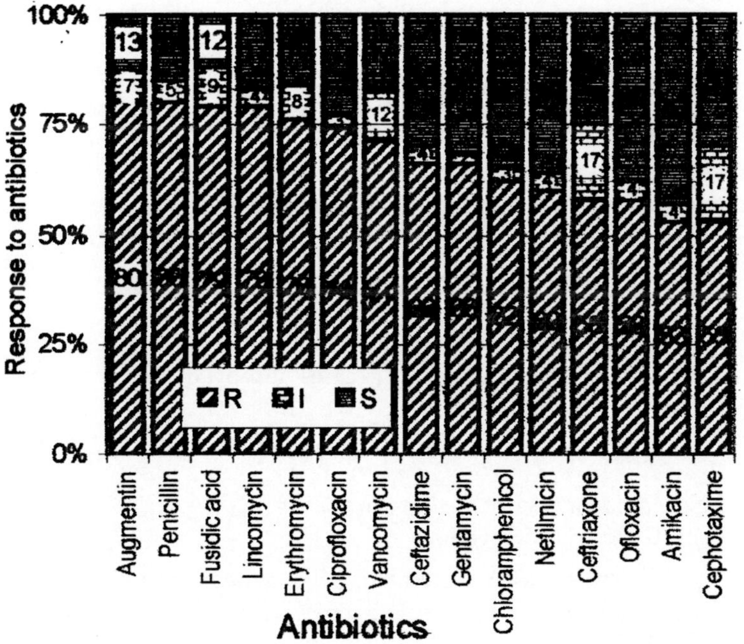

Figure 33.2: Antibiotic Response Against Bacterial Isolates

ciprofloxacin, gentamycin, ceftriaxone, ceftazidime, netilmicin, cephotaxime and ofloxacin. While *Citrobacter freundii* had the same MAR index against vancomycin and lincomycin. *Klebsiella pneumoniae* also had same MAR index (0.0013) against lincomycin, augmentin, ciprofloxacin, erythromycin, ofloxacin and netilmicin (Table 33.1).

Table 33.1: Bacterial Isolates and Antibiotic Sensitivity

Organisms	*R*	*I*	*S*
C. freundii	11	3	16
Ent. aerogenes	104	14	47
Esch. Coli	62	0	13
Kleb. pneumoniae	19	04	07
M. luteus	130	08	12
M. roseus	62	08	20
Pr. vulgaris	27	0	03
Ps. aeruginosa	377	39	154
S. aureus	98	05	47
S. epidermidis	93	14	43
S. saprophyticus	32	08	20
Total	**1015**	**103**	**382**

Ps. aeruginosa are maximally resistance due to the ability to produce a large number of extra cellular protective and toxic substances, and found resistance to commonly used antibiotics such as penicillin, augmentin, erythromycin, fusidic acid and lincomycin. The 80 per cent species of *Micrococcus luteus* were resistant to all 15 antibiotics tested. Thus, it deserves special attention among recently existing hospital resistant organisms. *Staphylococcus* spp. of general ward was sensitive while that of private room, out patient department and burn ward were found to be resistant to all antibiotics tested. Out of 60 isolates of *Pseudomonas* studied, more than 64 per cent isolates were resistant to more than six antibiotics tested. *Citrobacter freundii* was mostly resistant in indoor of private room whereas it shows sensitive pattern in outdoor environment of private room. *Escherichia coli* of surgery ward and *Enterobacter aerogenes* of general ward and X-ray, sonography sections were resistant to all antibiotics tested.

Thus study suggested that a number of factors influences the prevalence of antibiotic resistant in bacteria in hospital environment and sub-therapeutic and the therapeutic usage of antimicrobial drugs will result in increased propositions of multiple antibiotic resistant hospital pathogens. The periodic review of antibiotic usage is, therefore, of greatest importance to ensure that antibiotics are not used indiscriminately.

References

Aiblithnott, J.P., 1992. Staphylococcus. In: *Medical Microbiology*, 14th edn. (Eds.) Greenland, D., Slack, R.C.B., Peutherer, J.F. Churchill Livingston.

Basustaoglu, A.C., Gun, H., Saracli, M.A., Baysallar, M., Haznedaroglu, T., 1995. Development of resistance to imipnem among nosocomial isolates of *Pseudomonas aeruginosa*. *Eur. J. Clin. Microbiol. Infect. Dis.*, 11: 469–470.

Beggs, C.B., 2002. The use of engineering measures to control airborne pathogens in hospital buildings. Internet: (http://www.efm.leeds.ac.uk/CIVE/MTB/CBB-Nov8.pdf).

Beggs, C.B., 2003. The airborne transmission of infection in hospital buildings: Fact or fiction? *Indoor Built Environ.*, 12: 9–18.

Bennett, J.V., 1974. Nosocomial infection due to *Pseudomonas*. *J. Infect Dis.*, 230(Sppl): S4–7.

Brachman, P.S., 1970. Nosocomial infection: airborne or not? In: *Proceedings of the International Conference on Nosocomial Infections*, American Hospital Association, pp. 189–192.

Burden, D.W. and Whitby, J.L., 1967. Contamination of hospital disinfectants with *Pseudomonas* species. *Br. Med. J.*, 2: 153–155.

Cynthia, Friend, Norton, 1998. *Microbiology*, 2nd ed. Addison Wesley Publishing Company, USA, pp. 360–382.

Cotterill, S., 1996. An unusual source for an outbreak of methicillin-resistant *Staphylococcus aureus*. *J. Hospital Infection*, 32: 207–216.

Davis, W.W. and Stout, T.R., 1971. Disk plate method of microbiological antibiotic assay. *App. Microbiol.*, 22: 659–665.

Domale, V., Nanoty, V.D., Musaddique, M., 2005. Microbial studies on nosocomial infections due to environmental contamination in hospitals. *Nature Environment and Pollution Technology*, 4: 61–64.

Favero, M.S., Carson, L.A., Bond, W.W., Peterson, N.J., 1971. *Pseudomonas aeruginosa:* Growth in distilled water from hospitals. *Science*, 173: 836–838.

Fujita, K., Lilly, H.A., Kidson, A. and Ayliffe, G.A.J., 1981. Gentamicin-resistant *Pseudomonas aeruginosa:* Infection from mattresses in a burns unit. *Br. Med. J.*, 283: 219–220.

Garcia-dominquez, C., Martin, F., Santos Hurtado, I., Blanco, J., BIanco, M.T., Gomez Garcia, A.C., 1994. Susceptibilities of *Pseudomonas aeruginosa* to anti-pseudomonal antibiotics in a general hospital of Spain from 1989 to 1992. *J. Antimicrob. Chemother.*, 33: 1056–1059.

Grieble, H.G., Bird, T.J., Nidea, H.M. and Miller, C.A., 1974. Chute-hydro-pulping waste disposal system: A reservoir of enteric bacilli and *Pseudomonas* in a modern hospital. *J. Infectious Diseases*, 130: 602.

Kominos, S.D., Copeland, C.E., Gowsiak, B. and Postic, B., 1972. Introduction of *Pseudomonas* into a hospital via vegetables. *Appl. Microbiol.*, 24: 567–570.

Kurup, R., Wadher, B.J., Tumane, P.M. and Bhoosreddy, G.L., 2005. Isolation of MRSA plus VSSA strains causing hospital infection. *J. Microb. World*, 7: 93–96.

Lowbury, E.J.L., Thorn, B.T., Lilly, H.A., Babb, J.R. and Whitall, K., 1970. Sources of infection with *Pseudomonas aeruginosa* in patients with tracheotomy. *Jammed. Microbiol.*, 3: 39–56.

Mathias, A.J., Smacker, R.K., Sumatra, S. and Surabaya, S., 2000. An assessment of reservoirs of multi-resistant nosocomial pathogens in a secondary cares hospital. *Indian Journal of Microbiology*, 40: 183–190.

Mathias, A.J., Smacker, R.K. and Nandini, N., 2003. Multi-resistant *Pseudomonas aeruginosa* in an Indian secondary care hospital: Reservoirs and possible cross-transmission of the nosocomial pathogen. *Asian J. of Microbiol. Biotech. Env. Sci.*, 5: 71–75.

Nanoty, V.D., Musaddique, M. and Ahale, M.A., 2003. Microbiological studies on air in hospital environment. *J. Microb. World*, 5: 91–94.

Park, J.E. and Park, K., 1987. *Textbook of Preventive and Social Medicine*, 8th ed., pp. 372.

Sands, K.E.F. and Goldmann, D.A., 1998. Epidemiology of *Staphylococcus* and group *A. streptococci*. In: *Hospital Infections*, 4th ed., (Eds.) Bennett, J.V.. and Brachman, P.S. Lippincott-Raven Publishers.

Tambekar, D.H. and Gulhane, P.B., 2003. Studies on pathogenic air flora of hospitals in Amravati. In: 44th *Annual Conference of Association of Microbiologists of India*, Dharwad, November, 12–14, 2003.

Tambekar, D.H. 2004. An assessment of aeromicrobiology of hospitals in Amravati city. *Environmental Biology and Conservation*, 9: 91–93.

Tambekar, D.H., Kalbende, P.S. and Gulhane, P.B., 2005. Aeromicrobiology of hospitals in Amravati (India). *Indian J. Aerobiol.*, 18: 18–23.

Thomas, C.G.A., 1988. *Medical Microbiology*, 6th ed., pp. 183.

Wagenvoort, J., 1993. MRSA from air-exhaust channels. *Lancet.*, 341: 840–841.

Wells, W.F., 1995. *Airborne Contagion and Air Hygiene*. Harvard University Press, Cambridge, Massachusetts.

Whitby, J.L. and Rampling, A., 1972. *Pseudomonas aeruginosa* contamination in domestic and hospital environment. *Lancet*, 1: 15–17.

Chapter 34

Static Renewal Bioassay Test of Dimethoate on the Freshwater Fish *Cyprinus carpio*

H.B. Shivakumar and M. David*

Karnatak University's Research Laboratory, Toxicology Division, Department of Zoology, Karnatak Science College, Dharwad – 580 001, Karnataka, India

ABSTRACT

Static renewal bioassay test was conducted to determine the toxicity of insecticide dimethoate on fish *Cyprinus carpio. C. carpio* were exposed to different concentrations of organophosphate insecticide dimethoate 35 per cent EC for 96 hr and the per cent mortality was observed. The LC_{50} value was found to be 35 µl/l and one fifth of LC_{50} (7 µl/l) was selected as sub-lethal concentration for sub-acute studies. Behavioural patterns were observed in both lethal (1, 2, 3, 4 days) and sub-lethal concentration (1, 5, 10, 15 days). *C. carpio* in toxic media exhibited erratic and darting movements with imbalanced swimming activity, which might be due to the malfunctioning of neurotransmitters, followed by hyper and hypo opercular activity, loss of equilibrium, and mucus secretion all over the body were observed.

***Keywords**: Behaviour, Dimethoate toxicity, and Cyprinus carpio.*

Introduction

Millions of people around the world are dependent for protein and other nutritional requirements on the fisheries sector. It is now realized that natural Ocean and freshwater fisheries are being harvested to their near limit and the increased demands can only be met through diversification and application

* Correspondence author: Research Laboratory, Toxicology Division, Department of Zoology, Karnatak Science College Dharwad – 580 001, Karnataka, Mobile: 09448571295.

of modern tools. In India freshwater fish farming have been successfully developed to enhance fish production, but on the other hand toxicants originated from industrial, agricultural and municipal sources are directly or indirectly dumped into the aquatic ecosystem, which ultimately affects the non-target organisms like fish. The toxicity of pesticides in relation to mortality and impairment of metabolic activities in various organisms has been extensively studied (Choudhary and Shafi, 1985; Prasad and Shafi, 1989). Any change in the natural condition of aquatic medium causes behavioural responses in fish, ultimately leading to physiological adjustments (Rajamanickam and Karagaganapathy, 1988).

Dimethoate is one of the organophosphate insecticide extensively used in agriculture particularly on cotton, bengalgram, sunflower, citrus, groundnut, pulses, fruits and vegetables in Karnataka (India). The Environment Centre for National Toxicology says, there are a dozen highly dangerous chemicals including dimethoate. Dimethoate is a hard insecticide, which has become a matter of concern because of its potentiality and hazardous effect. Some detailed works have been published which indicate the need for giving a fresh look on the safe disposal of various toxicants into natural waters for the protections of fishes and aquatic ecosystem (Suparna Mullick and Konar, 1994). Hence in the present study an attempt bas been made to determine the LC_{50} value of dimethoate and its impact on behavioural aspects of the freshwater fish *Cyprinus carpio.*

Materials and Methods

Cyprinus carpio weighing 8±2 g and measuring an average length of 7 cm were collected from the State Fisheries Department, Dharwad and acclimatized to laboratory condition for 15 days in large cement tanks previously washed with potassium permanganate to free the walls from any fungal infections. Fish were maintained in dechlorinated tap water whose physico-chemical characteristics were analyzed following the methods mentioned in APHA (1998) and found as follows, Temperature 28±2°C, pH 7±0.2 at 27°C, Dissolved oxygen 5.7±0.9 mg/l, Carbon dioxide 8.3±0.4 mg/l, Calcium 16.04±1.2 mg/l, Magnesium 0.8±0.1 mg/l, Nitrate 9.14±0.13 µg/l, Phosphate 0.401±0.001 µg/l, Chlorinity 7.2±0.4 mg/l, Total hardness 43.4±3.4 mg/l, Total alkalinity 25.2±1.3 mg/1 and Specific gravity 1.00374. Fish were fed with oil cake and rice bran to keep them more or less in the same state of metabolic requirements.

Dimethoate 35 per cent purity was procured from Rallies India Ltd Mumbai. Required quantity of pesticide was added to the exposure container using variable micropipette. The fish were exposed in batches of ten to varying concentrations of dimethoate. After every 24 hr water was renewed to maintain the pesticide concentration. Fish killed in each concentrations of dimethoate was recorded. After 96hr of exposure the data obtained was subjected to per cent kill, probit analysis method (Finney 1971) and Dragstedt-Behrens equation (Carpenter, 1975) as mentioned (Shivakumar and David, 2004) to determine LC_{50} value. One fifth of LC_{50} value was selected for sub-lethal concentration for chronic study (1, 5, 10, 15 days). The control and the dimethoate-exposed fish were aerated frequently to prevent hypoxic condition of the medium. The control and dimethoate exposed fish were kept under continuous observation during the experiment period 24, 48, 72 and 96 h for lethal concentration and 1, 5, 10 and 15 days for sub lethal concentration.

Results and Discussion

The mortality data of *Cyprinus carpio* in different concentrations of dimethoate is presented. The graph plotted with per cent mortality against log concentration of dimethoate showed a sigmoid curve, and probit mortality against log concentration of dimethbate showed a linear curve. The LC_{50} values obtained through sigmoid curve and straight lines are 34.76 µl/l and 34.76 µl/1, respectively.

To evaluate the validity of graphic LC_{50} values, the data were subjected to Dragstedt and Behrens equation. The LC_{50} values obtained were 34.76 µl/l, 34.76 µl/l and 39.14 µl/l. for per cent mortality, probit mortality and Dragstedt and Bherens equation respectively. The mean of LC_{50} was found to be 35.76 µl/l, one fifth of the LC_{50} (7.15 µl/l) was chosen for sub-acute studies.

Fish in non toxic media were considered as control. In such environment fish maintained compact schooling and moved in co-ordinated manner. They were alert at the slightest disturbance and sensitive to light. No other extraordinary behaviour was observed.

Fish exposed to the lethal concentration of dimethoate, they migrated immediately to the bottom of the tank. The schooling behaviour was observed to be disrupted in the first day itself. Fish occupied twice the area than that of the control group. They were spread out and appeared to be swimming independent of one another. This was followed by irregular, erratic and darting movements with imbalanced swimming activity. The swimming behaviour was in a cock pattern rotating along horizontal axis and followed by 's' jerk, partial jerk, sudden, rapid, non-directed spurt of forward movement (burst swimming). Support to the present observation was reported eratic swimming in fish *Labeo rohita* exposed to fenvalerate (David *et al.*, 2002). The frequency of surfacing phenomenon was greater on the second day of exposure wherein the fish frequently come to the water surface.

Fish exposed to endosulfan surfaced occasionally, which was documented as to gulp maximum possible air to ease the pesticide stress (Devi, 1991). This fact is clearly evidenced in the present study. On day 3rd the fish were continuing in the surface area in a horizontal axis without showing escaping phenomenon (jumping). On the 4th day they lost their equilibrium and response, to external stimuli such as touch and light followed by drowning to the bottom. The fish eventually died with their mouth and operculum wide opened. Fish in sublethal concentration were found under stress, but that was not fatal.

The hyper excitability death of the animal might be due to the inhibition of acetylcholinestrease enzyme by disrupting the nervous activity by the accumulation of acetylcholine at nerve endings in central and pheriperal nervous system (O'Brien, 1966). A change in colour of the gill lamellae from reddish to light brown with coagulation of mucous on gill lamellae and abdomen swelling was seen in dead fish. This increase in the production of mucus may be to delimit the toxic effect of pesticides (David *et al.*, 2003). This implies that toxicants brought into contact with gill lamellae causes more damage to gill epithelium, which changes the colour (Piska Ravishankar *et al.*, 1992). The abdomen swelling was seen in the dead individuals of *C. carpio* exposed to pesticide which could be due to induction of intestinal acidosis, which in turn causes gas accumulation in the abdomen (Jagannatha Roa, 1983).

From the results arrived at during the study it can be concluded that dimethoate at very low concentration caused mortality and has grater impact on the fish behavior. Hence it is necessary, to select behaviour indices that relate to the organisms natural behaviour pattern in the field in order to derive a more accurate assessment of the hazards that a contamination may pose in natural systems.

References

Carpenter, P.L., 1975. *Immunology and Serology*, 3rd ed. W.B. Sunders Co., Philadelphia, USA.

Choudhary and Shafi, 1985. Occurrence of cancerous tumor in the kidney of catfish on intoxication in a phenolic waste. In: *Chemical Proc. 72. Indian Science Congress*, 3: 130–133.

David, M., H.B. Shivakumar, B.H. Ganti, S.B. Mushigerp, and R. Shivakumar, 2003. Toxicity evaluation of cypermethrin and its effect on oxygen consumption of the freshwater fish, *Tilapia mossambica. Indian J. Environ. Toxicol.*, 13(2): 99–102.

David, M., Mushigeri, S.B. and Prashanth, M.S., 2002. Toxicity of fenvalerate to the freshwater fish, *Labeo rohita* (Hamilton). *Geobios*, 29: 25–28.

Devi Swetharanyam, 1991. Studies on behavioural, physiological and biochemical changes in *Oreochromis mossambicaus* exposed to endosulfan. *Ph.D. Thesis*, Bharathidasan University, Tamil Nadu, India.

Environment Centre for National Toxicology, Media Release, Tuesday 26th October, 2004.

Finney, D.J., 1971. *Probit Analysis*, 3rd Edition, Cambridge University, Press, London.

Jagannatha Rao, K.S., 1983. Acidosis in the intestine of fish due to pesticide stress. In: *Proc. of National Conf. of Pollu. Res.*, Bhopal, India.

Mullick, Suparna & S.K. Konar, 1994. Pollution of water by mixture of metals, detergents, petroleum products and pesticide and its impact in fish. *Environment and Ecology*, 12(1): 189–194.

O'Brien, R.D., 1966. Mode of action of insecticides. *Ann. Rev. Entamol.*, 11: 369–402.

Piska Ravishankar, Waghray Sarala S. and Indira Devi. 1992. Sublethal concentration of synthetic pyrethriod cypermethrin to the common carp. *J. Environ. Biol.*, 13(2): 89–94.

Prasad, S. and Shafi, M.D., 1989. Toxicity of an industrial phenolic waste on the mortality of a hill stream fish *Lepidocephalus guntea. J. Ecobiol.*, 1(4): 308–311.

Rajamanickam and Karagaganapathy, P.R., 1988. Effect of Lindane (n-BHC) on the behavioural changes in freshwater fish *Tilapia mossambica.* In: *Proc. 2nd Nat. Symp. Ecotoxicol.*, pp. 143–145.

Shivakumar, R. and M. David, 2004. Endosulfan induced behavioural changes in the freshwater fish, *Cyprinus carpio. J. Ecotoxicol. Environ. Monit.*, 14(1): 65–69.

Chapter 35

Impact of Livestock Production Systems on the Environment

Ch. Vijaya[1] and R. Mallikarjuna Reddy[2]
[1]Department of Biotechnology, [2]Department of Dairy Science
Jawahar Bharathi Degree College, Kavali – 524 201, Nellore Dist., A.P.

ABSTRACT

Livestock rearing is carried out under three broad production systems namely grazing, mixed farming and industrial production system. Grazing and mixed farming systems depend largely on fodder available from common property resources thus eventually leading to their degradation. Industrial production system is found in urban dairy farms where the animals are maintained on purchased feeds and fodder causing less damage to the environment. The milk productivity of the cattle and buffaloes kept in grazing and mixed farming systems is much lower compared to that if animals under industrial production system. The lower milk producing ability of the animals in grazing and mixed farming systems could be attributed to the low availability of fodder which is the result of over grazing and eventual degradation of the grazing lands.

Introduction

One of the greatest challenges the world facing is to preserve its natural resources while at the same time producing sufficient food to satisfy the demands of a growing human population. World population is expected to grow from 6.0 billion now to about 8 billion by the year 2020. Incomes also continue to grow, especially in the developing world and in turn it will increase the demand for livestock products like milk and meat. In India also the demand for milk and meat are growing continuously exerting greater strain on the available resources. However, the production of livestock products like milk and meat depends upon the production system adapted by the farmers (Subramanyam and Reddy, 1999).

In India livestock rearing is carried out mostly with the feed and fodder available from village common lands. The common property resources (CPR's) of land consist of village forest, village common and pasture land, waste land, common threshing grounds, ponds, tanks etc. These resources help in sustaining a number of animals for draft and milk production, which would not have been permitted by the individual own land, especially for small farmers (Jodha, 1986).

Livestock rearing is carried out in many forms, usually referred to as production systems or agro-ecosystems. Production systems evolve as a result of agro-ecological interactions. Many production systems are currently at a sustainable equilibrium, with livestock being produced in harmony with nature and in environmentally sound systems. However, over the last few decades, several, production systems have lost this equilibrium because of the pressure caused by growing human population and increased demand for animal products (Haan, Stein Field and Black Burn, 1996). This articles provides an overview of the main production systems, the relative changes in the importance of these production systems with reference to milk yield and how these changes affects the resource base.

Data Source and Methodology

The study focuses on the interactions between livestock production systems and the environment. Thus, the study uses the primary data collected from the selected, households spread over in four districts of south coastal Andhra Pradesh. Data regarding the species and breed of the animals maintained, milk yield, feeds given, feeding practices etc is collected through the questionnaires during the year 2004 from about 200 house-holds coverings different agro-ecological zones. Primary data is collected in two rounds, *i.e.*, in summer and winter to avoid seasonal fluctuations. Secondary data related to livestock census, availability of grazing lands, forest lands etc. is collected from the related records available with the animal husbandry and agriculture departments.

Results and Discussion

Types of Production Systems

Based on the degree of integration with crops and its relation to land, the livestock sector has been classified into three board production systems (Haan, Stein Field and Black Burn, 1996). These are grazing, mixed farming and industrial production systems. For better understanding, the performance of these production systems are examined vis-à-vis difference agro-ecological zones *i.e.*, high irrigation, medium irrigation and low irrigation zones.

Production System vs. Milk Yields

Among different types of livestock production systems, grazing systems are found in low irrigation and medium irrigation zones (Table 35.1). In this system, the animals are largely maintained through the grass available from grazing lands. Animals are taken for grazing on common tending basis on all the days through out the year. Apart from grazing, little amounts of crop residues like Paddy straw, Jowar straw are fed to the animals at home. Further, the type of cows and buffaloes maintained in the areas (low irrigation) under grazing systems are of low quality and yields very little quantity of milk (Table 35.2). For instance, the average daily milk yield of graded buffalo is as low as 2.25 litres in grazing system as against 3.5 litres in mixed farming system and 6.5 litres in industrial system. (Urban dairy farms). The milk yield of crossbred cow is low in mixed farming system compared to industrial production system. The examination of production system vs. milk yields clearly reveals that extent of dependence on grazing influences the milk yields of dairy animals. The dependence on grazing is more in the case of grazing system and mixed farming systems and thus the milk yields of the animals maintained under these systems are lower compared to industrial production system. The lower milk yields of dairy animals maintained under grazing system could be attributed to the scarcity of feed and fodder. In the case of mixed farming system, the fodder available to the animals is more than that of grazing system because this system is largely adopted by upper strata of the farmers. The quality of the animals maintained by the farmers is good and therefore, their milk yields are higher than those maintained in grazing system. Urban dairy farms or Industrial production systems are observed in the urban areas or towns. It is observed that the animals in urban dairy farms are maintained on purchased feeds and fodders. The quality of the animals maintained is extremely good and their milk yields are higher than those in the other systems. Large number of graded Murrha buffaloes and Crossbred cows are maintained in Urban dairy farms.

Table 35.1: Feeding Practices of Dairy Animals Across Different Production Systems and Agro-climatic Zones

Production System	*Agro-ecological Zones*	*Resource Base*
Grazing	Low irrigation zone	1. Grazing and collection of fodder from communal lands.
	Medium irrigation zone	2. Feeding of crop residues.
Mixed Farming	All the zones	1. Grazing
		2. Feeding of crop residues
		3. Feeding of collected green fodder from own lands.
Industrial (Urban	Urban areas and High	1. Feeding of concentrates,
Dairy Farms)	irrigation zone	2. Purchased crop residues
		3. Green fodder from own lands
		4. No or little grazing

Table 35.2: Milk Production of Dairy Animals by Production System

Production System	*Breed*	*Milk Yield (Litres/day)*
Grazing	Local Cow	1.00
	Local Buffalo	1.75
	Graded Buffalo	2.25
Mixed	Local Cow	1.20
	Local Buffalo	1.80
	Graded Buffalo	3.50
	Crossbreed Cow	4.50
Industrial (Urban Dairy Farms)	Graded Buffalo	6.50
	Crossbred Cow	7.20

Grazing System

Environment vs. Production System

This system is extremely useful to the livestock keepers with little or no integration with crops. It is mainly based on native grassland or common lands of the village. Landless agricultural laborers and marginal farmers possess little land and thus, they depend largely on the common property resources to feed their animals. It is the common scenario in most of the villages that the farmers send their animals for grazing, on common tending basis during daytime. Thus, during grazing the livestock interact with the environment *i.e.*, with land, water and plant and animal biodiversity. In principle, grazing systems are closed systems, where the cattle manure is used within the system and thus helps the plant mass to grow abundantly. However, these interactions are strongly determined by the agro-ecological conditions of the area.

An effort has been made to work out the fodder availability of the common property resources used for grazing of cattle and buffaloes. The optimum land area required for grazing of bovines is 0.9 hectares per adult cattle unit (GOI, 1976). However, the CPR area available for grazing per adult cattle (ACU) unit is much lower in the high irrigation and medium irrigation zones while it is almost equal in low irrigation zone (Table 35.3). The fodder available from these grazing lands is abysmally low and negligible even in low irrigation zone where there is optimum grazing land for cattle to graze. This lower availability of green fodder to the grazing animals indicate the degraded condition of the CPR's and it could be due to higher bovine density. The bovine density of CPRs is higher in all the agro-climate zones compared to the optimum carrying capacity of 60 adult cattle units per 100 hectares (Dhaliwal *et al.*, 1998). In case we take horses, donkeys, small ruminates also into consideration, then the pressure on grazing lands would further aggregate (Pandey *et al.*, 2005). Thus, it is evident that the production system dependent on grazing would suffer resulting in the lower productivity of the animals maintained. Though the animals maintained are of low productivity in grazing systems people rear them for the sustained economic reasons of drought power supply and manure (Pandey, 2002). The sustained rearing of these low quality animals under grazing system and mixed farming systems causes degradation of CPR's.

Table 35.3: Fodder Availability from Grazing Lands

Agro-climatic Zone	*Area Required per Adult Cattle Unit (ha)*	*Area Available as per CPRs per ACU (hect)*	*Fodder Availability per ACU (kg.)*	*Bovine Density (No./100 hect)*
High Irrigation	0.90	0.08	1.58	833
Medium Irrigation	0.90	0.31	1.90	196
Low Irrigation	0.90	0.85	2.10	82

It is observed in the present study that the interactions between dairy animals and the grazing lands are negative and harmful to the biodiversity. Most of the dairy farmers send their animals to grazing on communal lands and forest areas present in the study area. Since the rainfall is low to moderate in the low and medium irrigation zones, the growth of the grass on communal lands is low. Dairy animals are found in the grazing lands on all the 365 days of the year for 5–7 hours a day. As soon as the grass starts growing after rains, the animals browse this grass and do not allow them to grow. As a result the productivity of grazing lands is declined and part of it is degraded due to over-grazing and animal trampling (Reddy, 2001).

Though the vegetation is extremely resilient an irreversible change in the plant biodiversity is observed in the present study. Due to selective grazing, a weed plant belonging to prosofis has grown extensively in the grazing lands thus resulting in lower availability of fodder to animals (Reddy, 2002).

Resource degradation, especially of land and biodiversity, is occurring in many of the grazing lands. For the most part of this is occurring where traditionally well managed common lands are becoming open access areas. In such open access situations, the interests of individual users conflict with those of the community causing what economists call "market failures". In these "free for all" situations, degradation is most severe. On the other hand, the grazing systems also offer potential for biodiversity enhancement. Identifying situation, incentives to correct the market failures and enhancing the livestock-environment synergies are thus of the biggest challenges in this system.

Mixed Farming System

About 80 per cent of the dairy animals in India are maintained in the mixed farming system. Globally, mixed farming system produce the largest share of total milk (90 per cent) and meat (54 per cent). In this system crops and livestock production are integrated on the same farm. The crop residues and green grass available from the fields of the farmers are used as feed for the animals. On the other hand, manure produced in the farm is used as fertilizer to the fields and the male progeny of cattle and buffaloes are used for traction. Therefore, resource use in mixed farming is often highly self-reliant as nutrients and energy flow from crops to livestock and back. Thus, such a closed system offers positive incentives to compensate for environmental effects (internalize the environmental costs), making them less damaging and more beneficial to natural resource base.

In mixed farming systems, livestock-environment interactions can be a focal-point. As human population pressure increase, some types of wild life or plant diversity can be lost while other types of plants may actually increase. The number of trees, types of trees, birds, insects and mammal species change as biodiversity is changed due to intensified human use of the resource base.

It is observed that in Irrigated mixed farming grazing is more likely to be contained within the system although the source of feed supply is likely to shift increasingly from non-arable grazing to grazing of crop residues and the use of fallow land. In low-irrigated mixed farming grazing is largely confirmed to communal grazing lands and thus the degradation of grazing lands is on increasing trend. Low availability of green fodder and the resultant lower milk yields of animals maintained under this system support this fact.

In addition, intensification of production, following increasing population pressure leads, especially in dairy production to the replacement of local livestock breeds by a small number of exotic breeds. The local cows and local buffaloes have been replaced by graded Murrah buffaloes in the study area and this could be attributed to the change in biodiversity due to over-grazing and increased demand for milk and milk products. These graded buffaloes and crossbreed cows are largely maintained at home and are personally taken for grazing to the fallow lands and communal lands by the owners.

Industrial Livestock Production System

Industrial production of milk, meat, pork and poultry is the fastest growing form of animal production. In India, industrial production of milk is practiced in urban areas. Large dairy farms are maintained in cities and towns with high yielding dairy animals, both crossbred cows and Murrah buffaloes in large numbers (in town like Nellore, Kavali, Gudur, Ongole, Guntur and Tirupathi). Since the milk consumers are looking for the supply of fresh milk, dairy farms in the cities and towns are increasing at a faster rate. Dairy plants belonging to A.P.D.D.C.F., Dodla Dairy, Tirumala Dairy etc. collect milk from the farmers situated at long distances, process the milk in aseptic conditions and supply the milk to the consumers on second or third day of its production. But urban consumers, especially those of rural back ground, do not relish this milk and thus they are encouraging the urban dairy farms for the fresh milk they needed.

Main Features of Urban Dairy Farms run under industrial production system, observed in the present study are the following:

1. Good quality and high yielding dairy animals are kept.
2. Depend largely on purchased concentrated feeds.
3. No grazing land is needed, and thus less damage to the environment.
4. Crop residues such as paddy straw is procured from villages.
5. When the animals become dry, they are sent to villages for maintenance. As and when they become pregnant and reach calving stage, they are taken back to city dairy farms. Since the animals sent to villages from city dairy farms are dry or less productive, they are generally maintained on grazing on communal lands leading over grazing and degradation of communal lands. It is found that the landless labour and marginal farmers procure this type of animals from city dairy farms under some agreement. Since their resource base is small their dependence on communal lands is more causing negative effect on the environment.
6. Calves are neglected because of high demand for milk resulting, in the loss of, progeny of good quality animals. In the long run, this phenomenon leads to the existence of low quality animals and thus affects biodiversity.

Environmental Benefits of Industrial Production System

The development of urban dairy farms alleviate pressures for deforestation and degradation of communal lands, thus saving land and preserving biodiversity. Second, the feed-saving technologies developed for this system can be effective at any scale and therefore, can be successfully transferred to mixed farming systems. The same holds true for waste prevention and treatment technologies which have been developed following regulations applied mainly to the industrial system. Therefore, the resource-saving and waste management technologies generated by the industrial systems bring benefits to the livestock sector as a whole.

On the other hand, the industrial system or urban dairy farming poses most of the environmental problems and offers possible solutions. The World's human population will increase from today's 5.5 billion to around 10 billion by the year 2030, *i.e.*, almost double. With increasing incomes, urbanization and aging populations, the world demand for milk and milk products and other livestock products is likely to triple, perhaps quadruple. Neither the grazing system nor the mixed farming system will be able to satisfy this increase in demand. The greatest part of the additional demand will have to be met by the urban dairy farms. For this happen, two requirements must be met.

1. Urban dairy farms and high potential cropping must be integrated in an area-wide crop-livestock integration. Mixed farming need to be practiced in a wider land use concept, where crop and livestock production are specialized but maintain their bio-physical links. This would allow for environmentally beneficial interaction, while giving enterprises the opportunity to specialize and to fully exploit economies of scale. It needs a new type of regional planning with proper breeding policies that takes into account nutrient balances, input and product markets, infrastructure and services.
2. The striving for increasing efficiencies must continue in order to alleviate the resource requirements for dairy animals. Cross-breeding of low yielding cows with high yielding

exotic bulls such as Jersey, H.F., Brown Swiss should continue with the required infrastructural support. Similarly, low quality buffaloes shall be upgraded by using well developed breeds like Murrah and Surti. Use of recent biotechnological applications would further facilitates the rapid development of low quality stock.

Conclusion

Three types of production systems *viz.*, grazing, mixed farming and industrial (urban dairy farming) systems are observed. Among these systems, grazing and mixed farming systems depend largely on grazing lands for feeding the dairy animals and therefore, cause negative effects on the environment due to overgrazing. The quality of the animals maintained in these systems is poor because of the lower availability of feed and thus, it resulted in the lower milk productivity of the animals. On the other hand, the high yielding graded Murrah buffaloes and crossbred cows are observed in urban dairy farms. The animals in the urban dairy farms are maintained by the purchased feeds and fodders and grazing is not practiced generally. Hence, it does not cause degradation of grazing lands. Therefore, the industrial production system (urban dairy farms) appears to be more efficient and eco-friendly production system to rear cattle and buffaloes for milk production.

Acknowledgement

Authors are thankful to UGC for the financial assistance provided for the Minor Research Project.

References

Dhaliwal, G.S., Sangha, G.S. and Ralham, P.K., 1998. *Fundamentals of Environmental Science*. Kalyani Publications, New Delhi.

GOI, 1976. Report of the National Commission on Agriculture, 1976, Part IX. Ministry of Agriculture and Irrigation, Government of India, New Delhi.

Hann de, Steinfeld, H. and Blackburn, H., 1996. *Livestock and Environment: Finding a Balance*. A publication of FAO and World Bank.

Jodha, N.S., 1986. Common property resources and rural poor in dry region of India. *Economic and Political Weekly*, 21(27): 1169–1181.

Pandey, U.K., 2002. Livestock in the household economy. *Agricultural Situation in India*, pp. 471–486.

Pandey, U.K., S.K. Goyal and Manget Ram, 2005. The bovine wealth of Haryana: Trends and constrains. *Indian Journal of Dairy Science*, 58(1): 1–11.

Reddy, R.M., 2001. Rural poverty, dairy development and environmental degradation: Need for alternate paradime. *Indian Journal of Environmental Protection*, 21(3): 274–281.

Reddy, R.M., 2002. Weed invasion in grazing lands of Andhra Pradesh. *Intensive Agriculture*, 40(5–6): 15–17.

Subramanyam, S. and Mallikarjuna Reddy, 1999. *Economics of Bovine Production in Andhra Pradesh*. Centre for Economic and Social Studies, Hyderabad, April.

Chapter 36

Agrobacterium rhizogenes Mediated Gus Gene Transformation

*S. Sethilkumar**

Centre of Advanced Study in Marine Biology, Annamalai University, Parangipettai – 608 502, T.N., India

ABSTRACT

Transfer of β-Glucuronidase (GUS) as a reporter gene in to *Nicotiana tobaccum* plants through *Agrobacterium rhizogenes* infection was attempted in the present study. Genetically altered strain *Agrobacterium rhizogenes* biovar 1 bearing Ri plasmid encoded β-glucuronidase (pAT AG3) was used as vector to infect tobacco leaf discs in co-cultivation method. Integration of the transfer DNA (T-DNA) of a root-inducing (Ri) plasmid of *A. rhizogenes* into the plant DNA was evident by microscopically and histochemical assay.

Keywords: *Agrobacterium rhizogenes, β-Glucoronidase, Histochemical assay.*

Introduction

The common soil bacteria *Agrobacterium rhizogenes* and *A. tumefacienes* are the two species, widely used to transform higher plants including tobacco plants. *A. rhizogenes* derived hairy roots and plants have application for many areas of research, including production of secondary metabolites and production of transgenic crop plants and crop improvement (Christey, 2001; Strycharz and Shetty, 2002). Hairy root disease symptoms involve gene transfer from a bacterial plasmid to the host plant genome. This unusual mechanism has led to the development of *A. rhizogenes* as vectors in genetic studies and crop improvement.

*E-mail: senthilk_s2003@yahoo.co.in.

The first major clue that *Agrobacterium* virulence, *i.e.*, the ability to form hairy roots was linked to the presence of tumor inducing (Ti) or root inducing (Ri) plasmid (Hamilton and Fall, 1971). Moor *et al.* (1979) demonstrated that hairy roots would not develop on plant cells unless Ri plasmid is present. Researchers have succeeded in transforming a wide range of plant species through *A. rhizogenes* by its ability of to transfer a segment of DNA (the T-DNA) from the Ri plasmid to the plant nuclear genome (Schell and Van Mantagu 1979; Mc Cormicket *et al.*, 1986; Klee *et al.*, 1987; Vahala *et al.*, 1989; Wurtele Bulka, 1989; Diaz *et al.*, 2000 and Dernier *et al.*, 2001). Targeted genes used to transform in to plants have come from variety of plant, animal and bacterial species (Bevan *et al.*, 1983; Elzen *et al.*, 1985). *Agrobacterium* mediated transformed plant cells synthesize opines, a group of amino acid derivatives, usually not found in plant cells (Bevan and Chilton, 1982). Plant tissues transformed by wild-type strain of *A. rhizogenes* can be readily distinguished by their oncogenic phenotype such as hairy roots (Tepfer, 1984). An advantage of using oncogeneic *A. rhizogenes* for plant transformation is that clonal transgenic tissue is readily obtained from hairy roots. The enzyme β-glucuronidase, popularly known as GUS breaks down the glucuronides giving a coloured reaction, so that its presence can be detected *in-situ* (Ouartsi *et al.*, 2004).

In the present article, we describe the transformation of GUS gene in to *Nicotiana tobaccum* plant through co-cultivation with *A. rhizogenes* and performed histochemical assay to confirm the presence of GUS activity.

Materials and Methods

Growth and Culture of *Agrobacterium rhizogenes*

A strain of *A. rhizogenes* biovar 1 bearing β-glucuronidase encoded Ri plasmid (pATAG3) was used for this study. Bacterial cultures was prepared by incubation of a loopful of bacteria in 20 ml of liquid YEB medium supplemented with Kanamycin (100 μg/ml) (Beach and Gresshoff, 1988) for 18 h at 28 C with reciprocatory shaking at 130 rpm.

Leaf Discs Infection

Young leaflets of *N. tobaccum* were collected and discs were prepared as described by McCormick *et al.* (1986). Leaf discs were surface-sterilized by dipping them in 50 per cent (v/v) ethanol for 30 seconds and then soaking them for 5 minutes in a 10 per cent (v/v) sodium hypochlorite solution. The sterilized leaf discs were rinsed with sterile water and co-cultivated with broth culture of *A. rhizogenes* (approximately 10^{-9} cells/ml). After incubation at 25 C (100 μmol/m^2/SPAR, 18 h photoperiod) for 2 days the leaf discs were agitated in liquid hairy root induction medium (MS salts with sucrose 1 g; vitamines (1X); Carbenicillin (250 mg/l); Cefotaxim (250 mg/l) and agar 0.8g; distilled water 100 ml and pH adjusted to 5.8) for 24 h, and then plated on to semi-solid MS medium (Murashige and Skoog, 1962) containing Cefotaxim to eliminate remaining other bacteria.

Histochemical Analysis

β-Glucuronidase (GUS) assays in cotransformed leaf discs was performed as described by Eede *et al.* (1992). Leaf discs were placed in different wells of the multi-well plate and 200 μl of GUS staining solution [Phosphate buffer solution 94 ml which constituted (KH_2PO_4 100 mM and K_2HPO_4 100 mM) Triton X-100/ethanol solution 1 ml (Triton X-100 10 ml, ethanol 40 ml and distilled water 50ml) and X-Glue 5 mg] was added to each wells including positive control well and incubated at 37 C for 2 to 3 hours in the dark.

Results and Discussion

A genetically modified strain of *A. rhizogenes* biovar 1 was used to study the expression of GUS transformation on to *N. tobaccum*. Here we described the co-cultivation method as effective infection procedure to transfer the GUS gene in to the *N. tobaccum* plant and subsequent *in vitro* callus induction on solid media to establish hairy root.

A significantly higher number of co-transformed leaf discs were obtained with co-cultivation method. This high efficiency of the leaf disc infection protocol is probably due to the longer incubation of the chimeric leaf discs in petridish with appropriate light physical conditions such as temperature and light intensity (Quandt *et al.*, 1993; Stiller *et al.*, 1997; Denier *et al.*, 2001 and Caturla *et al.*, 2002). Leaf disc protocol offered several practical advantages over other methods such as seedling infection protocol. First callus induction was generated much faster *i.e.* after 1–2 weeks of incubation. Second

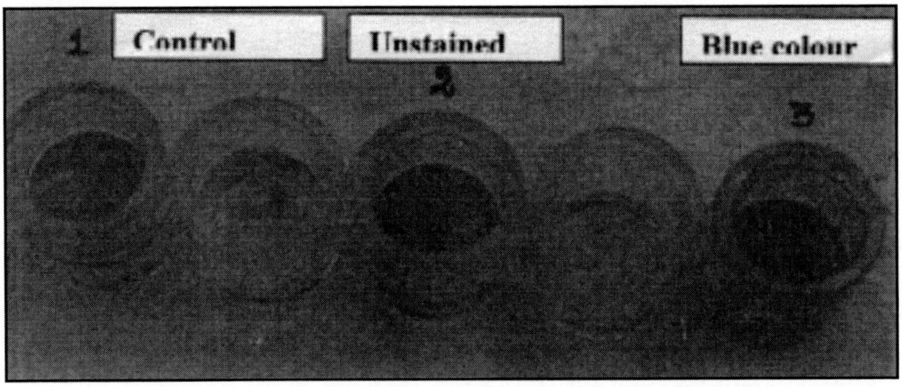

Figure 36.1: GUS Gene Activity in Leaf Discs after Histochemical Assay. (1) Control; (2) Unstained Leaf Tissue and (3) Deeply Stained Leaf Tissue

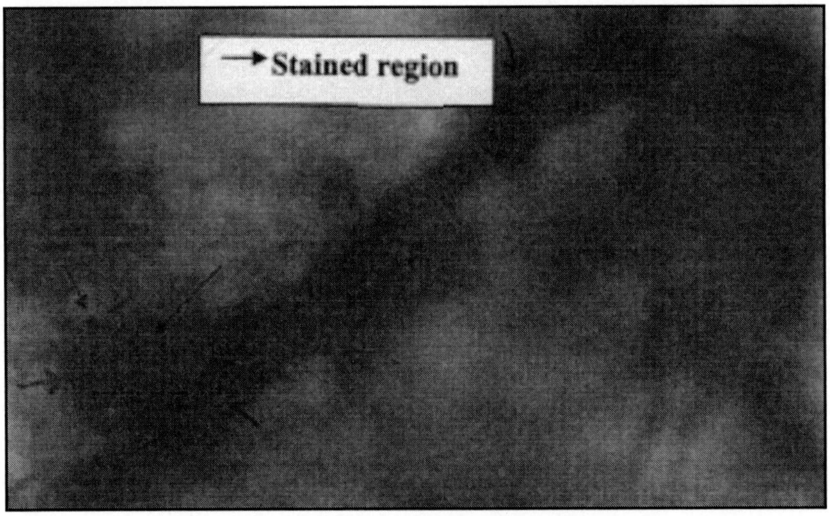

Figure 36.2: Microscopic Observation of Deeply Stained Transformed Leaf Tissue

upon leaf disc infection, the GUS activity could be easily visualized under light microscope. Third, *in vitro* growth on solid media greatly facilitated handling of callus for further histochemical analysis.

Excess bacterial growth on explants during co-cultivation is common and may stress the plant tissues. This was reduced by washing the explants in sterile distilled water. The regeneration medium was supplemented with carbenicillin to kill the bacteria and knamycin to prevent proliferation of non transformed cells (Velde *et al.*, 2003). Preconditioning of the explants prior to co-cultivation by culturing leaf discs in hormone containing medium reduces stress on the explants and promote cell division and transformations.

After 2 weeks of inoculation with *A. rhizogenes*, leaf discs were subjected to assay for GUS activity through histochemical analysis. Uninoculated leaf discs were served as control. GUS activity was seen on both micro well plates and microscopic visualization. The transformed GUS gene was detected using X-gluc staining solution which depicts the blue color formation in leaf discs as a result of GUS activity (Figure 36.1). In microscopic detection, the GUS activity was strong in the nodule parenchyma cells (Figure 36.2). The activity was lower in other central uninfected tissues (Figure 36.2). The GUS expression pattern in *N. tobaccum* corresponded with the patterns earlier published for other systems showing that it is primarily active in vascular tissues (Stiller *et al.*, 1997 and Martirani *et al.,*, 1999). β-glucuronidase was used for screening of cotransformed leaf discs. Analysis of GUS expression results in damage of tissue, making it impossible to monitor the transgenic growth in vivo. But, as it is stated by Chalfie *et al.* (1994) the GUS marker has the advantage that it does not require an exogenous substrate or cofactor except for oxygen to visualize.

References

Beach, K.H. and P.M. Gresshoff, 1988. Characterization and culture of *Agrobacterium rhizogenes* transformed roots of forage legumes. *Plant Sci.*, 57: 73–81.

Bevan, M., R. Flavell and M. Chilton, 1983. A chimeric antibiotic resistance gene as a selectable marker for plant cell transformation. *Nature*, 304: 184–187.

Bevan, M.J. and M. Chilton, 1982. T-DNA of the *Agrobacterium* Ti and Ri plasmids. *Ann. Rev. Genet.*, 16: 357–384.

Caturla, M., C. Chaparro, S. Herman, and M. Holsters, 2002. Suppression subtractive hybridization to emich low-abundance and submergence-enhanced transcripts of adventitious root primordia of *Sesbania rostrata. Plant Sci.*, 162: 915–921.

Chalfie, M., Y. Tu, G. Euskirchen, W.W. Ward, and D.C. Prasher, 1994. Green fluorescent protein as a marker for gene expression. *Science*, 263: 802–805.

Christey. M.C., 2001. Use of Ri-mediated transformation of production of transgenic plants. *In Vitro Cell. Dev. Biol. Plant*, 37: 687–700.

Dernier, A.B., M. Chabaud, F. Garcia, G. Becard, C. Rosenberg, and D.G. Barker, 2001. *Agrobacterium rhizogenes*–transformed roots of *Medicago truncatula* for the study of nitrogen-fixing and endomycorrhizal symbiotic associations. *Mol. Plant-Microbe Interact.*, 14: 695–700.

Diaz, C.L., H.P. Spaink, and J.W. Kijne. 2000. Heterologous rhizobial lipochitin oligosaccharides and chitin oligomers induce cortical cell divisions in red clover roots, transformed with the pea lectin gene. *Mol. Plant-Microbe Interact.*, 13: 268–276.

Eede, V.D., G.R. Deblaere, K. Goethals, M. Van Montagu, and M. Holsters. 1992. Broad host range and promoter selection vectors for bacteria that interact with plants. *Mol. Plant–Microbe Interact.*, 5: 228–234.

Elzen, V.D., P.J.M., J. Townsend, K.Y. Lee and J.R. Bedbrook, 1985. A chimeric hygromycin gene as a selectable marker in plant cells. *Plant. Mol. Biol. Rep.*, 5: 299–302.

Hamilton, R.H. and M.Z. Fall, 1971. The loss of tumor initiating ability in *Agrobacterium tumefaciens* by incubation at high temperature. *Experentia.*, 27: 229–230.

Klee, H., R. Horsch, M. Hinchee, M. Hein and N. Hoffmann, 1987. The effects of overproduction of two *Agrobacterium tumefaciens* T-DNA auxin biosynthetic gene products in transgenic petunia plants. *Genes Dev.*, 1: 186–196.

Martirani, L., J. Stiller, R. Mirabella, F. Alfano, A. Lamberti, S.E. Radutoiu, M. Iaccarino, P.M. Gresshoff, and M. Chiurazzi, 1999. T-DNA tagging of nodulation- and root-related genes in *Lotus japonicus:* expression patterns and potential for promoter trapping and insertional mutagenesis. *Mol. Plant-Microbe Interact.*, 12: 275–284.

McCormick, S., J. Niedermeyer, J. Frey, A. Barnason, R. Horsch and R. Fraley, 1986. Leaf disc transformation of cultivated tomato (*L. esculentum*) using *Arobacterium tumefaciens. Plant. Cell. Rep.*, 5: 81–84.

Moore, L., G. Warren and G. Stroebel, 1979. Involvement of a plasmid in the hairy root disease of plants caused by *Agrobacterium rhizogenes. Plasmid*, 2: 617–626.

Murashige, T. and F. Skoog, 1962. A revised medium for rapid growth and bio assays with tobacco tissue cultures, *Physiol. Plant*, 15: 473–497.

Ouartsi, A., D. Clerot, A.D. Meyer, Y. Dessaux, J. Brevet and M. Bonfill, 2004. The T-DNA ORF8 of the cucumopine-type *Agrobacterium rhizogenes* Ri plasmid is involved in auxin response in transgenic tobacco.

Quandt, H.J., A. Puhler, and I. Broer, 1993. Transgenic root nodules of *Vicia hirsuta:* A fast and efficient system for the study of gene expression in indeterminate-type nodules. *Mol. Plant-Microbe Interact.*, 6: 699–706.

Schell, J. and M.V. Montagu, 1979. The Ti plasmids of *Agrobacterium tumefaciens* and their role in crown gall fonnation. In: *Genome Organization and Expression in Plants,* (Ed.) C.J. Leaver. Plenum press, New York, pp. 453–470.

Stiller, J., L. Martirani, S. Tuppale, R.-J. Chian, M. Chiurazzi, and P.M. Gresshoff, 1997. High frequency transformation and regeneration of transgenic plants in the model legume *Lotus japonicus, J. Exp. Bot.*, 48: 1357–1365.

Strycharz, S. and K. Shetty, 2002. Effect of *Agrobacterium rhizogenes* on phenolic content of *Mentha pulegium* elite clonal line for phytoremediation applications. 38: 287–293.

Tepfer, D., 1984. Transformation of several species of higher plants by *Agrobacterium rhizogenes:* Sexual transmission of the transformed genotype and phenotype. *Cell*, 37: 959–967.

Vahala, T., S. Stabel and T. Eriksson, 1989. Genetic transformation of willows (*Salix* spp.) by *Agrobacterium tumefaciens. Plant. Cell. Rep.*, 8: 55–58.

Velde, W.V., J. Mergeay, M. Holsters and S. Goonnachtig, 2003. *Agrobacterium rhizogenes*–mediated transformation of *Sesbania rostrata. Plant Sci.*, 165: 1281–1288.

Wurter, E.S. and K. Bulka, 1989. A simple efficient method for the *Agrobacterium* mediated transformation of carrot callus cells. *Plant. Sci.*, 61: 253–262.

Chapter 37

Evaluation of Phytochemical and Antibacterial Properties of Medicinal Plants

***Archana Moon*[1], *Aqueel Khan*[1] *and B.J. Wadher*[2]**
[1]*PGTD Biochemistry,* [2]*PGTD Microbiology*
RTM Nagpur University, L.I. T Premises, Nagpur – 444 033

ABSTRACT

Extracts from 30 plant species were tested for antibacterial activity. These plant species were selected on account of their reported traditional uses in the treatment of various diseases caused by bacteria. The antibacterial activity was tested against *Escherichia coli, Staphylococcus aureus, Bacillus subtilis, Klebsiella aerogenes, Klebsiella pneumoneae, Pseudomonas fluorescens* and *Salmonella typhii.* Out of the thirty plants investigated, 26 plants exhibited antibacterial activity against the test organisms. Amongst these 26 plants, 5 plant extracts *viz., Alangium salvifolium, Centella asiatica, Ocimum sanctum, Sphaeranthus indicus, Soyameda febrifuga* were found to have a comparatively higher potential for antibacterial activity.

Keywords: *Antibacterial activity, Plant extracts.*

Introduction

The development of resistant microbial strains due to continuous exposure of bacteria to antibiotics and the side effects caused due to the continuous usage of antibiotics are the two important reasons for an upsurge in research to find substitutes for antibiotics. In a quest to find substitutes for antibiotics, the medicinal plants which are widely used by the traditional medical practitioners for the treatment of infectious diseases are put to systematic scientific investigation (Prakash and Gupta, 2005). In the present study 30 plant species were tested for antibacterial activity on account of their reported traditional uses in the treatment of diseases caused by bacteria (Jain, 1968; Dastur, 1976; Nadkarni, 1976; Kirtikar and Basu, 1996; NISCAIR, 2003).

Materials and Methods

Plant Material

Fresh plants were selected for antibacterial assays on the basis of their mention in traditional systems of medicines and also used by the local people of rural areas of Nagpur District. 30 plants as mentioned in Table 37.1 were collected by undertaking ethnomedicinal investigations in local, rural and urban fringe areas. The collected plants were used within 24 hours of their being collected or stored at 4 C until use.

Preparation of Plant Extract

The plant material/parts of plants were cleaned free of soil and dirt and chopped into small pieces of approximately 1 g. 10 g of these were crushed in a mortar and transferred to 100 ml conical flask. The powdered plant material was extracted with 40 ml distilled water. The flasks were stoppered and kept at room temperature for 24 hours with occasional shaking. The contents were carefully filtered. This crude aqueous extract was tested for antibacterial activity against listed microorganisms. Fresh leaves were crushed and filtered through muslin cloth and extract tested for antibacterial activity (Perez *et al.*, 1990).

Test Organisms

The microorganisms utilized in this study were clinical isolates of *Escherichia coli, Staphylococcus aureus, Bacillus subtilis, Klebsiella aerogenes. Klebsiella pnuemoneae, Pseudomonas fluorescens* and *Salmonella typhii.* The bacterial cultures were maintained on Nutrient Agar (Himedia, Mumbai) at 4°C and subcultured every two weeks. *P. fluorescens* was subcultured weekly.

Media

Nutrient Agar (M001), Agar Agar Type I (RM666), Mueller Hinton Agar No. 2 (M1084) and Nutrient broth (M002) were procured from Hi-Media, Mumbai. The preparation of media was done strictly according to the manufacturer's instructions.

Antibacterial Testing

Agar well Diffusion Technique was used in this study (Tumane P., Wadher B *et al.*, 2000).

Inoculum Preparation

A bacterial inoculum was prepared by inoculating a loopful of test organism into a 3 ml sterile nutrient broth tube and incubated at 37 C for 18 hours. The turbidity was matched with 0.5 Mc Farland's Nephelometer Standard (WHO Drug Information, 1993). Dilutions to the tube were done with sterile nutrient broth to get a cell density corresponding to 6 10^8 CFU/ml.

Antibacterial Susceptibility Test

A suspension (0.1 ml) of the test organisms from the 18 hour cultures was thoroughly mixed with 20 ml of sterile Mueller Hinton Agar maintained at 45–50 C. The seeded M.H. Agar is poured in presterilized petri plates and set aside. After solidification, the seeded agar was punched with a flamed (sterile) 10 mm cork borer in order to obtain a well of 10 mm diameter in the center of the petri plate. 100 μl of the aqueous plant extract is loaded into the well accurately with a micropipette (with

Table 37.1: Antibacterial Activity of Plant Extracts

Sl.No.	Plant (Botanical Name)	Plant Part	Antibacterial Activity	Test Organisms, Zone of Inhibition in mm Diameter						
				E. coli	*S. aureus*	*B. subtilis*	*K. aerogenes*	*K. pneumoniae*	*P. fluorescens*	*S. typhii*
1.	*Aegle marmelos*	Root bark	+	16	14	16	–	–	–	16
2.	*Alangium salvifolium*	Leaf	+	16	18	20	–	–	22	20
		Root bark	+	18	20	24	–	–	26	24
3.	*Andrographis paniculata*	Leaf	–	–	–	–	–	–	–	–
		Root bark	+	12	10	–	–	–	–	–
4.	*Astercantha longifolia*	Leaf, Stem	+	12	18	16	–	12	18	16
5.	*Benincasa hispida*	Leaf	–	–	–	–	–	–	–	–
6.	*Bixa orellana*	Leaf	+	18	20	21	22	22	18	–
		Root bark	+	16	18	19	17	16	14	–
7.	*Bridelia retusa*	Stem	+	–	18	–	–	–	18	–
8.	*Calotropis gigantia*	Flower	–	–	–	–	–	–	–	–
		Leaves	+	16	18	16	–	–	14	–
9.	*Centella asiatica*	Aerial parts	+	22	21	18	14	12	13	28
10.	Citrus *limon*	Fruit juice	+	24	28	30	22	24	28	24
11.	*Datura stromanium*	Fruit	+	–	21	–	–	–	–	–
12.	*Emblica officinales*	Fruit	+	21	20	–	15	13	14	14
13.	*Eupatorium odoratum*	Leaf	–	–	–	–	–	–	–	–
14.	*Euphorbia hirta*	Leaf, Stem	+	–	16	10+/–	–	–	–	–
15.	*Gardenia resiniflora*	Leaf	+	–	20	–	14	16	20	–

Contd...

Table 37.1–Contd...

Sl.No.	Plant (Botanical Name)	Plant Part	Antibacterial Activity	Test Organisms, Zone of Inhibition in mm Diameter						
				E. coli	*S. aureus*	*B. subtilis*	*K. aerogenes*	*K. pneumoniae*	*P. fluorescens*	*S. typhii*
16.	*Helianthus annus*	Leaf, Flower, Root	–	–	–	–	–	–	–	–
		Stem bark	+	–	16	14	–	–	–	–
17.	*Morus alba*	Root bark	+	14	22	–	14	17	–	–
18.	*Myristica fragrans*	Seeds	+	–	16	14	–	–	–	–
19.	*Ochna gamblei*	Stem	+	–	22	18	–	–	14	18
20.	*Ocimum sanctum*	Leaves	+	20	23	20	22	20	19	24
21.	*Pongamia pinata*	Leaf, Stem bark	–	–	–	–	–	–	–	–
22.	*Psoralea corylifolia*	Seeds	+	–	–	–	–	–	–	26
23.	*Sphaeranthus indica*	Aerial parts, Stem	+	–	20	22	–	–	–	18
24.	*Schleichera oleosa*	Stem bark	+	–	20	16	16	16	16	16
25.	*Solanum trilobatum*	Aerial parts	+	–	18	–	13	15	21	13
26.	*Soyameda febrifuga*	Stem bark	+	–	28	24	–	–	–	–
27.	*Termanalia arjuna*	Stem bark	–	–	–	–	–	–	–	–
28.	*Termanalia bellerica*	Stem bark	+	–	16	14	12	12	14	–
		Fruit	+	–	14	14	12	12	12	–
29.	*Thespesia populnea*	Aerial parts	+	19	18	–	20	20	13	–
30.	*Vitis vinifera*	Fruit	+	–	19	–	–	–	–	–

presterilized tips). The petri plates were delicately handled and kept in refrigerator for 30 minutes and then at room temperature for 30 minutes which facilitated diffusion of the plant extract. The petri-plates were then incubated at 37 C for 24 hours (Perez *et al.*, 1990). The zone of inhibition was measured with HiAntibiotic ZoneScale (PW096), HiMedia, Mumbai.

Soxhlet Extraction and Phytochemical Analysis

The short listed five plants *viz.*, *Alangium salvifolium*, *Centella asiatica*, *Ocimum sanctum*, *Sphaeranthus indicus* and *Soyameda febrifuga* were subjected to Petroleum ether, Chloroform, Acetone, and Methanol Soxhlet extraction and were tested for sterols, alkaloids, saponins, cardiac glycosides, cyanogenetic glycosides, anthroquinones, tannins, phenols, amino acids and carbohydrates. The concentration of phytochemical was rated as +: Low, ++: Moderate, +++: Good, ++++: High, –: Negative, ±: Not clear.

Minimum Inhibitory Concentration

The determination of M.I.C. was done by the Agar Dilution Method (NCCLS, 1990). Stock solution of 100 mg/ml of the methanolic extract of selected five plants were prepared in DMSO-Tris buffer (3 : 7). 100 µl to 3000 µl of this stock solution was added to 20 ml sterile M.H. Agar to achieve final concentration of 0.5 mg/ml, 1 mg/ml, 2 mg/ml, 5 mg/ml, 8 mg/ml, 10 mg/ml and 15 mg/ml. These were poured in petridishes and allowed to solidify. The reverse side of the plate was divided into 10 checker board blocks by glass marker to accommodate bacterial cultures. A bacterial inoculum of all the test organisms was prepared as discussed in antibacterial testing section. All the plates including control plates without plant extracts were spot inoculated with 10 µl of bacterial inoculum using sterile micropipette tips. The petriplates were incubated at 37°C for 24 hours. The result was read as presence or absence of bacterial growth. Complete suppression of growth was required for an extract to be declared active. MIC was determined as the least concentration of extract inhibiting the growth of the test organisms.

Results and Discussion

The results of preliminary screening for antibacterial activity are shown in Table 37.1. Of the 30 plants selected, 26 plants exhibited antibacterial property. *S. aureus* showed susceptibility to 28 parts of plant extracts out of 36 tested. *K. aerogenes* and *S. typhii* showed least susceptibility. Remarkable antibacterial activity was observed in *Alangium salvifolium*, *Bixa orellana* and Citrus *limon* both possess exceptional antibacterial activity. The antibacterial activity of *Bixa orellana* was reported earlier in 2002. (Castello, Phatak *et al.*, 2002) Citrus *limon* fruit juice was found to be highly active against *B. subtilis*. The high activity of citric acid against multi-drug resistant *P. aeruginosa* has also been reported (Nagoba, Gandhi *et al.*, 1998). Citrus *limon* contains 6–8 per cent citric acid. Commendable activity was also seen in *Sphaeranthus indica*, *Soyameda febrifuga*, *Centella asiatica* and *O. sanctum* leaf extract. The antibacterial activity of *O. sanctum* leaves against *S. aureus* was reported in 1989. Shrivastav and Gupta have validated *O. sanctum* as a medicinal plant in 2002. Also, Prakash and Gupta have provided therapeutic uses of *O. sanctum* and *Eugenol* in 2005.

The phytochemical analysis was performed qualitatively (Harborne, 1984; Trease and Evans, 1989; Kokate *et al.*, 2004). *A. salvifolium* showed presence of sterols in its ether and chloroform extracts and presence of tannins and alkaloids in its acetone and methanol extracts. *S. febrifuga* showed presence of tannins, saponins, alkaloids in its acetone and methanolic extracts whereas small amounts of sterols were detected in petroleum ether and chloroform extracts of *S. indicus* and *C. asiatica* and alkaloids in its chloroform extracts. *O. sanctum* showed presence of sterols in ether extracts, alkaloids in chloroform extracts and tannins in acetone and methanolic extracts (Tables 37.2 and 37.3).

Table 37.2: Preliminary Phytochemical Analysis of Plant Extracts

Tests	*Alangium salvifolium*				*Soyameda febrifuga*				*Ocimum sanctum*			
	Ether	*Chloroform*	*Acetone*	*Methanol*	*Ether*	*Chloroform*	*Acetone*	*Methanol*	*Ether*	*Chloroform*	*Acetone*	*Methanol*
Sterols					++	+	–	–	+++	++	–	–
Salkowski's test	++++	++	–	–	++	+	–	–	+++	++	–	–
Liberman test	++++	++	–	–	++	+	–	–	+++	++	–	–
Liberman Burchard test	++++	++	–	–								
Alkaloids			++++	++++	–	–	+	+				
Dragendorf's reagent	–	+	++++	+++	–	–	+	+	–	++	–	–
Mayer's reagent	–	+	++++	+++	–	–	+	+	–	++	–	–
Wagner's test	–	+	++++	+++	–	–	+	–	–	++	–	–
Hager's test	–	+	++++	++++	–	–	+	+	–	++	–	–
Tannic acid test	–	+	++++	++++	–	–	+	+	–	++	–	–
Scheibler's test	–	+	++++	++++					–	++	–	–
Saponins					–	–	–	++				
Foam test	–	–	++++	++++				–	–	–	–	–
Flavinoids	–	–	–	–	–				–		+	++
Cardiac Glycosides					–	–	–	–				
Keller–Killani test	–	–	+/–	–	–	–	–	–	–	–	–	–
Legal's test	–	–	+/–	–				–	–	–	–	–
Cyanogenetic glycosides					–	–	–	–				
Grignard'st test	–	–	–	–				–	–	–	–	–
Anthroquinones					–	–	–					
Bortranger's test	–	–	–	–					–	–	–	–
Tannins					–	–	+	++++				
Ferric chloride test	–	–	+	++	–	–	+	++++	–	–	++	+++

Contd...

Table 37.2–Contd...

Tests	*Alangium salvifolium*				*Soyameda febrifuga*				*Ocimum sanctum*			
	Ether	*Chloroform*	*Acetone*	*Methanol*	*Ether*	*Chloroform*	*Acetone*	*Methanol*	*Ether*	*Chloroform*	*Acetone*	*Methanol*
Lead acetate test	–	–	+	++	–	–	+	+++	–	–	++	+++
Potassium dichromate test	–	–	+	++	–	–	–	++++	–	–	++	+++
Gelatin solution test	–	–	–	–/+	–	–	+	++	–	–	++	+++
Bromine water test	–	–	+	+						–	++	+++
Phenols					–	–	+	+				
Ferric chloride test	–	–	++	+	–	–	+	–	–	–	++	–
Nitric acid test	–	–	–	–	–	–	–	–	–	–	–	–
Phthalic acid test	–	–	–	–					–	–	–	–
Proteins					–	–	–	–	–	–	–	–
Biuret test	–	–	–	–	–	–	–	–	–	–	–	–
Xanothoproteic test	–	–	–	–	–	–	–	–	–	–	–	–
Milton's test	–	–	–	–								
Amino acids					–	–	–	–	–	–	–	
Ninhydrin test	–	–	–	–								
Carbohydrates					–	–	+++	+++				
Mollisch test	–	–	–	–	–	–	+	–	–	–	–	++
Barfoed's test	–	–	–	–	–	–	–	–	–	–	–	–
Fehling's test	–	–	+++	+++					–	–	–	–

The investigation of petroleum ether, chloroform and acetone extracts did not show any antibacterial activity for the five plants selected but methanolic and aqueous extracts showed a high potential for antibacterial activity (Table 37.5). Aqueous extracts showed less antibacterial activity than the alcoholic counterparts due to decrease in the activity of water soluble inhibitors/ phytochemicals. So, the nature of solubility of phytochemical maybe the factor in showing increased or decreased antibacterial activity.

Table 37.3: Preliminary Phytochemical Analysis of *S. indicus* and *Centella asiatica*

Tests	*S. indicus*				*Centella asiatica*			
	Ether	*Chloroform*	*Acetone*	*Methanol*	*Ether*	*Chloroform*	*Acetone*	*Methanol*
Sterols								
Salkowski's test	+++	++	–	–	++	+	–	–
Liberman test	+++	++	–	–	++	+	–	–
Liberman Burchard test	+++	++	–	–	++	+	–	–
Alkaloids								
Dragendorf's reagent	–	+++	–	–	–	–	++	+++
Mayer's reagent	–	+++	–	–	–	–	++	+++
Wagner's test	–	+++	–	–	–	–	++	+++
Hager's test	–	+++	–	–	–	–	++	+++
Tannic acid test	–	+++	–	–	–	–	++	+++
Scheibler's test	–	+++	–	–	–	–	++	+++
Saponins								
Foam test	–	–	–	–	–	–	+	+
Flavinoids	–	–	–	–	–	–	–	–
Cardiac Glycosides								
Keller–Killani test	–	–	–	–	–	–	–	–
Legal's test	–	–	–	–	–	–	–	–
Cyanogenetic glycosides								
Grignard's test	–	–	–	–	–	–	–	–
Anthroquinones								
Bortranger's test	–	–	–	–	–	–	–	–
Tannins								
Ferric chloride test	–	–	–	+	–	–	++	+++
Lead acetate test	–	–	–	+	–	–	++	+++
Potassium dichromate test	–	–	–	+/–	–	–	++	+++
Gelatin solution test	–	–	–	+/–	–	–	++	+++
Bromine water test	–	–	–	+/–	–	–	++	+++
Phenols								
Ferric chloride test	–	–	–	+	–	–	–	–
Nitric acid test	–	–	–	–	–	–	–	–
Phthalic acid test	–	–	–	–	–	–	–	–

Contd...

Table 37.3–Contd...

Tests	*S. indicus*				*Centella asiatica*			
	Ether	*Chloroform*	*Acetone*	*Methanol*	*Ether*	*Chloroform*	*Acetone*	*Methanol*
Proteins								
Biurel test	–	–	–	+	–	–	–	–
Xanothoproteic test	–	–	–	+	–	–	–	–
Milton's test	–	–	–	+	–	–	–	–
Amino acids								
Ninhydrin test	–	–	–	–	–	–	–	–
Carbohydrates								
Mollisch test	–	–	–	+	–	–	+	++
Barfoed's test	–	–	–	–	–	–	–	–
Fehling's test	–	–	–	–	–	–	–	–

The present results agree with the other reports that secondary metabolites like tannins, flavonoids, steroids etc showed antibacterial activities. (Munekazu *et al.*, 1994; Lutete *et al.*, 1999; Van teigh *et al.*, 2004; Hideka *et al.*, 2005).

The Minimum Inhibitory Concentrations of the selected 5 plants are mentioned in Table 37.4. The MIC results also prove that the plant extracts exhibit bacteriostatic activity at high dilution rates. The overall results reveal that the plants or plant extracts can be effectively used as therapeutic agents.

Table 37.4: Minimum Inhibitory Concentration of Methanolic Extracts of Plants

Sl.No.	*Microorganism*	*MIC in mg/ml of Methanolic Extracts*				
		A.s	*S.i.*	*S.f*	*C.a*	*O.s*
1.	*E. coli*	> 15	5	2	10	–
2.	*S. aureus*	1	5	0.5	2	8
3.	*P. fluorescens*	8	5	1	10	15
4.	*K. pneumoniae*	> 15	5	2	> 15	10
5.	*S. typhii*	10	8	1	10	5

Conclusion

The present study supports the view that several ethnomedicinal plants might be useful as antibacterial agents resulting in the development of novel drugs. The application of such results involves the comparing of the MIC with dilutions in order to work out the mechanism of action which could be either bacteriostatic or bactericidal. In general, antibiotics require in vivo concentration in tissue that is equal to or greater than the inhibition concentration in vitro. The reason for the inhibitory effect of these extracts is presumably due to the presence of bioactive phytochemicals which inhibit growth of bacteria. This study has highlighted some plants which are worthy of further investigation for their antibacterial activities to assess the *in vitro* and *in vivo* activities of these extracts against pathogenic strains of microorganisms.

Table 37.5: Antibacterial Activity of Selected Plant Extracts

Sl.No.	Test Organisms	Control 1 2% Methanol	Control 2 Distilled Water	Type of Extracts	Diameter of Zone of Inhibiont in mm. Conc. in mg/ml																			
					Alangium salvifolium				Sphaeranthus indicus				Soyameda febrifuga				Centella asiatica				Ocimum sanctum			
					10	30	50	100	10	30	50	100	10	30	50	100	10	30	50	100	10	30	50	100
1.	E. coli	–	–	Methanol	12	13	14	24	–	–	12	14	–	14	22	40	–	–	8	14	+/–	11	12	16
		–	–	Aqueous	–	–	–	–	–	–	–	+/–	–	13	15	28	–	–	–	+/–	–	–	–	+/–
2.	S. aureas	–	–	Methanol	20	26	28	30	26	32	38	41	26	32	39	> 45	12	17	18	20	16	20	22	25
		–	–	Aqueous	14	20	19	34	20	26	28	33	28	29	38	> 45	–	–	–	+/–	+/–	–	–	–
3.	K. pneumoniae	–	–	Methanol	14	20	21	27	12	14	15	17	12	14	15	17	13	15	18	19	13	15	18	20
		–	–	Aqueous	–	–	–	–	–	–	–	–	–	–	–	13	–	–	–	–	–	12	14	16
4.	P. fluorescens	–	–	Methanol	17	19	21	27	13	14	15	17	13	14	15	16	12	14	15	17	12	14	16	22
		–	–	Aqueous	–	–	–	–	–	–	–	–	–	–	13	22	–	–	–	–	+/–	11	12	17
5.	S. typhii	–	–	Methanol	20	22	23	29	13	20	18	24	13	20	25	42	20	25	28	30	–	16	20	25
		–	–	Aqueous	–	14	20	26	–	–	–	–	12	14	20	36	18	20	21	28	+/–	12	14	19

References

_____2003. *The Wealth of India*, Vols. I, II, III, IV, V. NISCAIR.

_____1996. *Indian Medicinal Plants: A Compendium of 500 Species*, Vols. I, II, III. Orient Longman.

Castello, Marie-Claire, Anita Pathak, Naresh Chandra and Madhuri Sharon, 2002. Antimicrobial activity of crude extracts from plant parts and corresponding calli of Bixa; *orellana* L. *Indian J. of Exp. Biol.*, 40: 1378–1381.

Dastur, J.K. *Medicinal Plants of India and Pakistan*. D.B. Taraporewala Sons and Co., Bombay.

Harborne, J.B. 1984–85. *Phytochemical Methods*, 2nd Edition. Chapman and Hall, New York, London.

Hnuma, Munekazu, Yasu Toshi Okliwa, Toshiyuki Tanaka, Feng Chicho, Yusako Kobayashi and Ken-Ichi, Migau Chi, 1994. Antioral microbial activity of isoflavonoids in root bark of *Ormosia monosperma*. *Phytochemistry*, 37L 889.

Jain, S.K., 1968. *Medicinal Plants*. National Book Trust, New Delhi.

Kirtikar, K.R. and B.D. Basu. *Indian Medicinal Plants*, 2nd edn., Vols. I, II, III and IV. Lalit Mohan Basu, Allahabad.

Kobayashi, Hideka and Elvira de Mejia, 2005. The genus *Ardisia*: A novel source of health promoting compounds and phytopharmaceuticals. *Journal of Ethnopharmacology*, 96: 347–354.

Kokate, C.K., A.P. Purohit and S.B. Gokhale, 2004. *Pharmaconosy*, 27th edn. Nirali Publication.

Lutete, Tona, K. Kambu, D. Ntondde and K.C. Manga, 1999. Antimicrobial activity of tannins. *Fitoterapia*, pp. 279.

Nadkarni, K.M., 1976. *Indian Materia Medica*, Vols. I and II. Popular Prakashan Pvt. Ltd., Bombay.

Nagoba, B., Gandhi, S.P., Wadher, B.J. and Deshmukh, S.R., 1998. Citric acid treatment of a severe electric burn complicated by multiple antibiotic resistant *Pseudomonas aeruginosa* case report." *Burns*, London, 24(5): 581–583.

Perez, C., Paul, M. and Bazerque, 1990. An antibiotic assay by the agar well diffusion method. *Acta Biologiae et medicine Experimentalis*, 15: 113–115.

Prakash and Gupta, 2005. Therapeutic uses of *O. sanctum* with a note on eugenol and its pharmacological actions: A short review. *Indian J. Physiol. Pharmacol.*, 49(2): 125–131.

Tarfa, Florence D., Obiageri Obodozie, Emanuel Mshelia, Kolo Ibrahim and Temple V.J., 2004. Evaluation of phytochemical and antimicrobial properties leaf extracts of *Tapinanthus sessifolius* (P. Beauv) van Teigh. *Indian Journal of Experimental Biology*, 42: 326–329.

Trease, G.E. and Evans, W.C., 1989. *Pharmacognosy*, 13th edn., Tindall, London, pp. 171, 219, 270.

Tumane, P., Wadher, B., Khan, A., Gomashe, A. and Ingle, A., 2000. Antibacterial activity of plant extracts. *J. of Microb. World*, 2(2): 47–5.

WHO Scientific Working Group, 1983. Antimicrobial Resistance Bulletin of the WHO, 61: 383–391.

WHO, 1993. Drug Information.

Chapter 38

Impact of Organic Mulberry Cultivation Practices on Soil Microbial and Fertility of M_5 Mulberry Garden

K. Rashmi[1], M.A. Shankar[2], T.K. Narayanaswamy[1], K.R. Sreeramulu[3] and Rajegowda[1]

[1]Department of Sericulture, [2]Dryland Agriculture and [3]Agricultural Microbiology University of Agricultural Sciences, GKVK, Bangalore – 560 065

ABSTRACT

The effect of organic mulberry cultivation practices on the microbial development and fertility status of soil was studied in M_5 mulberry garden during 2004–05 at Main Research Station, Hebbal, University of Agricultural Sciences, Bangalore. The results revealed that, application of biofertilizers 10 kg each of *Azospirillum brasilense* + *Aspergillus awamori* + 20 per cent each of recommended N through compost, green manure (*Glyricidia maculata*), castor cake, vermicompost, fertilizer + remaining P and K through fertilizers (T_{11}) significantly increased the available macro and micronutrient contents in the soil. Whereas, the increased number of beneficial microbes *viz.*, *A. awamori* and *A. brasilense* population (15.00 10^3 and 15.00 10^6 cfu/g of soil, respectively) were found at 10 kg each of *A. brasilense* + *A. awamori* + 25 per cent each of recommended N through compost, green manure (*Glyricidia maculata*), castor cake, vermicompost (T_{10}) compared to control.

Keywords: *Mulberry, Organic manures, Soil microbial, Soil fertility.*

Introduction

Quality of mulberry leaf plays an important role in the nutrition of silkworm. Production of quality leaf depends on several factors including the native soil fertility. The fertility of soil not only depends on its chemical composition but also on the nature of microorganisms living in it. However, organic manures are known to improve soil fertility, physical, chemical and biological properties of soil. In addition, the inorganic fertilizers are also recognized as a key component for improved crop productivity and sustainable fertility management. However excess use of chemical fertilizers in an unbalanced manner has a deleterious effect on soil structure and soil properties such as alkalinity, acidity and nutrient deficiencies which leads to total loss of soil health. In order to balance this situation, the different organic manures *viz.,* farm yard manure (FYM), compost, green manure, vermicompost, oil cakes *etc.* acts as a better source of nutrients either alone or in combination with fertilizers. Hence, the present investigation was undertaken to study the impact of organic farming practices for mulberry cultivation and their influence on microbial and fertility status of soil.

Materials and Method

A field experiment was carried out during 2004–2005 in established irrigated M_5 mulberry garden planted at a spacing of 60 60 cm. The experiment was laid out in randomized complete block design (RCBD) with 13 treatments of 3 replications each. Before the commencement of experiment the soil analysis report revealed the presence of carbon content (10.6 per cent), nitrogen (200 kg/ha), phosphorous (50 kg/ha), potassium (86 kg/ha), soil pH 6.8 and the microbial population such as bacteria (7 10^6 cfu/g soil), fungi (2 10^3 cfu/g of soil), actinomycetes (1 10^4 cfu/g of soil), *Azospirillum brasilense* (4 10^6 cfu/g of soil) and PSB (1 10^3 cfu/g of soil). The cultivation practices for mulberry were followed as suggested by Dandin *et al.* (2003). The treatments were as follows.

T_1: 100 per cent recommended N through compost; T_2: 50 per cent recommended 'N' through compost + 50 per cent recommended 'N' and remaining P, K through fertilizer; T_3: 100 per cent recommended N through green manure (*Glyricidia maculata*); T_4: 50 per cent recommended N through green manure + 50 per cent recommended N and remaining P and K through fertilizer; T_5: 100 per cent recommended N through castor cake; T_6: 50 per cent recommended N-castor cake + 50 per cent recommended N and remaining P and K through fertilizer; T_7: 35 per cent recommended N through compost + 30 per cent recommended N through castor cake + 35 per cent recommended N through green Manure; T_8: 100 per cent recommended N through vermicompost; T_9: 50 per cent recommended N through vermicompost + 50 per cent recommended N and remaining P and K through fertilizer; T_{10}: Biofertilizer 10 kg each of *Azospirillum brasilense* + *Aspergillus awamori* + 25 per cent recommended N through compost, green manure, castor cake, vermicompost; T_{11}: Biofertilizer 10 kg each of *Azospirillum brasilense* + *Aspergillus awamori* + 20 per cent each recommended N through compost green manure, castor cake, vermicompost. + 20 per cent recommended N and remaining P and K through fertilizers; T_{12}: Recommended 20 tonnes FYM + 300 : 120 : 120 NPK kg/ha/year through fertilizer (control); T_{13}: Only fertilizer 300 : 120 : 120 NPK kg/ha/year (control).

The soil fertility and microbial population was estimated after the harvest of third crop. The available nitrogen in soil was determined by alkaline potassium permanganate method. Available phosphorous was estimated by Brays No. 1 method, available potassium by Flame photometer method. Calcium and available magnesium were determined by Versanate titration method and available sulphur by Turbidometric method as per the procedure outlined by Jackson, 1973. The available copper, manganese and Zinc were estimated by Atomic absorption spectrophotometer method. The data were analysed statistically by using simple RCBD as outlined by Cochran and Cox (2000).

Results and Discussion

The fertility status and microbial population of M_5 mulberry garden was significantly superior when treated with biofertilizers at 10 kg each of *A. brasilense* and *A. awamori* + 20 per cent each of recommended N through compost, green manure, castor cake, vermicompost, fertilizers + remaining P and K through fertilizers (T_{11}).

After harvest of third crop the soil analysis indicated that, among the different treatments, T_{11} has recorded significantly higher available nitrogen (274 kg/ha), phosphorous (80.60 kg/ha), potassium (259 kg/ha), calcium (4.85 c.mol/ kg), magnesium (4.10 c.mol/kg), sulphur (38 ppm), copper (1.40 ppm), manganese (22.00 ppm) and Zinc (2.00 ppm) contents (Table 38.1), Availability of nutrients from both organic and inorganic sources was greatly influenced by nitrogen fixation by *A. brasilense* and solubalization of P by *A. awamori.*

Table 38.1: Macro, Secondary and Micro Nutrient Contents in Soil Grown with M_5 Mulberry as Influenced by Different Organic Manures and Inorganic Fertilizers

Treatments	*Available N (kg/ha)*	*Available P (kg/ha)*	*Available K (kg/ha)*	*Available Ca (C.mol\Kg)*	*Available Mg (C.mol\Kg)*	*Available S (ppm)*	*Available Cu (ppm)*	*Available Mn (ppm)*	*Available Zn (ppm)*
T_1	232	243	64.30	4.39	3.45	26.66	1.00	18.00	1.30
T_2	201	216	60.80,	4.20	3.00	30.00	0.90	17.00	1.30
T_3	185	224	62.20	4.12	2.88	37.00	1.00	16.00	1.20
T_4	239	232	70.90	4.35	3.33	29.00	1.10	17.00	1.50
T_5	195	239	69.80	4.34	3.30	27.00	0.80	19.00	1.40
T_6	245	240	75.40	4.52	3.70	33.00	1.30	20.00	1.70
T_7	191	236	71.60	4.00	2.22	27.00	1.00	17.00	1.60
T_8	183	226	72.90	3.90	2.98	31.00	0.90	18.00	1.40
T_9	242	249	74.80	4.50	3.60	33.00	1.30	17.00	1.70
T_{10}	212	240	73.56	4.38	3.45	32.00	1.20	18.00	1.50
T_{11}	274	259	80.60	4.85	4.10	38.00	1.40	22.00	2.00
T_{12}	268	255	78.70	4.69	3.90	35.00	1.30	20.00	1.90
T_{13}	188	214	60.40	3.70	2.26	20.00	0.80	14.00	1.20
F test	*	*	NS	*	*	*	*	*	*
SEm ±	5.94	5.88	–	0.06	0.05	3.08	0.13	1.25	0.15
CD at 5%	17.35	17.18	–	0.17	0.15	9.01	0.39	3.67	0.46

Significantly higher bacterial (65,00 x 10^6 cfu/g of soil), fungal (30.66 10^3 cfu/g of soil) and actinomycetes population (7.00 10^4 cfu/g of soil) were recorded in T_{11}. Whereas, maximum *A. awamori* (15.00 10^3 cfu/g of soil) and *A. brasilense* population (15.00 10^6 cfu/g of soil) were recorded in the plots manured with bio fertilizers 10 kg each of *A. brasilense* and *A. awamori* + 25 per cent each of recommended N through compost, green manure, castor cake, vermicompost and fertilizers (T_{10}) (Table 38.2). Prabhuraj *et al.* (2005) reported that, combination of FYM, sericulture wastes compost, green manure and biofertilizers with NPK recorded significantly higher population of inoculated Phosphate solubalizing microorganisms (6.1 10^4 cfu/g of soil) and Nitrogen fixing bacteria (5.1 10^4 cfu/g of soil).

Table 38.2: Microbial Population (cfu/g) in Soil of M_5 Mulberry Garden as Influenced by Different Organic Manures and Inorganic Fertilizers

Treatments	*Bacterial (No. x 10^6)*	*Fungi (No. x 10^3)*	*Actimycetes (No. x 10^4)*	*Aspergillus awamori (No. x 10^3)*	*Azospirillum brasilense (No. x 10^6)*
T_1	33.00	10.00	03.00	10.00	08.00
T_2	32.00	10.33	03.00	06.00	07.00
T_3	31.33	18.33	03.00	09.00	06.00
T_4	33.00	21.66	04.00	10.00	08.00
T_5	25.66	21.66	04.33	09.00	07.00
T_6	42.66	30.33	06.00	10.00	07.00
T_7	34.66	12.66	05.00	11.00	10.00
T_8	33.66	19.00	05.33	08.00	08.00
T_9	38.00	27.00	06.00	11.00	11.00
T_{10}	32.00	24.00	05.66	20.00	16.00
T_{11}	65.00	30.66	07.00	15.00	15.00
T_{12}	51.00	30.66	06.66	13.00	12.00
T_{13}	24.33	09.00	02.33	06.00	07.00
F test	*	*	*	*	*
SEm ±	2.15	1.39	0.89	1.32	1.19
CD at 5%	6.28	4.05	2.61	3.86	3.48

From the above results it is concluded that, the integrated use of organic manures, biofertilizers with lower dose of inorganic fertilizers not only economized the N requirement of crop but also helped in sustained availability of nutrients for crop production. The application of organic manures in turn provides adequate biomass as a feed for the microbes and helps in increasing microbial population in soil.

Acknowledgement

The research investigation is part of the Department of Science and Technology (DST) New Delhi funded project entitled "Significance of organically grown mulberry on bio-chemical and reproductive biology of silkworm seed cocoons (*Bombyx mori* L.). The financial help received from the DST is greatly acknowledged.

References

Cochran and Cox, 2000. *Experimental Design*, Second edition. Wiley Classics Publishers, p. 95–145.

Dandin, S.B., Jayanth Jayaswal and Giridhar, K., 2003. *Handbook of Sericulture Technologies.* CSB, Bangalore, pp. 259

Jackson, M.L., 1973. *Soil Chemical Analysis.* Prentice Hall of India (Pvt.) Ltd., New Delhi, p. 259–260.

Prabhuraj, D.K., Bongale, U.D., Sukumar, J., Sanaulla, H. and Thimma Reddy, H., 2005. Comparative study on the organic and integrated nutrient management in mulberry. In: *Progress of Research in Organic Sericulture and Seri-Byproducts Utilization*, 2005, pp. 146–148.

Chapter 39

Sustainability through Environmental Conservation on Agricultural Land Resources

S.R. Singh and Th. Manimala Devi
Post Graduate Studies Centre, HRDRI, Canchipur, Imphal – 795 003

ABSTRACT

Sustainability of Thoubal district based on agricultural land explored about the planning development of the area spontaneously. Agricultural land resource, a unit of sustainable development accorded with 21496.50 hectares in 1991 and 33215.0 hectares in 2001 highlighted the provision view. In depth analysis, an ecologically productive land per capita in hectares (EPLC) compute 0.073127 in 1991 and 0.09066 in 2001; ecologically deficit land per capita (EDLC) in hectare as 1.926872 and 1.90933 in 1991 and 2001 respectively; ecologically deficit land per capita (EDLC) in per cent available in decades and in annum compute 2634.96 and 263.496 in 1991 and 2106.03 and 210.603 in 2001 respectively scrutinize the ways of strengthening the upshot planning and programmes in substantiation with full conservation of nature and environment.

Keywords: *Sustainability, Agricultural land, Ecologically productive, Ecologically deficit land per capita, Conservation of nature and environment.*

Introduction

The World Commission on Environment and Development, the Brundtland Commission in 1987 recommended sustainable development is the way to achieve global environmental security. More empathetically the term itself had been used before the Brundtland commission endorsed it, notably in the "World Conservation Strategy', jointly prepared by the World Conservation Union (IUCN), the United Nations Environment Programme (UNEP), the World Wide Fund for Nature (WWF) in 1980. (Ruth and William, 2000).

The World Commission on Environment and Development in 1993 was created, published "Our Common Future Report" in which it stated that sustainable development is a process of change in which it defines sustainable development as "Development that meet the needs of the present without compromising the ability of future generations to meet their own needs."

Moreover, for practical approach "Sustainable development is a process of change in which the exploitation of resource, the direction of investments, the orientation of technological and institutional change are all in harmony and enhance both current and future potentials to meet human needs and aspirations".

Agenda 21 emphasizes that sustainable development will reverse poverty and environment degradation thus provides a plans for combating degradation of land, air, water and for conserving forests and biological diversity (Tewari, 1994).

The Earth Summit held in Rio de Janeiro in 1992 also targeted towards the sustainable development. However, no such investigation have been worked out in Manipur. The present study implies on the multifacet meaning of the sustainable development based on the following objective to determine the sustainable development on the capabilities of agricultural area based on EPLC in hectares, EDLC in hectares, EDLC in percentage available in decades and annum etc.

Materials and Methods

Surveys being to produce information to desirable, compare and predict attitudes, opinions, values and behaviour based on what people say or see and what is contained in records about them and their activities the present survey's aim is fixed to the goal (Fink, 1995; Yadava, 1995). During the present work samples in surveys was adopted according to convenient for the data mainly to probability sampling like simple random sampling, stratified random sampling, systematic sampling and cluster sampling whereas non-probability sampling like convenience sampling, snowball sampling, quote sampling and focus groups. Then measure survey and their reliability and validity was checked adopting reliability *i.e.* test-retest, alternate-form, internal consistency, interobserver and validity *i.e.* face, content, criterion and construct. All the criterion include concurrent and predictive. The construct also include convergent and divergent.

According to P.V. Young, data has categorised as documentary sources and field sources. Documentary sources include material already collected whether published or unpublished. Among documentary sources the first place may be given to the books written by different experts. And reports of surveys, memories, accounts of travels, historical accounts, official published data and other unpublished record are also belongs to documentary sources. Field sources including living persons, scholars, scientists, research workers, leaders who have worked with the social group or have studied the problem.

For the determination of sustainable development on the capabilities of agricultural land of the Thoubal district under mention formulae have adopted following Rees, 1992.

1. Ecologically productive land per capita, EPLC (in Ha) = $\frac{\text{TA (in Ha)}}{\text{TP (in nos.)}}$

where,

EPLC: Ecologically productive land per capita in hectares.

TA: Total area in hectares.

TP: Population in numbers.

2. Ecologically deficit land per capita (EDLC) in hectares

   EDLC = FP – EPLC in hectares

   FP: Foot print = 2 Ha since India belongs to (2–3) Ha foot print.

   EPLC: Ecologically productive land per capita in hectares.

3. Ecologically deficit land per capita (EDLC) in per cent available in decades.

$$\text{EDLC in \% available in decades} = \frac{\text{EDLC (in Ha)}}{\text{EPLC (in Ha)}} \times 100$$

4. Ecologically deficit land per capita (EDLC) in per cent available in annum.

$$\text{EDLC in \% available in annum} = \frac{\text{EDLC (in Ha)}}{\text{EPLC (in Ha)}} \times 10$$

Results and Discussion

The population of Thoubal district in 1991 recorded 293958 persons and 21496.50 hectares in agricultural area. The calculated ecologically productive land per capita accord 0.073127 and ecologically deficit per capita as 1.926872. The percentage available of ecologically deficit per capita in decades and annum was 2634.96 and 263.496 respectively. The district recorded a population of 366341 persons and 33215 hectares in agricultural area during 2001. The calculated value of ecologically productive land per capita accord 0.09066 hectares and ecologically deficit land per capita in hectares compute as 1.90933. The ecologically deficit land per capita in percentage available in decades and annual was computed as 2106.03 and 210.603 respectively (Table 39.1). The illustrated data were plotted in Figure 39.1.

Table 39.1: Determination of Sustainability on Agricultural Area for Developmental Planning of Thoubal District and Manipur State

District/ State	*Year*	*Total Population (TA) in Nos.*	*Total Agricultural Area (TAA) in Ha*	*Ecologically Productive Land per Capita (EPLC) in Ha*	*Ecologically Deficit Land per Capita (EDLC)*			*Remarks*
					in Hectares	*in % Available in Decades*	*in% Available in Annum*	
Thoubal	1991	293958	21496.50	0.073127	1.926872	2634.96	263.496	
	2001*	366341	33215.0	0.09066	1.90933	2106.03	210.603	Assuming 2 Ha as footprint
Manipur	1991	1837149	148572.62	0.080871	1.919128	2373.06	237.306	
	2001	2388634	155232.0	0.06498	1.93502	2977.87	297.787	

Sources: 1. Statistical abstract of Manipur, 2001; 2. Directorate of Remote Sensing, Manipur; 3. Department of Agriculture, Manipur; 4. Provisional Population.

Manipur state recorded 1837149 persons of population and 148572.62 hectares of agricultural area during 1991. The calculated value of ecologically productive land per capita in hectare accord 0.080871 and 1.919128 hectares as ecologically deficit land per capita. The population of Manipur state in 2001 recorded 2388634 persons and 155232 hectares in agricultural land. The calculated

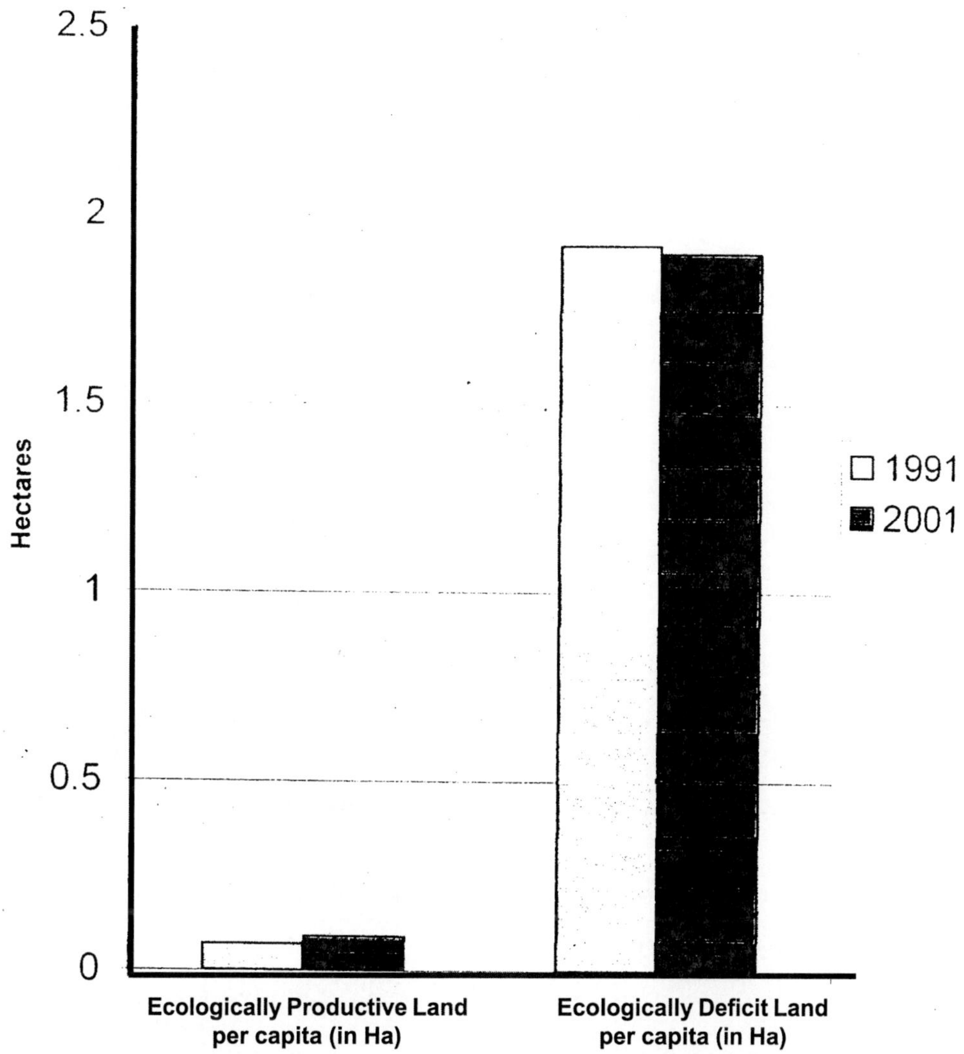

Figure 39.1: Sustainability Development on Agricultural Area of Thoubal District

value of ecologically productive land per capita in hectares accord 0.06498 and ecologically deficit land per capita in hectare as 1.93502. The percentage available of ecologically deficit land per capita in decades and in annum accord 2977.87 and 297.787 respectively (Table 39.1). The exemplified data were plotted in Figure 39.2.

Table 39.1 revealed the increased area of cultivable land in the district with an area of 11718.5 hectares in year 1991 year than that of the year 2001 and increased area of agricultural land in the Manipur state. The district accord the area of cultivable land 21496.50 hectares, the ecologically productive land per capita in hectares 0.073127, ecologically deficit land per capita in hectare 1.926872, ecologically deficit land per capita in percentage available in decade 2634.96, and ecologically deficit

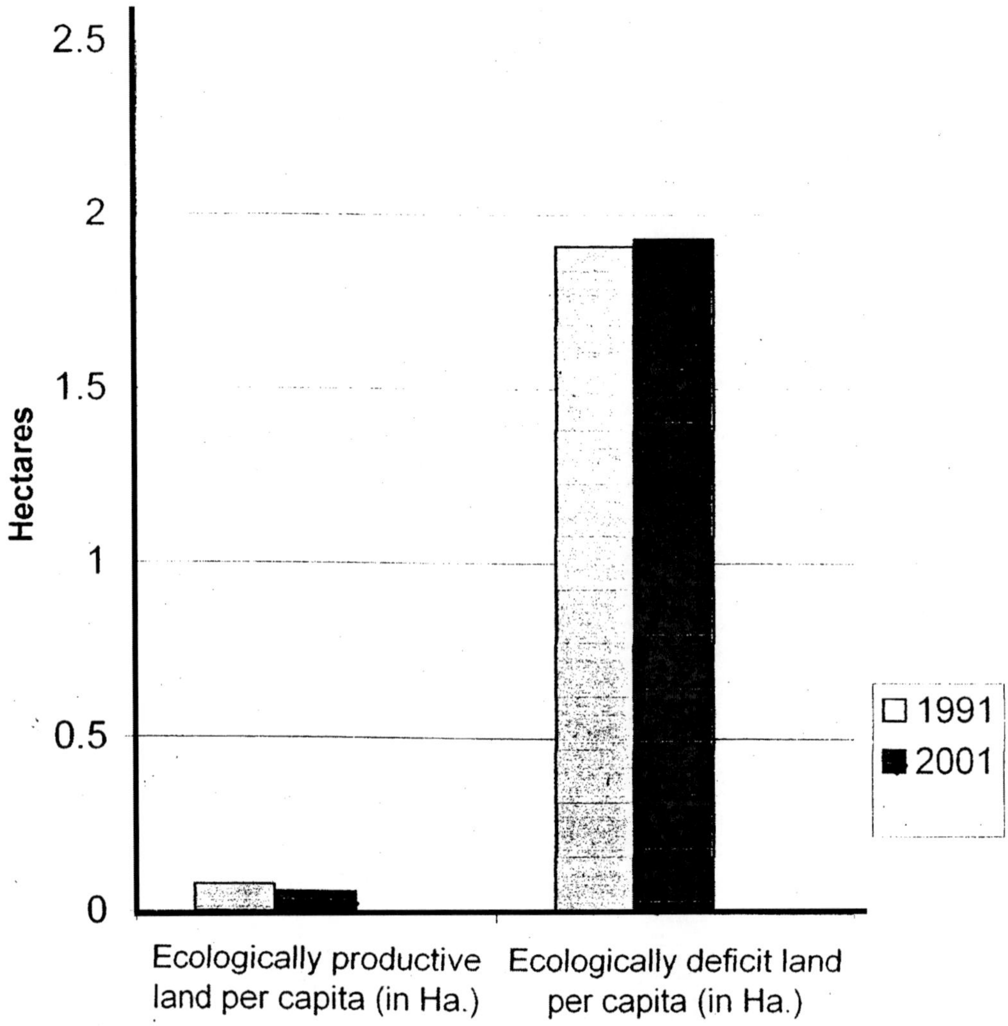

Figure 39.2: Sustainability Development on Agricultural Area of Manipur State

land per capita in percentage available in annum as 263.496 during 1991. During 2001, the district observed the total area of cultivable land of 33215.0 hectares, ecologically productive land per capita 0.09066 hectares, ecologically deficit land per capita in hectare 1.90933, ecologically deficit land per capita in percentage available in decade 2106.03 and ecologically deficit land per capita in percentage available in annum 210.603.

Manipur observed 148572.62 hectares of cultivable land in 1991 with a calculated value of ecologically productive land per capita in hectare as 0.080871, ecologically deficit land per capita in hectare as 1.919128, ecologically deficit land per capita in percentage available in decade as 2373.06 and ecologically deficit land per capita in percentage available in annum as 237.306.

During 2001, the state record the population of 2388634 persons and agricultural land 155232.0 hectares. The computed value of ecologically productive land per capita in hectares as 0.06498, ecologically deficit land per capita in hectares as 1.93502, ecologically deficit land per capita in percentage available in decades as 2977.87 and ecologically deficit land per capita in percentage availab1e in annum as 297.787.

The increase area in agricultural land naturally deduct from the natural land area of the district mainly from forest, scrub, wasteland etc. The situation is compelled by human being for producing more food production for feeding the ever increasing population. The present analysis clearly showed that the population pressure have great impact on the environment as nature keep a vast area of natural vegetation or resources to maintain the stability of ecological balance in an area. It maintains the system of all ecological components in a very well manage living system in that particular area for restoration of lifelines of any disturbance comes to that system nature suffers a lot of problems to maintain them and took many years in restoring the system sometimes it may not be in such a situation that enable to maintain them in that state the ecological balancing system fails and caused the eco-disturbances and consequence to unwanted environment. Thus, invite the special needs of human kind to check the consistent increase of disturbing ecological balance and destroying the natural resources to a state of newer restoring back. The investigating area highlighted the consciousness of mankind towards the right use of natural resources and immediate action of maintenance. The finding was in collaboration with the results of other workers in different places on earth.

Kircher *et al.* (1984) suggested land tenure problems on the lowland plains and technical difficulties/prejudices in the extensive low hilly areas have directed the in-migrants on the steep, often mountainous land to practice their "Kaingin" or shifting cultivation, resulting in accelerating destruction of the forests and an inevitable degradation of natural resources in all parts of the island and its adjoining marine waters (the 'present trends' scenario).

Sustainable development is underlined by an ethical norm focussing on a new relationship of co-operative and co-existence of humans with nature as compared to the prevailing conquest and exploitation of nature by humans (Batie, 1989). For the current practitioners of agricultural science, including agricultural economists, many of the concepts and tools employed are rooted in technically efficient resource development (or utilization). But this quality of agricultural systems, although definable and amenable to measurement, has been incriminated as inadequate or inappropriate for a sustainable perspectives that embodies a sense of ecological connections between action of humans and the consequences for them.

Acknowledgements

The authors thank to the Head, Department of Economics and Statistics, Remote Sensing and Agriculture, Government of Manipur, for supplying the required data.

References

Arlene, F., 1995. *How to Ask Survey Questions*? Sage Publications India Pvt. Ltd.

Batie, S., 1989. Sustainable development: Challenges to the profession of agricultural economics. *Amer. J. Agric. Econ.*, 71: 1083–1101.

Kirchner, J., Leduc, G., Goodland, R. and Drake, J., 1984. *Carrying Capacity Population Growth and Sustainable Development*. Population and Development Series. Washington, D.C., The World Bank.

Rees, W., 1992. Ecological footprints and appropriated carrying capacity: What urban economics leaves out. *Environment and Urbanization*, 4(2): 121–130.

Ruth, A.E. and William, R.E., 2000. *The Environment Encyclopedia,* Vol. 9. Salmonella Tropical Dry Forest.

Tewari, D.N., 1994. ICFRE and sustainable development. *Indian Forester,* November.

Yadava, S.S. and Yadava, K.N.S., 1995. *Statistical Analysis for Social Sciences.* Manak Publications.

Chapter 40

Sustainability through Agrochemicals on the Conservation of Natural Resources

S.R. Singh and Th. Manimala Devi

Post Graduate Studies Centre, HRDRI, Canchipur, Imphal – 795 003

ABSTRACT

Sustainability through agrochemicals, the indispensable component of food production, for the ever increasing population and hitherto the hungry mouth enforced the compulsive adoption of judicious use of need base application has been computed with the need of 970.08 tonnes, 1248.56 tonnes, 1061.60 tonnes, 1071.04 tonnes of $CaCO_3$ to neutralize the acidity formed by the urea applied for the year 2000–01, 2001–02, 2002–03 and 2003–04 respectively. Total environmental hazardous chemicals of fertilizers, pesticides, herbicides etc. not to cook with them through foods but to conserve the precious resources of lives on earth have been analysed.

***Keywords**: Sustainability, Agrochemicals, Acidity, Natural resources, Environmental degradation.*

Introduction

Even though the soil is a dynamic natural body and biochemically weathered and synthesized products of nature, there is no more to replace it, as being a natural habitat for plants and other living organisms. Further, soil is associated with the growth nutrition, yield and production of the plants. Consequently, life on earth depends directly on the living soil and the aquatic ecosystem of rivers. Without fertile soil and the microbial fauna that inhabit it, food would not grow, dead living things would not decay and nutrients would not be recycled. Yet the earth 's soils are being stripped away, rendered sterile and contaminated with toxic chemicals at a rate that can not be sustained. So far in

Manipur such trend of investigation have not yet been undertaken. Henceforth, the present investigation have been undertaken for determination of sustainability of chemical fertilizers, pesticides, herbicides etc. on the conservation of natural resources, with the following objectives:

1. To determine the acidity percentage in terms of neutralizing agents.
2. To determine the affect of environment through natural resources.
3. To investigate the sources of sustainability.

Materials and Methods

Surveys being to produce information to describe, compare and predict attitudes, opinions, values and behaviour based on what people say or see and what is contained in records about them and their activities the present survey's aim fixed to the goal (Fink, 1995; Yadava, 1995).

Samples in surveys during the present work was adopted mainly for probability sampling like simple random, stratified random, systematic and cluster sampling. Whereas non-probability sampling like convenience, snowball, quote and focus groups sampling was practiced.

Then measure survey and their reliability was checked expediting reliability *i.e.* test-retest, alternate form, internal consistency, inter-observer and validity *i.e.* face, content, criterion and construct. All the criterion include concurrent and predictive. The construct also include convergent and divergent other required data accords from documentary and field sources.

Documentary sources includes books, report of surveys, memories, accounts of travels, historical accounts, official published data, other unpublished record etc. Field sources includes living persons, scholars, scientists, research workers, leaders, people who have worked with the social group or have studied the problem. Under this method the data which contain necessary information was accorded by observing the person concern *viz.*, direct observation and in certain case if the observer is not possible, information from the informants were adopted. For such data collection, more than 300 farmers were selected randomly and adopted the methods and practices (Fink, 1993; Converse, 1987; Fowler, 1993 and Frey, 1989).

Results and Discussion

Thoubal district consumed chemical fertilizers like urea, DAP, MOP and Murate of Potash as high as 12126, 1378, 651 and 838 tonnes respectively in 2000–01. In the next year, *i.e.*, 2001–02, the district consumed 11604 tonnes of urea, 1404 tonnes of DAP, 585 tonnes of MOP and 651 tonnes of SSP; in 2002–03, 13270, 1490, 810, 910 tonnes of urea, DAP, MOP and SSP respectively were consumed by the district. In 2003–04, Thoubal district consumed 13388 tonnes of urea, 880 tonnes of DAP, 350 tonnes of MOP, 140 tonnes of SSP and 655 tonnes of Murate of Potash (Table 40.1).

Manipur state consumed 38319, 4353, 2215 and 1860 tonnes of urea, DAP, MOP and Murate of Potash respectively in 2000–01. In 2001–02, the state consumed 38708 tonnes of urea, 5443 tonnes of DAP, 1879 tonnes of MOP, 1338 tonnes of SSP and in 2002–03, 45677, 5000, 2860, 3163 tonnes of urea, DAP, MOP and SSP respectively were utilized as a chemical fertilizers by the state as a whole. Manipur state recorded 47246 tonnes of urea, 5370 tonnes of DAP, 2235 tonnes of MOP, 690 tonnes of SSP and 3415 tonnes of Murate of Potash (Table 40.1).

Manipur state consumed pesticides of Monocrotophos upto 4500 litres, Endosulphan upto 5080 litres, Quininephos upto 3500 litres, Cypermethrin upto 6440 litres and 7000 litres in other pesticides in 2000–01. During 2001–02, the state as a whole consumed about 5000 litres, 6020 litres, 4500 litres,

7000 litres and 8500 litres of Monocrotophos, Endosulphan, Quininephos, Cypermethrin and others respectively. In the next year *i.e.* 2002-03, the state utilized 5400 litres of Monocrotophos, 7500 litres of Endosulphan, 5050 litres of Quininephos, 7050 litres of Cypermethrin and 8550 litres of other pesticides. Manipur state consumed a total of 6500, 8550, 7000, 7500 and 10000 litres of Monocrotophos, Endosulphan, Quininephos, Cypermethrin and others pesticides in 2003–04 respectively (Table 40.2).

Table 40.1: Consumption of Chemical Fertilizers in Thoubal District and Manipur State

Year	*Urea (Tonnes)*	*DAP (Tonnes)*	*MOP (Tonnes)*	*SSP (Tonnes)*	*Murate of Potash (Tonnes)*	*Total (Tonnes)*
Thoubal District						
2000–01	12126	1378	651	–	838	14993
2001–02	11604	1404	585	651	–	14244
2002–03	13270	1490	810	910	–	16480
2003–04	13388	880	350	140	655	15413
Manipur State						
2000–01	38319	4353	2215	–	1860	46747
2001–02	38708	4553	1879	1338	–	46478
2002–03	45677	5000	2860	3163	–	56700
2003–04	47246	5370	2235	690	3415	59046

Source: Directorate of Economics and Statistics, Govt. of Manipur.

Table 40.2: Consumption of Pesticides in Manipur State

Year	*Monocrotophos (Litre)*	*Endosulphan (Litre)*	*Quininephos (Litre)*	*Cypermethrin (Litre)*	*Others (Litre)*	*Total*
2000–01	4500	5080	3500	6440	7000	26520
2001–02	5000	6020	4500	7000	8500	31020
2002–03	5400	7500	5050	7050	8550	33550
2003–04	6500	8550	7000	7500	10000	39550

Source: Delbros India, Khoyathong, Imphal.

Manipur state consumed herbicide of 2,4-D (liquid) upto the capacity of 13 kilolitres, 17 kilolitre, 22 kilolitre and 25 kilolitre in 2000–01, 2001–02, 2002–03 and 2003–04 respectively. The state utilized 2,4-D (solid) upto 24 metric tonnes in 2000–01, 29 metric tonnes in 2001–02, 34 metric tonnes in 2002–03 and 40 metric tonnes in 2003–04. In case of herbicides of Butachlor (liquid), the consumption recorded 1.1 kilolitre, 1.2 kilolitre, 1.8 kilolitre and 2.0 kilolitres in 2000–01, 2000–02, 2000–03 and 2003–04 respectively. Regarding the Butachlor (solid) and Glyphosphate (liquid), the state consumed 10 metric tonnes and 0.2 kilolitres in 2000–01, 13 metric tonnes and 0.5 kilolitres in 2001–02, 17 metric tonnes and 0.8 kilolitres in 2002-03 and 20 metric tonnes and 1.0 kilolitre in 2003–04 respectively (Table 40.3).

Thoubal district consumed 13388 tonnes of urea which indicate the required amount of $CaCO_3$ to neutralize the acidity upto 970.80 tonnes which is about 7.24 per cent in 2000–01. The district consumed

13270 tonnes of urea in 2001–02 and about 124.856 tonnes of $CaCO_3$ to neutralize the acidity and occupied 9.40 in per cent. Similarly, 11604 and 12126 tonnes of urea was consumed in 2002–03 and 2003–04 respectively. The amount of $CaCO_3$ to neutralize the acidity also raised up to 1061.60 and 1071.04 tonnes with a percentage of 9.14 and 8.83 for 2002–03 for 2002–03 and 2003–04 respectively. Manipur state as a whole consumed urea upto 38319 tonnes in 2000–01, 38708 tonnes in 2001–02, 45677 tonnes in 2002–03 and 47246 tonnes in 2003–04. The amount of $CaCO_3$ to neutralize the acidity is 3064.52, 3096.64, 3654.16, 3779.68 tonnes in 2000–01, 2001–02, 2002–03 and 2003–04 respectively. The percentage amount of $CaCO_3$ to neutralize acidity for the year from 2000–01 to 2003–04 accords 8.0 for each year (Table 40.4). The illustrated data of the district and state were plotted in Figure 40.1 (a) and (b).

Table 40.3: Consumption of Herbicides in Manipur State

Year	*2,4-D Liquid (Kilolitre)*	*2,4-D Solid (Metric Tonnes)*	*Butachlor Liquid (Kilolitre)*	*Butachlor Solid (Metric Tonnes)*	*Glyphosphate Liquid (Kilolitre)*
2000–01	13	24	1.1	10	0.2
2001–02	17	29	1.3	13	0.5
2002–03	22	34	1.8	17	0.8
2003–04	25	40	2	20	1

Source: Delbros India, Khoyathong, Imphal.

Table 40.4: Urea Consumption and the Amount of $CaCO_3$ to Neutralize the Acidity for Thoubal District and Manipur State

Year	*Thoubal District*			*Manipur State*			*Contribution Share to Whole in %*
	Urea Consumption (in tonnes)	*Amount of $CaCO_3$ to Neutralize Acidity (Tonnes)*	*Amount of $CaCO_3$ to Neutralize Acidity (%)*	*Urea Consumption on Tonnes*	*Amount of $CaCO_3$ to Neutralize Acidity (Tonnes)*	*Amount of $CaCO_3$ to Neutralize Acidity (%)*	
2000	13388	970.08	7.24	38319	3065.52	8.0	34.93
2001	13270	1248.56	9.40	38708	3096.64	8.0	34.28
2002	11604	1061.60	9.14	45677	6354.16	8.0	25.40
2003	12126	1071.04	8.83	47246	3779.68	8.0	25.66

Table 40.1 revealed that Thoubal district uses different chemicals in agriculture and allieds, fertilizers like urea accord 12, 126 tonnes in 2000–01, 11604 tonnes in 2001–02, 13270 tonnes in 2002–03 and 13388 tonnes in 2003–04. In DAP, the district consumed 1378, 1404, 1490 and 800 tonnes in 2000–01, 2001–02, 2002–03 and 2003–04 respectively and 651 tonnes in 2000–01, 585 tonnes in 2001–02, 810 tonnes in 2002–03 and 350 tonnes in 2003–04 in MOP fertilizers. Similarly in case of SSP fertilizers, the district consumed 651 tonnes, 910 tonnes, 140 tonnes in 2001–02, 2002–03 and 2003–04 respectively. Again 838 tonnes in 2000–01 and 655 tonnes in 2003–04 of Murate of Potash were consumed for agricultural purposes. All these chemical fertilizers have special impact to the environment directly to soil, water and indirectly to air. Impact in the sense of deterioration of the resources and in broad view to the environment. Regarding the intensity of consumption of these

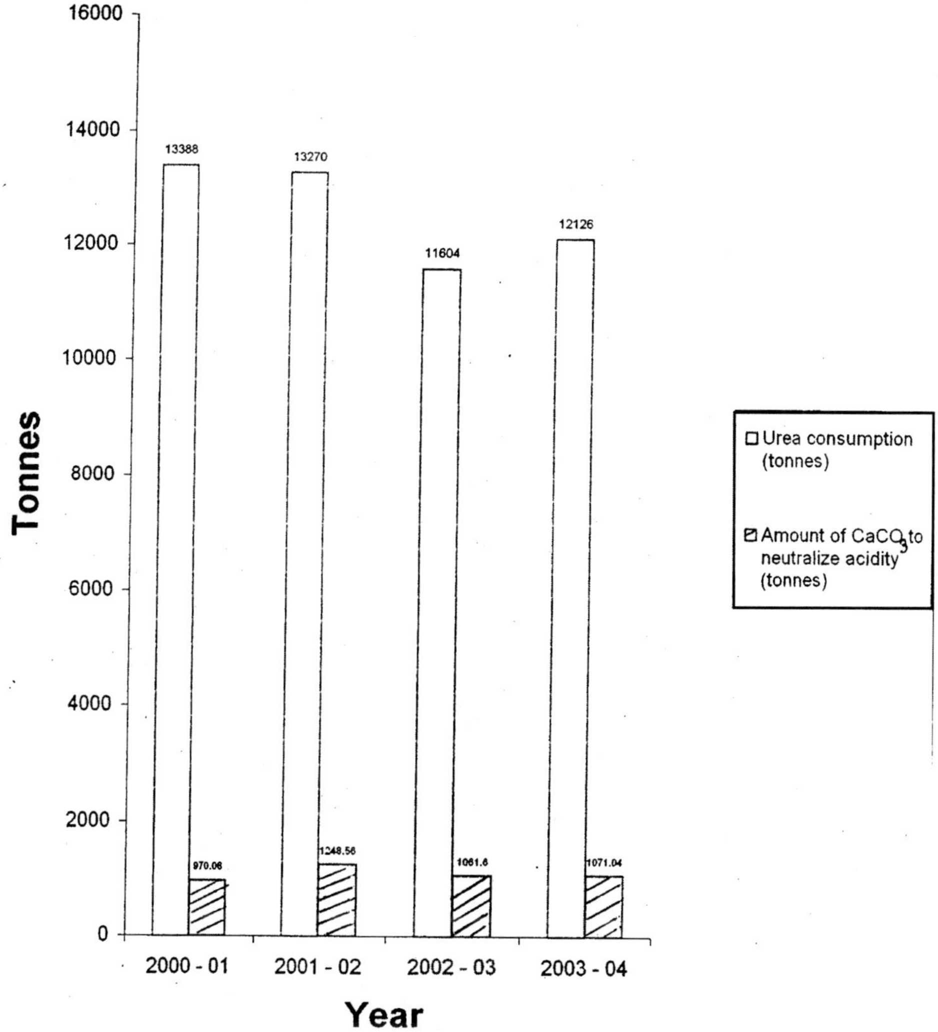

Figure 40.1(a): The Amount of Urea Consumption and the Amount of $CaCO_3$ to Neutralize Acidity for Thoubal District During 2000–01 to 2003–04

chemical fertilizers increases upto 14993 tonnes in 2000–01. The trend continues and attained 14244, 16480, 15413 tonnes in 2001–02, 2002–03 and 2003–04 respectively. The consumption of chemicals by Thoubal district in agriculture and allied ranked first in comparison with other districts of the state and shared 29.72 per cent to the state as a whole. The finding clearly showed the highest rank in addition of chemicals to soil and unjudicious utilization of resources and advancement towards the deterioration of environment in part or a piece to the whole. The trend is seriously supportive to the national environmental degradation. During the period of investigation, Manipur uses different chemical fertilizers like urea, DAP, MOP, SSP and Murate of Potash etc. thus highlight the judicious need base dose recommendation so as to have sustainability of agro-chemical fertilizers with least affect to natural resources.

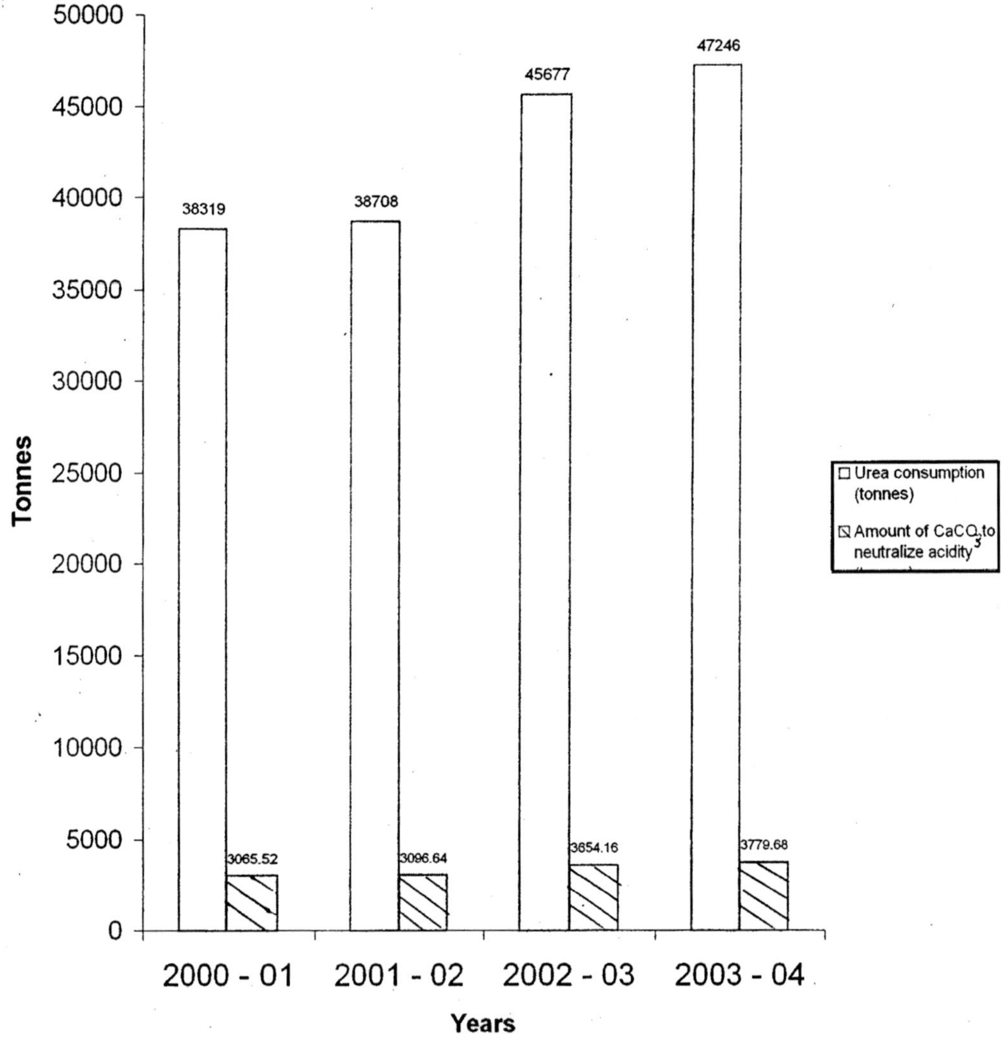

Figure 40.1(b): The Amount of Urea Consumption and the Amount of $CaCO_3$ to Neutralize Acidity for Manipur State During 2000–01 to 2003–04

Due advancement of science and technology, food and agricultural production rises upto the peak in past few decades. With heavy input of chemicals which have short and long term affect from insecticides, pesticides, fungicides, acricides, nematicides, herbicides, weedicides etc. are freely used by the farmers for agricultural, horticultural and floricultural production. Directly and indirectly, all these pesticides, herbicides, nematicides etc. harm to soil, air, water starting from immediate effect to long term effect of biomagnification. Food production is sufficient or even surplus in the developed countries due to advanced agricultural technologies. However, it is not so in the under developed or developing countries. Further, more than half of the population virtually go half feed. Sen (1965) had remarked 22 years ago that these half fed are in reality 'half alive' and it is no better even today. The peasant population mostly depend on their own cultivated products or else go hungry. The situation becomes more depressing as the various types of pests snatch away a substantial quantity of their foodstuff.

Table 40.2 revealed the consumption of pesticides in Manipur state during 2000–01 to 2003–04 with a heavy pressure to them for more production for ever increasing mouths.

Table 40.3 revealed the amount of consumption of different herbicides including 2,4-D (liquid), 2,4-D (solid), Butachlor (liquid), Butachlor (solid) and Glyphosphate (liquid) etc. in Manipur state. Thus question naturally comes that where has gone these chemicals and what would be the fate of destiny.

Table 40.4 depict the consumption of urea and amount of $CaCO_3$ to neutralize the acidity of Thoubal district and Manipur state. The district not only handout a major shared in food grain production but also contribute acidity to soil, natural resources *i.e.* hampering the environment. The finding emphasize the judicious use of fertilizers both in amount and way of application a compulsive, prerequisite so as to conserve natural resources. The finding was in corroboration with that of Pandey (1976), Edwards (1974), Sahu (1997). Similar findings of pesticides persistence and degradation were reported by Smith *et al.* (1955), Wallis *et al.* (1957), Nath *et al.* (1974), Sarode and Lal (1982) and Agnihotri *et al.* (1981). Further in this connection it may be mentioned that Meena (2005) stated that "I ruined my soil with chemicals, now I am trying to save it. I am a certified organic farmer." However, the intensive use with utmost care of chemical fertilizers, pesticides both injudiciously and need base must be foremost priority and compulsory.

Acknowledgements

The authors are thankful to the Director of Economics and Statistics, Government of Manipur and also to the Delbros India, Khoyathong, Imphal, for supplying various relevant data.

References

Agnihotri, N.P., Pandey, S.Y., Jain, H.K. and Dewan, R.S. 1981. Rapid single step extraction and clean up method for estimation of residues of insecticides in soil. Paper presented by symposium on *"The insecticide act and pesticide residue control"*. CFTRI, Mysore, Karnataka.

Converse, J., 1987. *Survey Research in the United States.* University of California Press, Berkeley.

Edwards, C.A., 1974. *Persistent Pesticides in the Environment,* 2nd ed. CRC Monoscience Series, Butter Worths, London.

Fink, A., 1995. *How to Ask Survey Questions?* Sage Publications India Pvt. Ltd.

Fink, A., 1993. *Evaluation Fundamentals: Guiding Health Programs, Research and Policy.* Newbury Park, CA, Sage.

Fowler, F.J., 1993. *Survey Research Methods.* Newbury Park, CA: Sage.

Prey, J.H., 1989. *Survey Research by Telephone.* Newbury Park, CA: Sage.

Meena, Menon, 2005. *Organic Revolution.* The Hindu Survey of the Environment, 136 pp.

Nath, G., Kathpal, T.S and Srivastava, B.P., 1974. Persistence of endosulfan residues in Bhendi fruits. *Indian J. Pl. Prot.,* 2: 45–51.

Pandey, S.Y., 1976. Adsorption, desorption and degradation of dicrotophos, monocrotophos, disulfoton and phorate in soil. *Ph.D. Thesis,* Kanpur University, Kanpur.

Sahu, Dibakar, 1997. Fate of Catbofuran and Cytrolane on maize crop and soil. *Ph.D. Thesis*. P.G. School, IARI, New Delhi.

Sarode, S.V. and Lal Rattan, 1982. Persistence of residues of fenitrothion in or on Okra. *Indian J. Agric. Sci.*, 52: 135–138.

Sen, 1965. In the Bulletin "Facts about concerns on pesticides in relation to ecology". *K. Dhari.*

Smith, F.F., Giang, P. and Taylor, E.A., 1995. Reduction of malathion residues on vegetables by washing. *J. Econ. Ent.*, 48: 209–210.

Wallis, R.L., Smith, F.F., Wheeler, H.G. and Taylor, F.A., 1957. Malathion residues on vegetables berry and tobacco crops. *J. Econ. Ent.*, 50: 362–363.

Yadava, Surendra S. and Yadava, K.N.S., 1995. *Statistical Analysis for Social Sciences*. Manak Publications.

Chapter 41

Bioremedial Approach with Reference to Degradation of Hydrocarbon Compounds: A Study

Limpon Bora* and M.C. Kalita
Department of Biotechnology
Guahati University, Guwahati, Assam, India

ABSTRACT

All together 12 samples were collected from Tezpur and adjoining areas. In a whole 54 strains were isolated in the nutrient agar medium from the samples collected. Out of the 54 strains, 28 strains were found to be positive and have shown n-Hexadecane degradation activity.

Keywords: *Bioremediation, n-Hexadecane degradation, Microbes.*

Introduction

The development of biotechnology has lead to a revolution in the field of environmental management. There is an rapid development of chemical industries that has produced a great variety of chemical compounds that have led to modernization of our lifestyles during the past century. Contamination of soils, groundwater, sediments, surface water, and air with hazardous and toxic chemicals are serious problems, which have been faced by our world today (Boopathy, 2000). The ecological impact of spilled crude oil and its products in aquatic ecosystem is diverse and well

* Corresponding Author: E-mail: limponbioera@rediffmail.com.

documented and generally severe toxic effects are associated with low boiling points volatile aromatic hydrocarbons then heavier fuel oils. Research is going on various ways to develop methods for the prevention and reduction, reuse and recycling of materials for the management of the environmental problems. With the advancement in biotechnology the way of the management has been changed. The environmental biotechnologies are going new paths, resulting in processes within *"clean technologies"* with maximum production and less residues. Bioremediation has finds its applications in the field of soil contamination with reference to pesticides, herbicides, oil *(*Hydrocarbons*) (Dojika *et al.*, 1998), food waste and heavy metals etc. It is a naturally occurring process by which microorganisms either immobilize or transform environmental contaminants to innocuous end products. In this paper we have reported the results of hydrocarbon degradation of the isolated strains from the samples collected from Tezpur and adjoining areas.

Material and Methods

Reagents

Peptone and Yeast extract, Glucose, $MgSO_4$, KH_2PO_4, $CaCl_2$, $FeSO_4$ and Trace elements (Sarabhai chemicals Baroda, Glaxo industries).

Sample Collection and Processing

The samples were collected from various oil spillage area and garages of sonitpur district of Assam, India. All together 12 samples were collected from Tezpur and adjoining areas. pH of the samples was recorded at the time of collection. Samples were collected asceptically and transported to laboratory in the sterile polythene bags and glass vials (Autoclaved) and processed later on. Serially diluted samples were centrifuged at 5000 rpm for 15 minutes as pretreatment to make the extraneous matter to settle.

Culture Medium

Nutrient agar medium (Himedia) was used to isolate the bacterial strains (The medium was sterilized by autoclaving at 15 lbs pressure (121 C) for 15 minutes.

Plating and Incubation

Serially diluted samples were inoculated by spread plate method. The plates were incubated at 38±2°C for the isolation of the strains.

Isolation of Strains

The plates were screened for bacterial colonies after 48 hours of incubation. The bacterial colonies appeared in agar plates were selected based on their morphological characteristics. After isolation of colonies in pure culture in slants, they were preserved in the refrigerator for the further study.

Cultures were maintained on YEPG agar (gl^{-1} yeast extract 5; peptone 5; Glucose 4; and agar 10) slants.

Growth of Organism

The strains were cultivated in a medium containing Ammonium sulphate–0.4 gm, Na_2HPO_4–0.36 gm, KH_2PO_4–0.17 gm, $MgSO_4.7H_2O$–0.02 gm, $CaCl_2$–0.05 gm, $FeSO_4.7H_2O$–0.001 gm, Trace element–0.1 ml. The pH of the medium was adjusted to 7.2. 50 µl of n-Hexadecane was used as a hydrocarbon substrate.

Results and Discussions

Microorganisms responsible for the degradation of Hydrocarbon compounds are widely distributed in nature such as in marine, fresh water and soil habitats, including bacteria fungi and yeasts. There are special organism species, which can better degrade a certain oil components, such as aromatics, poly aromatic hydrocarbons which degradation depends upon certain factors such as the molecular weight of the compound. During this study it is found that out of 54 strains isolated 28 strains have shown positive activity by showing the degradation of hydrocarbon source n-Hexadecane (Table 41.1). According to Balba *et al.*, many technology are available to do the excavation and containment in safe land fills, biological stabilization in situ, soil washing and solvent extraction etc. bioremediation appears to be one of the most promising technologies for dealing with environmental contaminants. Bioremediation stimulates the natural process and are less expensive and ecofriendly to the environment.

Table 41.1

Sl.No.	Strains	n-Hexadecane Degradation	Sl.No.	Strains	n-Hexadecane Degradation
1.	B243	+	28.	C356	+
2.	B43	+	29.	C367	+
3.	B245	+	30.	C52	+
4.	B252	–	31.	B277	–
5.	B104	+	32.	B-212	–
6.	B40	+	33.	B-188	+
7.	B-179	+	34.	B-243	–
8.	B270	+	35.	B90	–
9.	B63	+	36.	C341	–
10.	B-140	+	37.	B66	+
11.	B250	–	38.	B119	–
12.	B48	+	39.	B98	+
13.	B256	–	40.	B37	+
14.	B50	+	41.	B54	–
15.	B139	–	42.	B156	–
16.	B-237	–	43.	B-184	–
17.	B-129	+	44.	B-207	+
18.	B-172	+	45.	300	–
19.	B70	+	46.	B-189	+
20.	B42	+	47.	B274	+
21.	C360	–	48.	C403	–
22.	C368	–	49.	B10	–
23.	C385	+	50.	B182	–
24.	C359	+	51.	C335	+
25.	C354	+	52.	B153	–
26.	C365	–	53.	C401	–
27.	C368	–	54.	B333	–

References

Balba, M.T., Al-Awadhi, N. and Al-Daher, R., 1999. Bioremediation of oil contaminated soil: Microbiological method for feasibility assessment and field evaluation. *J. Microbial. Meth.*, 32: 155.

Boopathy, R., 2000. Factors limiting bioremediation technologies. *Bioresource Technology*, 74: 63.

Dojka, M.A., Hugenholtz, P., Haack, S.K. and Pace, N.R., 1998. Microbial diversity in a hydrocarbon and chlorinated-solvent contaminated aquifer undergoing intrinsic bioremediation. *Appl. Environ. Microbial.*, 64: 3869.

Nelson-Smith, A., 1970. The problem of oil pollution of the sea. *Adv. Mar. Biol.*, 8: 215–306.

Chapter 42

Biomonitoring Approach with Benthic Macro-invertebrates for Water Quality Assessment in a Medium Reservoir

Ch. Srinivas and Ravi Shankar Piska
Department of Zoology, University College of Science, Osmania University, Hyderabad – 500 007

ABSTRACT

The present study deals with the benthic macro-invertebrates as a tool for biomonitoring for water quality assessment in a medium reservoir, Edulabad, Ranga Reddy District, Andhra Pradesh, India. The present work carried for three years from 2000–01 to 2002–03. The water spread area is 247.31 ha and water used for irrigation and fisheries. Polluted river Musi water enter into reservoir through a canal called Narayanarao Kathwa. Oligochaets, Hemipterans, dipterans, Coleopterans and molluscans found in the reservoir, out of which dipterans dominated. The biological data of various pollution indices like saprobic score and diversity scores calculated to quantify the water quality of the reservoir. Biomonitoring studies indicated that the reservoir was heavily polluted.

Keywords: *Biomonitoring, Benthic macro-invertebrates, Water quality, Pollution, Reservoir.*

Introduction

Monitoring of environmental components is an important pre-requisite fur pollution control activities. It is difficult to monitor water quality only by using physico-chemical methods due to large number of pollutants and their low concentrations. To overcome this difficulty environmental scientists, all over the world, are exploring the possibility of using biomonitoring techniques in addition to physico-chemical monitoring.

Aquatic organisms have preferred habitats, defined by physical, chemical and biological features. Variation in one or more of these can lead to stress on individuals and possibly a reduction in the total number of species or organisms that are present. In extreme situations of environmental change, certain species will be unable to tolerate the changes in their environment and will disappear completely from the area concerned, either as a result of death or migration. Thus the presence or absence of certain species or family groups or the total species number of abundance, have been exploited as a means of measuring environmental degradation.

Biomonitoring is the introduction of biological variables for assessment of the structural and functional aspects of aquatic ecosystems. Through biomonitoring, the cumulative effect of all the pollutants can be determined and the overall health of the aquatic ecosystems could be properly assessed. Biomonitoring can be used as a cost-effective mean for supplementing the physico-chemical techniques. In the present study, benthic macro-invertebrates used to assess the pollution levels in the medium reservoir.

Materials and Methods

The present study conducted in a medium reservoir, Laxminarayana reservoir, Edulabad, Rangareddy District, Andhra Pradesh, India during three years from 2000–01 to 2002–03. For the Bio-monitoring study of the reservoir, four sites were selected and the benthic macro-invertebrates samples collected in the same sites to minimize seasonal variation. Samples collected during biologically active period (October–May), the collected sediment sample has been sieved for separating animals from detritus. US standard Seive No. 30 with pore size 0.6 mm used to retain organisms. The other way was to pick up organisms from sieve with forceps and placed in collection bottle.

The collected organisms preserved with 4 per cent formalin in wide mouth polyethylene bottles for further identification, verification and analysis. The criteria followed for water quality evaluation was based on the comparison of diversity score and saprobic score as per deZwart (1993) and CPCB (2002).

The bottom fauna collected at a sampling station for evaluating saprobity is randomized by gently shaking the collection bottle. The contents poured in to a flat white enamel or plastic tray with vertical and horizontal lines on the bottom. Clumps of organisms dispersed by pouring water river them. Sequential comparison Index (SCI) method involved a comparison of sequentially encountered individuals and the difference of two specimens which can easily be observed up to the species level. The diversity was the ratio of the total number of runs and the total number of organisms encountered. However, diversity alone can't indicate overall health of water body that is why the diversity of benthic animals was always compared with saprobity score.

Saprobic Score method involves inventory of the presence of local benthic macro-invertebrate forms up to species or family level. All possible families having saprobic indicator value were classified on a score-scale of 1 to 10 according to their preference for saprobic (oxygen availability) water quality. The saprobity scores of all families registered are averaged to produce BMWP (Biomonitoring Working party). The pollution levels concluded after comparison with the help of Table 42.1 (deZwart, 1993; CPCB, 2002).

Table 42.1: Biological Criteria for Water Quality Evaluation

Sl.No.	*Taxonomic Groups*	*Range of Saprobic Score (BMWP)*	*Range of Diversity Score*	*Water Quality Characteristics and Class*
1.	Ephemeroptera, Plecoptera, Trichoptera, Hemiptera, Diptera	7 and more	0.2–1	Clean A
2.	Ephemeroptera, Plecoptera, Trichoptera, Hemiptera, Odonata, Diptera	6–7	0.5–1	Slight Pollution B
3.	Ephemeroptera, Plecoptera, Trichoptera, Hemiptera, Odonata, Crustacea, Mollusca, Polychaeta, Coleoptera, Diptera, Hirudinea, Oligochaeta	5–6	0.4–0.9	Moderate Pollution C
4.	Mollusca, Hemiptera, Coleoptera, Diptera, Oligochaeta	2–5	0.4 and less	Heavy Pollution D
5.	Diptera, Oligochaeta or No macroinvertebrates	0–2	0–0.2	Severe Pollution E

Results

The Biomonitoring water quality evaluation of Laxminarayana reservoir for three years from 2000–01 to 2002–03 indicated that this reservoir was showing D class as per CPCB Criteria (Table 42.2). The water body classified as Heavily Polluted. During present study Seven out of 24 months the water quality showed class C which was moderately polluted. This was due to the inflow of fresh water from catchment area after heavy showers. Where as 17 out of 24 months the water quality showed class D, *i.e.* class D was dominated. The mean diversity score was 0.37 and saprobic score was 4.62. The mean scores indicate class D.

Table 42.2: Biomonitoring Water Quality Status of Lakshminarayana Reservoir from 2000–01 to 2002–03

Year	*Indices and Class*	*Oct*	*Nov*	*Dec*	*Jan*	*Feb*	*Mar*	*Apr*	*May*
2000–01	Diversity Index	0.5	0.38	0.4	0.4	0.35	0.39	0.32	0.4
	Saprobity Index	4.4	4.5	4.1	4.3	5.0	4.9	4.5	4.3
	Water Quality Class	C	C	C	D	D	D	D	D
2001–02	Diversity Index	0.39	0.42	0.3	0.35	0.4	0.33	0.34	0.38
	Saprobity Index	5.0	4.4	4.3	4.9	4.5	5.0	4.6	4.6
	Water Quality Class	D	C	C	D	D	D	D	D
2002–03	Diversity Index	0.31	0.28	0.33	0.29	0.36	0.45	0.39	0.4
	Saprobity Index	4.2	5.0	4.4	4.7	5.0	4.8	5.0	4.6
	Water Quality Class	C	D	C	D	D	D	D	D

Class 'C': Moderate pollution; Class 'D': Heavy pollution.

In the reservoir, oligochaeta, hemiptera, diptera, coleoptera and mollusca animals found abundantly. Apart from these animals polychaeta, ephimeroptera, plecoptera, trichoptera, odoneta and hirudenea found rarely in the reservoir (Table 42.3). Out of the above orders, diptera dominated;

followed by hemiptera and coleoptera. Molluscans and oligochaets also found in good number. Based on different organisms of various groups the saprobic and diversity scores analysed to conclude the water quality of the reservoir.

Table 42.3: The Benthic Macro-invertebrates in Laxminarayana Reservoir

Orders	*Benthic Macro-invertebrates*
Hemiptera	*Notonecta glauco*
	Anisops sardea, A. nivens
	Lemnometra fluviorum
	L. anadyomene
	Ranetra elongata
	Laccotrephes maculates
	L. rubber
	Corixa ima
	Lithoceros indicum
Coleoptera	*Cybister limbatus, C. rugulosus*
	Hydaticus fabricii
	Lacuphilus chirensis
	Hydrophilus indicus
Diptera	*Larvae of Culex* sp., *Anapheles* sp. and *Aedes* sp.
	Chryptochiranomus sp.
	Chiranomus reparius sp.
	Harischia sp., *Polypeedium* sp.
	Tendipes sp.
Odoneta	Dragon and Damsel fly nymphs
Epimeroptera	May fly nymphs
Trichoptera	Caddis fly larvae
Pleoptera	Stone fly nymphs
Mollusca	*Planorbis exustus, Anisus* sp.
	Vivipora sp., *Alocinma orcula*
	Eulota scalpturita, Glessula sp.
	Pila sp., *Carbicula* sp.
Annelida (Oligochaeta)	*Tubifex tubifex, Glycera* sp.

Discussion

Biomonitoring is the use of biological response to assess changes in the environment due to anthropogenic causes. In the present study biomonitoring used to assess the pollution levels in the reservoir. Biological monitoring of water quality could be useful for assessing the overall health of water body because:

1. The number of chemicals to be monitored are unending.
2. Biological effects often occur at concentrations below analytical capabilities. Many of the pollutants are present in such low concentrations that the influent sensitivity is too poor to determine the micro-quantity of such pollutants.
3. The toxicants may act quite differently in mixture than individually. Such toxicants affect the ecosystem in a synergistic manner which cannot be detected by chemical analysis alone.
4. Chemical nature of the toxicant is highly dynamic in environment with time and space, whereas biological system can integrate all environmental variables are a long period of time in terms of effects, which can be easily measured and quantified.

Environmental pollutants from anthropogenic sources affect the aquatic ecosystem in a synergistic manner, which cannot be detected comprehensively by determination of selected physico-chemical parameters alone. Whereas, biological system can integrate all environmental variables over a long period of time in terms of effects, which can be easily measured and quantified.

According to CPCB (2002) during the development and validation of biomonitoring methodology, established the fact that among all the biological components of the aquatic ecosystem, benthic macro-invertebrates are best suitable for evaluation of water quality. Most of these macro-invertebrates particularly insect larvae, which are aquatic, share their biological life in fresh water, while their adults fly over for miles together in search of suitable fresh water environment for reproduction, breeding and laying eggs in suitable habitats and establish biological communities.

According to Kennish (1992) biomonitoring was a valuable assessment tool that is receiving increased use in water quality monitoring programmes of all types. deZwart *et al.* (1993) and CPCB (2002) reported that the animals such as aquatic and semi-aquatic insects like ephemeroptera, plecoptera, hemiptera, coleoptera, trichoptera, odoneta and diptera, mulluscans, annelids like oligochaetes and leaches can be used in monitoring studies to assess the pollution levels of a water body. Ramakrishnan (2003) reported that several biological communities including phytoplankton, periphyton, micro-phytobenthos and aquatic macrophytes can be used in the biomonitoring studies to assess the pollution levels of a water body. He also stated that the level of degradation of water reservoirs can be quantified by biomonitoring technology more than conventional chemical methods. Phytoplankton, periphyton and microphytobenthos used in biomonitoring by him. Cairns (1979), Gunale (1991), Pati and Sahu (1993), Acharjee *et al.* (1995), Baruah *et al.* (1997) and Sultana and Sharief (2004) reported that the planktons serve as an index of water quality in respect of pollution. They also stated that plankton population observation can be used as reliable tool for bio-monitoring studies to assess the pollution status of aquatic biota. Trivedi (1999) reported that different organisms like bacteria, algae, zooplankton, benthic macro-invertebrates and fish can be used for biomonitoring studies.

Palmer (1969) presented the list of most pollution tolerant algae species. According to Hynes (1960) heavy pollution affects whole taxonomic group of macro-invertebrates. Mild pollution can bring changes in individual species. Mason (1981) reported the pollution tolerant benthic macro-invertebrates. There were *Dinocras cephalotes* (Plecoptera), *Ecdyonusus venosus* (Ephemeroptera), *Gammărus perlex* (Amphipoda), *Asellus aquatica* (Isopoda), *Chironomus reparius* (Diptera) and *Tubifex tubifex* (Oligochaeta). Trivedi (1999) reported that *Tubifex* found most abundantly near sewage outfalls and can grow in very low oxygen conditions. *Gammarus* cannot survive in low oxygen condition. Stoneflies, mayflies and caddisflies disappear with increasing pollution.

The water of Laxminarayana reservoir was highly polluted and its biodiversity of phyto and zooplankton, benthic macro-invertebrate and ichthyofauna also declined. Major fish kills also noticed during the present study. Bioaccumulation of heavy metals found in the tissues of fishes. The fish growth rates and fish yields also decreased significantly. All the above changes occurred in the reservoir due to heavy pollution, which was due to the entry of polluted river Musi water into the reservoir. River Musi water severely polluted due to sewage, effluents from industries, agriculture, etc.

Acknowledgement

We are thankful to Prof. B. Raghavendra Rao and Prof. P. Umamaheswara Reddy for their encouragement and suggestions during the work.

References

Acharjee, B.A., M. Dutta, M. Chaudhary and V. Pathak, 1995. Phytoplankton species diversity indices in Dighate Jeel, Assam, India. *Environment and Ecology,* 13(3): 660–662.

Baruah, B.K., M. Chaudhary and M. Das, 1997. Plankton as index of water quality with reference to paper mill pollution. *Poll. Res.,* 16(4): 259–263.

Cairns, J. Jr., 1979. Final summary. In: *Biological Indicators and Water Quality,* (Eds.) A. James and L. Evison. John Wiley and Sons, New York, p. 226–227.

deZwart, D., 1995. Manual on integrated water quality evaluation Appendix 6: *Taxonomical key for biological water quality determination.*

Gunale, V.R., 1991. Algal communities as indicators of pollution. *J. Eniron. Biol.,* p. 223–232.

Hynes, H.B.N., 1960. *The Biology of Polluted Waters.* Liverpool Univ. Press, Liverpool.

Kennish, M.J., 1992. *Ecology of Estuaries: Anthropogenic Effects.* CRC Press, Boca Raton.

Mason, C.F., 1981. *Biology of Freshwater Pollution,* Longman.

Palmer, C.M. 1969. A composite rating of algae tolerating organic pollution. *Br. Phycol. Bul.,* 15: 78–82.

Ramakrishnan, N., 2003. Biomonitoring approaches for water quality assessment in two waterbodies at Tiruvannamalai, Tamil Nadu India. In: *Proceedings of the Third International Conference on Environment and Health,* Chennai, India, (Eds.) Martin J. Bunch, V. Madha Suresh and T. Vasantha Kumaran. 15–17 December, 2003, Chennai.

Pati, S. and B.K. Sahu, 1993. Biomonitoring of water pollution in Rengali reservoir, Orissa. *Environment and Ecology,* 11(4): 982–984.

Sultana, M. and Sharief, D., 2004. Water pollution studies in the Double lake (Erretai eri) with relation to Phytoplankton. *J. Aqua. Biol.,* 19(1): 15–18.

Trivedy, R.K., 1999. *Encyclopedia of Environmental Pollution and Control,* Vol. 1, Enviro Media, India.

Chapter 43

Variations in Heavy Metal Concentrations Before and After Tsunami in Nagapattinam Coasts, South East Coast of India

***P. Martin Deva Prasath*[1], *M. Palanisamy*[2] *and R. Kunjitham*[2]**

[1]*Department of Chemistry, TBML College, Porayar – 609 307, Nagapattinam District, Tamil Nadu, South India*

[2]*Department of Chemistry, Poompuhar College, Melaiyur – 609 107*

ABSTRACT

Assessment of heavy metals (Zn, Cu, Fe, Mn, Ni, Cd, Co, and Pb) in water and sediment at six different stations of Nagapattinam coasts *viz.*, Velankanni (Station 1), Nagapattinam (Station 2), Vanjur (Station 3), Karaikal (Station 4), Tharangampadi (Station 5) and Poombuhar (Station 6) lying along the south east coast of India were studied before and after tsunami attack by using atomic absorption spectrophotometer. Significant variations were observed in all the metals at all the six stations after tsunami. Accumulation of heavy metal was observed in the order of Sediments > Water. In water, the order of heavy metals was found to be Mn > Fe > Zn > Cu > Ni > Cd > Co > Pb and in sediments, the order was Mn > Fe > Cu > Zn > other metals. In water uniformly all the metals recorded higher values after tsunami. In sediments, Zn and Cu recorded higher values whereas Fe and Mn registered lower values after tsunami. After tsunami, in water, the highest recorded value among heavy metals was Mn; the lowest recorded value among metals was Pb and it varied from 419.64 µg/l to 496.76 µg/l and 0.005 µg/l to 0.009 µg/l respectively. After tsunami, in sediments, the highest recorded value among heavy metals was Mn; the lowest recorded value among metals was Zn and it varied from 694.9 µg/g to 786.3 µg/g and 7.3 µg/g to 11.1 µg/g respectively. The concentration of Ni, Cd, Co and Pb were not found before and after tsunami in sediments. Consistently, in all the six stations studied, in the two samples, metal concentrations showed significant variations after tsunami in Nagapattinam coasts.

Keywords: *Tsunami, Water, Sediments, Heavy metals, Nagapattinam coasts, India.*

Introduction

A study on distribution of heavy metals in seawater is very important to understand the role in various biogeochemical process of the sea. Significant contributions have been made with reference to oceanic and coastal distribution of various heavy metals (Govindasamy and Azariah, 1999). Further studies reveal that heavy metals are biologically nondegradable and through food chain they may finally pass on to man (Thomas and Jaquet, 1976). Assessment of metals in water and sediment has become necessary enabling us to draw certain baseline information for monitoring, metals like Fe, Zn, Cu and Mn which are common toxic pollutants for fish (Srivastava, 1985). Further, the tsunami that struck the Indian coasts on 26th December 2004 with a speed of more than 700 km/h caused considerable changes in the marine ecosystem by polluting the marine environment with high turbidity and also breaking the coastal vegetations (Kannan *et al.*, 2005). Considering these facts, the present study is aimed to determine the accumulation of heavy metals in water and sediments from six coastal stations of Nagapattinam district of the Bay of Bengal, south east coast of India and to compare these values with pre tsunami data.

Description of the Study Areas

The study areas are Station 1 (Velankanni) situated 10 kilometers south of Nagapattinam, Station 2 is Nagapattinam, Station 3 (Vanjur) is situated 12 kilometers away from Nagapattinam in the north, Station 4 is Karaikal located 10 kilometers towards north of Station 3, Station 5 (Tharangampadi) is situated 13 kilometers north from Station 4 and Station 6 is Poombuhar situated 15 kilometers away from Station 5 towards the north.

Materials and Methods

Samples of water and sediments were collected from six stations of Nagapattinam coasts on 28th November 2004 (Pre-tsunami) and 26th January 2005 (Post-tsunami) for the estimation of various heavy metals.

Water (Brooks *et at.*, 1967)

Surface water samples were collected in pre cleaned and acid washed polypropylene bottles and they were filtered in Millipore filter paper (mesh size 0.45µ). The samples were preconcentrated with APDC-MIBK extraction procedure (Brooks *et al.*, 1967). The resulting solution was aspirated to the flame atomic absorption Spectrophotometer (Perkin-Elmer Model 373) for the determination of Zn, Cu, Fe, Mn, Co, Pb, Cd and Ni.

Sediment (Chester and Hughes, 1967)

Sediment samples were collected in pre cleaned, acid washed PVC corer, transferred to clean poly bags and transported to the laboratory. These samples were then washed with metal free double distilled water and dried in hot air oven at 110 C for 5 to 6 hours and ground to powder in a glass mortar and stored in pre cleaned polythene bags. 500 mg of the sample was taken and digested with a mixture of 1 ml of con. H_2SO_4, 5 ml of con. HNO_3 and 2 ml of $HClO_4$. A few drops of HP was added to achieve complete digestion and the sample was filtered and made up to 25 ml with metal free double distilled water for the estimation of Zn, Cu, Fe, Mn, Co, Pb, Cd and Ni using flame atomic absorption spectrophotometer (Chester and Hughes, 1967).

Results

Variation in the accumulation of Zn, Cu, Fe, Mn, Co, Pb, Cd and Ni in water and sediments recorded at the six stations of the Nagapattinam coasts before and after tsunami are given in Tables 43.1 and 43.2 and the Figures are shown in 43.1 to 43.12.

Table 43.1: Heavy Metal Concentration (µg.l^{-1}) Recorded in Water Samples of the Six Stations of Nagapattinam Coasts Before and After Tsunami

Heavy Metals	*Station–1*		*Station–2*		*Station–3*		*Station–4*		*Station–5*		*Station–6*	
	BT	*AT*	*BT*	*AT*	*BT*	*AT*	*BT*	*AT*	*BT*	*AT*	*BT*	*AT*
Zn (µg.l^{-1})	55.36	57.45	49.56	50.78	48.76	55.45	48.79	50.48	49.21	52.48	49.66	54.25
Cu (µg.l^{-1})	14.58	15.67	12.89	13.69	11.03	13.45	11.09	12.49	12.42	14.56	13.01	15.65
Fe (µg.l^{-1})	236.48	248.69	313.45	325.64	298.32	315.48	274.51	289.78	296.3	314.25	307.71	325.96
Mn (µg.l^{-1})	416.26	425.78	456.26	468.23	432.85	486.52	402.4	419.64	514.23	496.78	506.9	489.52
Co (µg.l^{-1})	0.06	0.07	0.05	0.06	0.06	0.08	0.04	0.06	0.03	0.06	0.04	0.07
Pb (µg.l^{-1})	0.002	0.007	0.001	0.005	0.002	0.005	0.002	0.004	0.002	0.005	0.006	0.009
Cd (µg.l^{-1})	0.05	0.06	0.04	0.09	0.04	0.06	0.03	0.09	0.09	0.1	0.06	0.08
Ni (µg.l^{-1})	0.26	0.34	0.15	0.27	0.11	0.29	0.26	0.29	0.14	0.28	0.16	0.24

BT: Before Tsunami–28th November 2004; AT: After Tsunami–26th January 2005

Station 1: Velankanni; Station 2: Nagapattinam; Station 3: Vanjur; Station 4: Karaikal; Station 5: Tharangampadi; Station 6: Poombuhar.

Table 43.2: Heavy Metal Concentration (µg.l^{-1}) Recorded in Sediment Samples of the Six Stations of Nagapattinam Coasts Before and After Tsunami

Heavy Metals	*Station–1*		*Station–2*		*Station–3*		*Station–4*		*Station–5*		*Station–6*	
	BT	*AT*	*BT*	*AT*	*BT*	*AT*	*BT*	*AT*	*BT*	*AT*	*BT*	*AT*
Zn (µg.l^{-1})	5	10.6	8.1	9.7	7.8	8.4	4	11.1	8	9.6	5.9	7.3
CU(µg.l^{-1})	15	51.2	41	49.8	18	48.6	20	52.8	30	48.4	31.4	47.9
Fe (µg.l^{-1})	277	130.09	397	129.5	230	135.7	2t5	124.49	270	128.5	257.4	145.6
Mn (µg.l^{-1})	998	694.9	957	745.6	926	786.3	401	704.9	820	724.9	851.1	771.6
CO (µg.l^{-1})	ND	ND	ND	ND	ND	ND	ND	ND	ND	ND	ND	ND
Pb (µg.l^{-1})	ND	ND	ND	ND	ND	ND	ND	ND	ND	ND	ND	ND
Cd (µg.l^{-1})	ND	ND	ND	ND	ND	ND	ND	ND	ND	ND	ND	ND
Ni (µg.l^{-1})	ND	ND	ND	ND	ND	ND	ND	ND	ND	ND	ND	ND

BT: Before Tsunami–28th November 2004; AT: After Tsunami–26th January 2005.

ND: Not Detectable

Station 1: Velankanni; Station 2: Nagapattinam; Station 3: Vanjur; Station 4: Karaikal; Station 5: Tharangampadi; Station 6: Poombuhar.

Discussion

In water, the accumulation of heavy metals was observed in the order of Mn > Fe > Zn > Cu > Ni > Cd > Co > Pb. After tsunami, the highest recorded value among heavy metals was Mn; the lowest recorded value among metals was Pb and it varied from 419.64 µg/l to 496.76 µg/l and 0.005 µg/l to

WATER

Heavy Metal Concentration (µg.l^{-1}) Recorded in Water Samples of the Six Stations of Nagapattinam Coasts Before and After Tsunami

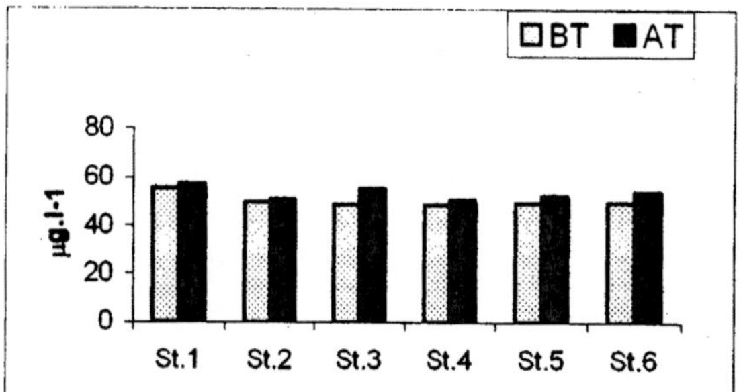

Figure 43.1: Zn Concentration in Water

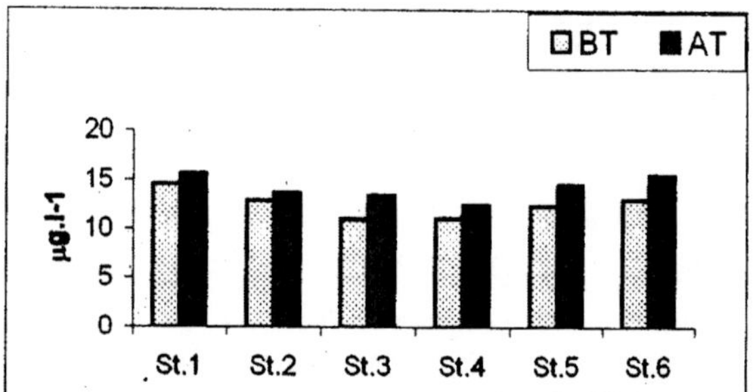

Figure 43.2: Cu Concentration in Water

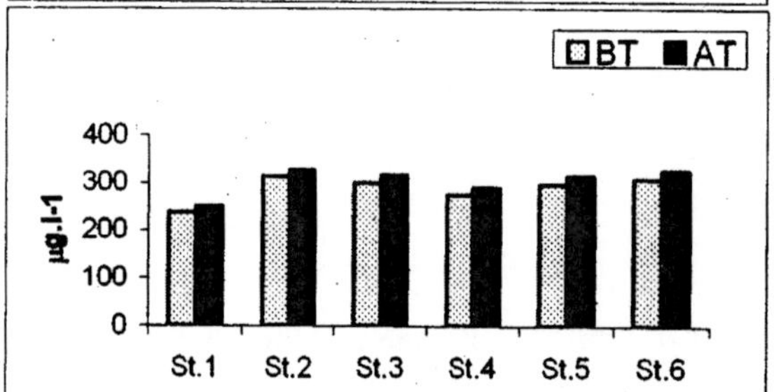

Figure 43.3: Fe Concentration in Water

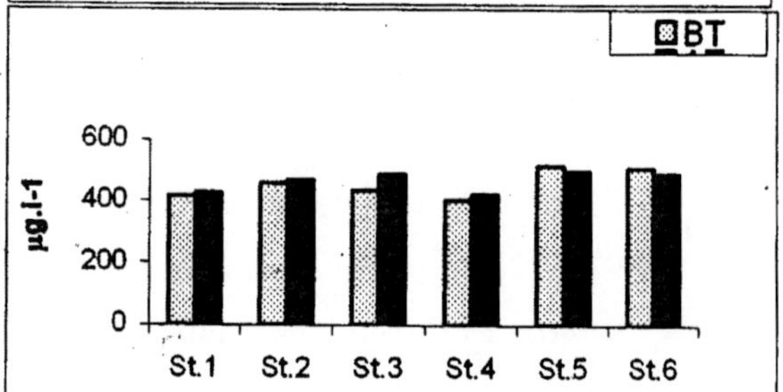

Figure 43.4: Mn Concentration in Water

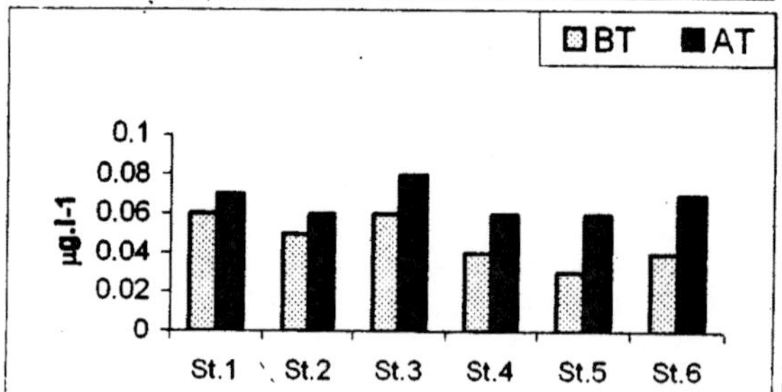

Figure 43.5: Co Concentration in Water

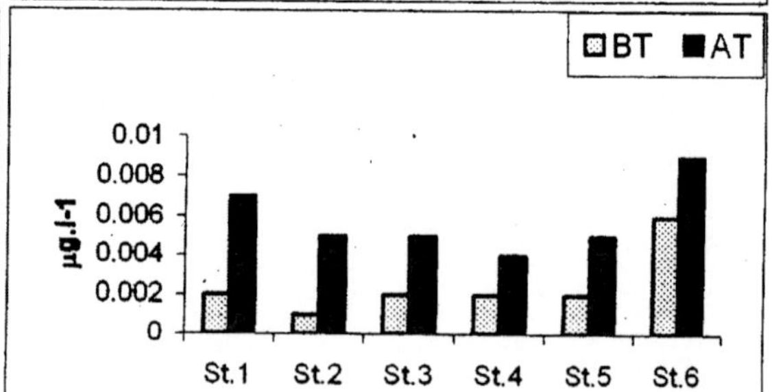

Figure 43.6: Pb Concentration in Water

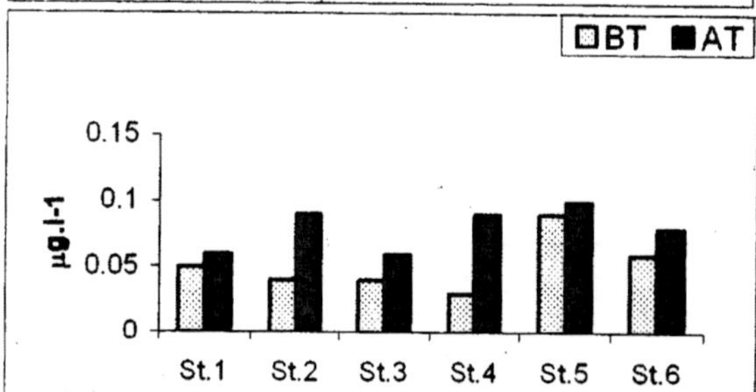

Figure 43.7: Cd Concentration in Water

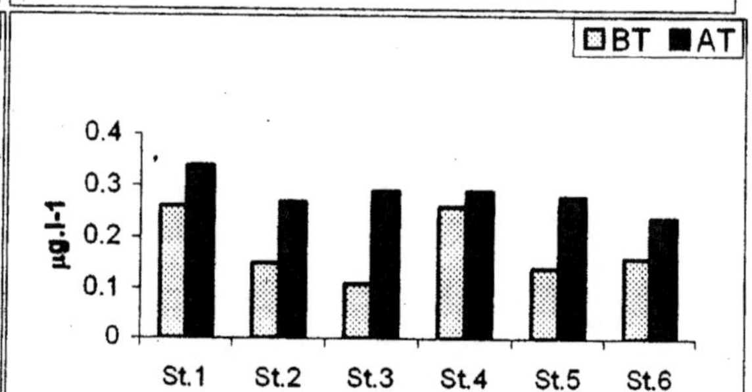

Figure 43.8: Ni Concentration in Water

SEDIMENTS

Heavy Metal Concentration (µg.l⁻¹) Recorded in Water Samples of the Six Stations of Nagapattinam Coasts Before and After Tsunami

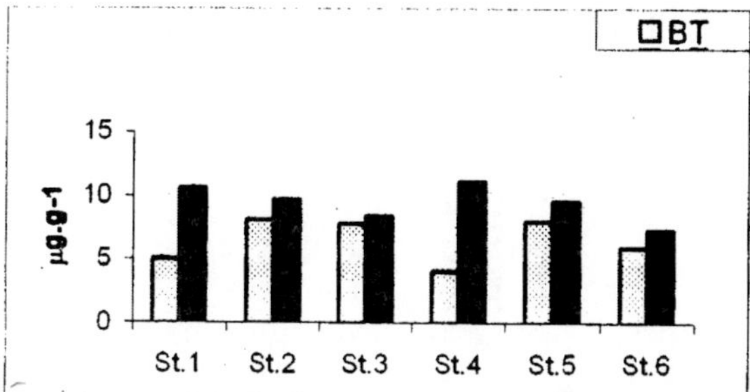

Figure 43.9: Zn Concentration in Sediment

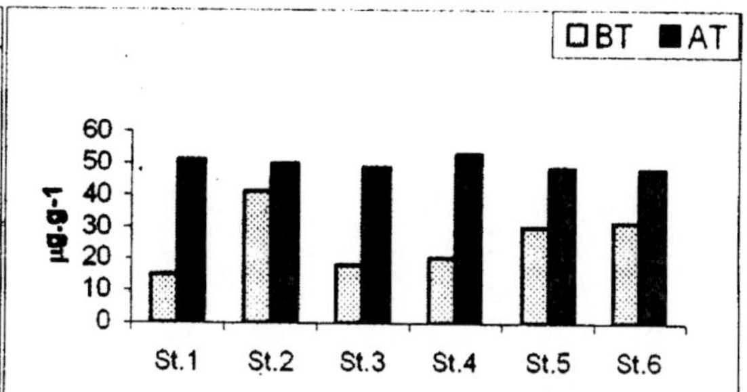

Figure 43.10: Cu Concentration in Sediment

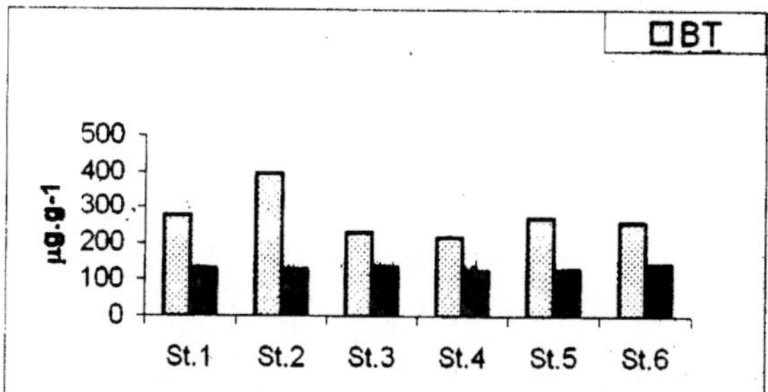

Figure 43.11: Fe Concentration in Sediment

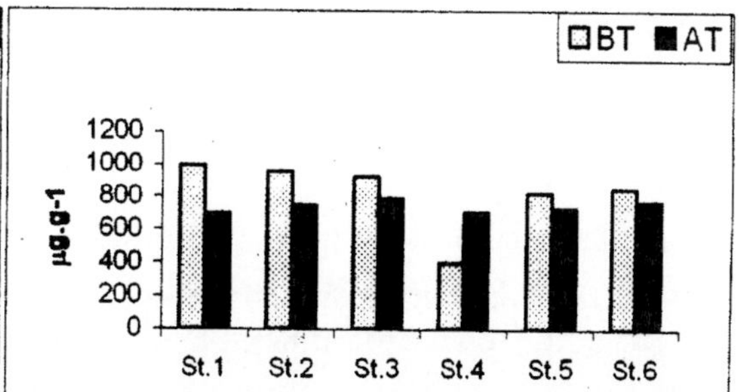

Figure 43.12: Mn Concentration in Sediment

0.009 µg/l respectively. Except Mn all the other metals (Zn, Cu, Fe, Cd, Cr, and Pb) recorded higher concentrations after tsunami. Stations 1, 2, 3 and 4 recorded higher concentrations whereas stations 5 and 6 recorded lower concentrations after tsunami. This variation in metal concentrations is due to the impact of tsunami that caused large scale seawater inundation and the receding tidal waves carried into the sea, debris, anthropogenic wastes, adjacent terrestrial parts including plastic materials, domestic and shrimp farm disposals from the nearby lands. Further, the heavy metals were higher in coastal waters due to the presence of major sources of metal pollution, intensive human activity and discharge of municipal waste and industrial effluents (Govindasamy and Azariah, 1999).

In sediments, the accumulation of heavy metals was observed in the order of Mn > Fe > Cu >, Zn > other metals. Zn and Cu recorded higher values whereas Fe and Mn registered lower-values after tsunami. After tsunami, in sediments, the highest recorded value among heavy metals was Mn; the lowest recorded value among metals was Zn and it varied from 694.9 µg/g to 786.3 µg/g and 7.3 µg/g to 11.1 µg/g respectively. The concentration of Ni, Cd, Co and Pb were not found before and after tsunami. In sediments, metal concentrations are mainly controlled by river inflow, sediment particle size, type and organic matter content which influence the absorption and accumulation of metals (Armstrong *et al.*, 1976; Leland *et al.*, 1973 and Loring. 1978). Further, the variation in the heavy metal concentrations in sediments in this study was due to the continuous stirring of sediments by the high speed of tsunami tidal actions.

Conclusion

In general, the present study clearly showed significant variations in the accumulation of heavy metals in water and sediments after tsunami in all the six stations. These variations in the marine environment is certainly brought about by the recent tsunami as similar type of variations in the physico-chemical characters were observed in the coastal water quality off Dakshina Kannada coast (Reddy *et al.*, 2005). The tsunami brought large-scale seawater inundation and created significant variations in the marine environment. The receding waves after the tsunami washed into the sea debris, adjacent terrestrial parts including plastic materials, sediment deposition, industrial effluents, metals and. anthropogenic wastes from the land into the ocean and this could have brought about the considerable variations in the metals observed. It may be expected that the high concentrations of metals in water can be gradually accumulated on the sediments and in due course it may get transferred to biota and thereby it may result in some major changes in the coastal bio diversity in the near future.

References

Anandan, G. 1994. Plankton ecology and heavy metals studies in the marine environs of Pondicherry, India. *Ph.D. Thesis*, Annamalai University, India, pp. 125.

Armstrong, P.B., G.M. Hanson and H.E. Gandette, 1976. Minor Elements of Great Bay estuary, New Hampshire. *Environ. Geol.*, 1: 207–214.

Brooks, R.R., B.J. Presley and I.R. Kalplan, 1967. APDC-M/BK extraction system for the determination of trace metal in saline water by atomic absorption spectroscopy, Talanta, 14: 809–816.

Brulhet, J., 1975. Observations on the biology of Mugil *cephalus ashentensis* and the possibility of its aquaculture on the Mauritania coast. *Aquaculture*, 5: 271–281.

Chester, R. and M.J. Hughes, 1967. A chemical technique for the separation of ferro manganese materials, carbonate minerals and absorbed trace elements from pelagic sediments. *Chem. Geol.*, 2: 249–263.

Cooper, J.J., 1983. Total mercury in fishes and the selected biota in Lohotan Reservoir, Navada. *Bull. Environ. Conton. Toxicol.*, 31: 9–17.

Govindasamy, C. and J. Azariah, 1999. Seasonal variation of heavy metals in coastal water of the Coromandal coast, Bay of Bengal, India. *Indian. J. Mar. Sci.*, 28: 249–256.

Kannan, L.T. Thangaradjou and A. Duraisamy, 2005. Effects of tsunami on the coastal and marine biodiversity of India. In: *Proceedings of the Brain Storming Session on Tsunami Mitigation Strategies*, held at Bharathidasan University, 25–26th February, 2005.

Leland, H.V., S.S. Shukla and N.F. Shrimp, 1973. Factors affecting distribution of lead and other trace elements in sediments of southern lake Michigan. In: *Trace Metals and Metal Organic Interactions in Natural Waters*, (Ed.) P.C. Singei, pp. 89–129.

Loring, V.H. 1978. In: *Environmental Biogeochemistry and Geo-microbiology, Methods, Metals and Assessment*. In: Proceeding of third international symposium on environmental biogeochemistry, (Ed.) W.E. Krumbein. Wolfenbuttal, W. Germany, and Ann Arbor. Science Publication. Inc., Ann Arbor. Michigan, pp. 1025–1040.

Reddy, H.R.V., Katti, R.J., Raveesha, K.P., Chandrashekara., Vikas, S.J., Nagendra Babu, R.S. and Shivakumar, K.S., 2005. Coastal water quality of Dakshina Kannada before and after tsunami. *Current Science*, 88(7): 1026–1027.

Srivastava, C.B.L., 1985. *A Textbook of Fishery Science and Indian Fisheries*. Kitab Mahal Publications, Allahabad, pp. 307.

Thomas, R.L., Jacquet, J.M., 1976. Distribution of heavy metals on the fish Mugil cephalus. *Fish Res. Board*. Canada, 33: 404.

Topping, G., 1973. Heavy metals in fish from Scottish waters. *Aquaculture*, 1: 373–377.

Chapter 44

A General Survey of Groundwater for Physico-chemical Investigations in Churu City (Rajasthan)

R.V. Singh[1], Dinesh Kumar[1] and Mukta Jain[2]
[1]Department of Chemistry, University of Rajasthan, Jaipur – 302 004, India, E-mail: kudiwal@datainfosys.net
[2]Department of Chemistry, Seth G.B. Podar College, Navalgarh

ABSTRACT

The present analyses were carried out to evaluate the magnitude of physio-chemical parameters of the groundwater of the Churu district and a comparative study has been made for the monsoon and pre monsoon seasons–2004. The sampling points were selected on the basis of their importance. The groundwater samples were collected at eight different sites. The analyses were carried out for the parameters such as temperature, pH, DO, BOD, COD, total alkalinity, total hardness, calcium, magnesium, carbonate, non-carbonate hardness, chloride, fluoride, nitrate, TS, TDS and TSS. From the observations, it may be inferred that fluoride and nitrate levels are unacceptable in drinking water samples taken from Churu city.

Keywords: *General survey, Drinking water, Monsoon and Pre-monsoon season, Parameters, Churu city and Water pollution.*

Introduction

Water is a vital factor in life, which constitutes 80 per cent of protoplasm 'a physical basis of life'. The human body contains 70 per cent of body weight in the form of water. The history of man has always been marked by an intimate relationship with water. It is well known that the earliest paleolithic implements of human existance have been traced in the river gravels. Water is the basis of all life forms and absolute necessity comprising domestic, agriculture and industrial activities. Even in this modern scientific era mankind has to depend on water for countless direct and indirect ways.

India is today one of the first ten industrial countries of the World. This is the result of industrial transformation in the post independence period. Today, we have a good infrastructure in the case of industries like metals, chemicals, fertilizers, automobiles, textiles, petroleum and food. India's utilisation quantum of water in the surface water sources is estimated to be approximately 70 million hectare metres (Tyagi *et al.*, 2003).

Materials and Methods

Study Area

Churu district is located in the Middle East. Churu is situated on the northeastern corner of the Rajasthan state and covers an area approximately 16,830 km^2, extending between latitudes 72 24'N and 29 0'N and longitudes 73 40'E and 75 41'E. Like most parts of the rural shield, a desert type of environment with dry hot weather characterizes the country. Very hot dry summers and cool to mild rainy winters characterize the climate of Churu. For most of the year, the weather in Churu is hot and dry. However, in summer, average temperature ranges between 30 and 50 C which in the winter fall below 10 C. Now-a-days this city is attracting different people from India and Abroad for tourism purposes. However, the quality of water in this city seems to be of lower quality. Therefore, a general survey has been carried out to point out the negative points of the water quality of this city.

Samples were collected in polythene bottles of good quality of two litre capacity and brought to the laboratory. Some parameters like: temperature, pH, and DO, were measured on the spot at the time of samples collection using water portable kit. For the analysis of other parameters, samples were kept by adding suitable reagents.

Temperature was measured by thermometer, pH was determined using digital pH meter, DO was fixed and measured by titrimetre method. Total solid and total dissolved solid were determined by evaporation of water sample in known volume. BOD was measured by titrimetric method after 5 days incubation in BOD incubator at 20 C temperature and COD was measured by reflux method. Total alkalinity, total hardness and chloride by titrimetric method. Anions fluoride and nitrate were estimated by ion selective electrode method. Reagents used for the present analysis were AR grade and double distilled water was used for preparing various solutions.

Results and Discussion

The results obtained from the analyses of physico-chemical parameters of pre-monsoon, 2004 groundwater collected from Churu district is shown in the Table 44.1. The results obtained from the analysis of physico-chemical parameters of during monsoon, 2004 groundwater collected from same sites of Churu districts in the Table 44.2.

Temperature

The maximum temperature (32.6°C) was found during pre-monsoon and minimum temperature (27.6 C) was found during monsoon.

pH

The pH of pre-monsoon of all groundwater samples was less than during monsoon groundwater samples. It shows slightly alkalinity of water of during the monsoon season.

Table 44.1: The Physico-chemical Parameters of Various Sampling Stations During Pre-monsoon (16 June, 2004)

Parameter	*Govt. ENT Hosp. (BW)*	*P. Chaudhary (TW)*	*Govindka School (TW)*	*Gadh (OW)*	*Bharatia Wala Kulm (TW)*	*Railway Platform (BW)*	*Civil Colony (TW)*	*Hanuman Gadhi (TW)*
Temperature °C	32.1	32.3	32.4	32.2	32.5	32.6	32.4	32.5
pH	7.6	7.4	7.5	7.4	7.6	7.8	7.9	7.8
Total Alkalinity (mg/L)	610	600	610	210	320	500	560	450
Total Hardness (mg/L)	290	340	330	720	350	380	320	550
Calcium Hardness (mg/L)	80	130	120	270	140	150	130	210
Magnesium Hardness (mg/L)	210	210	210	450	210	230	190	340
Carbonate Hardness (mg/L)	290	340	330	210	320	380	320	450
Non-carbonate Hardness (mg/L)	Nil	Nil	Nil	510	30	Nil	Nil	100
Chloride (mg/L)	720	900	820	800	780	850	900	1360
Nitrate (mg/L)	72	66	304	676	242	156	70	112
Fluoride (mg/L)	4.5	3.5	4.0	1.5	4.0	3.8	3.2	2.0
DO (mg/L)	6.9	6.4	2.9	4.6	7.2	5.9	6.1	6.8
BOD (mg/L)	0.4	Nil	0.2	Nil	Nil	Nil	Nil	Nil
COD (mg/L)	5.6	3.2	1.6	14.4	Nil	Nil	Nil	14.4
Total Dissolved Solid (mg/L)	2620	2960	3360	3200	3040	2880	3040	3680
Total Solid (mg/L)	2823	3220	3650	3455	3248	3100	3258	4000
Total Suspended Solid (mg/L)	203	260	290	255	208	220	218	320

Dissolve Oxygen (DO)

DO is one of the most important parameters in assessing water quality and reflects the physical and biological process prevailing in the drinking water. A good water should have the solubility of oxygen 7.6 and 7.0 mg/L at 30 C and 35 C, respectively proposed by Kudesia (1985).

DO of the during monsoon vary slightly and found more than in pre-monsoon of groundwater samples. These high values indicated that the solubility of oxygen increases with the decrease of temperature. But site numbers 3 and 4 have low values of DO. It may be due to biological impurities. Comparative studies have been shown in Figure 44.1.

Biochemical Oxygen Demand (BOD)

The BOD of all groundwater samples are under considerable range in both the seasons.

Chemical Oxygen Demand (COD)

The values of this parameter for all the groundwater samples were beyond the permissible value except sample site 4 and site 8 in both seasons. These exceptional high values show that groundwater samples were rich either with respect to organic compounds or oxidisable substances.

Total Alkalinity

Total alkalinity of the during monsoon samples is less than pre-monsoon of groundwater samples except samples site 4 and site 8. These exceptional high values show water samples contain high concentration of bicarbonate and hydroxyl ions and microbial decomposition of organic matter in the water body observed by Abbasi *et al.* (1999).

Table 44.2: The Physico-chemical Parameters of Various Sampling Stations During Monsoon (21 October, 2004)

Parameter	*Govt. ENT Hosp. (BW)*	*P. Chaudhary (TW)*	*Govindka School (TW)*	*Gadh (OW)*	*Bharatia Wala Kulm (TW)*	*Railway Platform (BW)*	*Civil Colony (TW)*	*Hanuman Gadhi (TW)*
Temperature °C	27.8	27.6	27.9	27.6	27.8	27.7	27.9	27.6
pH	8.1	8.0	7.9	7.7	8.0	82	8.2	8.4
Total Alkalinity (mg/L)	250	350	400	220	220	300	300	650
Total Hardness (mg/L)	340	360	400	1030	480	520	380	640
Calcium Hardness (mg/L)	200	130	140	240	200	290	100	190
Magnesium Hardness (mg/L)	140	230	260	790	280	230	280	450
Carbonate Hardness (mg/L)	250	350	400	220	220	300	300	640
Non Carbonate Hardness (mg/L)	90	10	Nil	810	260	220	80	Nil
Chloride (mg/L)	790	670	870	820	720	880	900	700
Nitrate (mg/L)	140	132	340	830	295	246	160	194
Fluoride (mg/L)	3.35	3.40	3.85	0.321	3.30	3.47	3.40	1.76
DO (mg/L)	7.2	6.8	3.2	4.9	7.8	6.5	6.5	7.2
BOD (mg/L)	0.6	Nil	0,3	Nil	Nil	Nil	Nil	Nil
COD (mg/L)	4.2	2.1	1.9	9.6	Nil	Nil	Nil	10.5
Total Dissolved Solid (mg/L)	1975	1675	2175	2050	1800	2200	2250	1750
Total Solid (mg/L)	2080	1755	2310	2170	1895	2350	2415	1839
Total Suspended Solid (mg/L)	105	80	135	120	95	150	165	89

Figure 44.1: DO in mg/L for the Water Samples from Churu District

Total Hardness

The total hardness of the during monsoon samples is more than pre-monsoon of groundwater samples except site 3. The high values of total hardness at site 4 and site 8 is due to the cations like calcium and magnesium. The presence of toxic heavy metals in the water may also cause high hardness in water. However, there is no firm evidence suggesting drinking of hard water cause any adverse effect on health (Doctor *et al.*, 1998).

Chloride

The chloride concentration was at toxic level in all the points in pre-monsoon and monsoon seasons samples. Comparative study of both seasons have been shown in Figure 44.2.

Nitrate

Nitrate pollution is a major problem in Rajasthan at present. Nitrate in groundwater samples ranged between 66 mg/L to 676 mg/L during pre-monsoon season and ranged from 132 mg/L to 830 mg/L during monsoon. The obtained values of nitrate were found higher than the permissible limit (100 mg/L) prescribed by HIS. Higher value of nitrate in the study area may be due to excessive use of nitrogen containing fertilizers according to Majumdar *et al.* (2000).

The observed values of nitrate and comparative studies during pre-monsoon and monsoon seasons have been shown in Figure 44.3.

Fluoride

The recorded values of fluoride were vary between 1.5 mg/L to 4.5 mg/L during pre-monsoon season and ranged from 0.32 mg/L to 3.47 mg/L during monsoon season in the groundwater samples. The fluoride concentration in all the groundwater samples (except site No. 4) were observed above the permissible limits (1.5 mg/L). Due to high concentration of fluoride in groundwater, skeletal and dental fluorosis, deformation of knee, joints and other parts of body occurs in persons (Madhavan and Chouhisa *et al.*, 2001) who live in these areas. Comparative study of both the seasons has been shown in Figure 44.4.

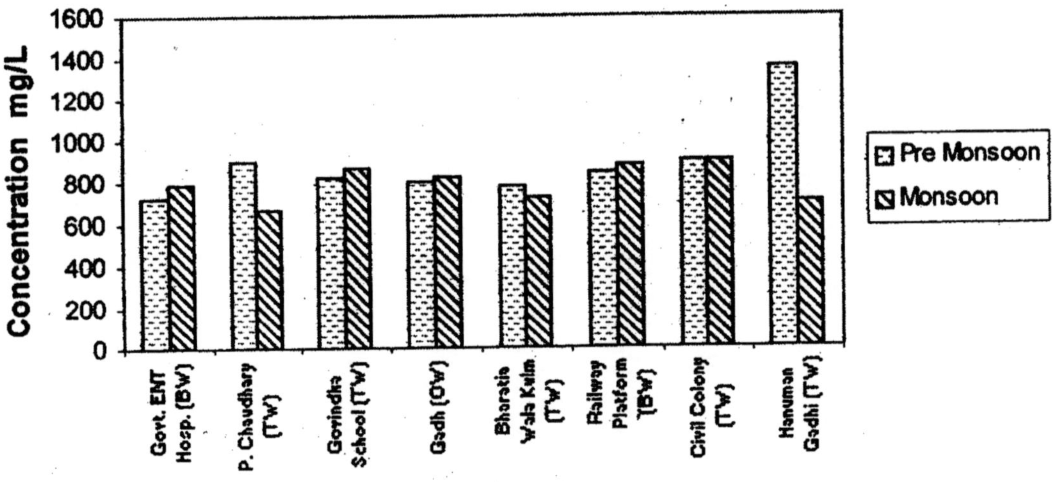

Figure 44.2: Chloride in mg/L for the Water Samples from Churu District

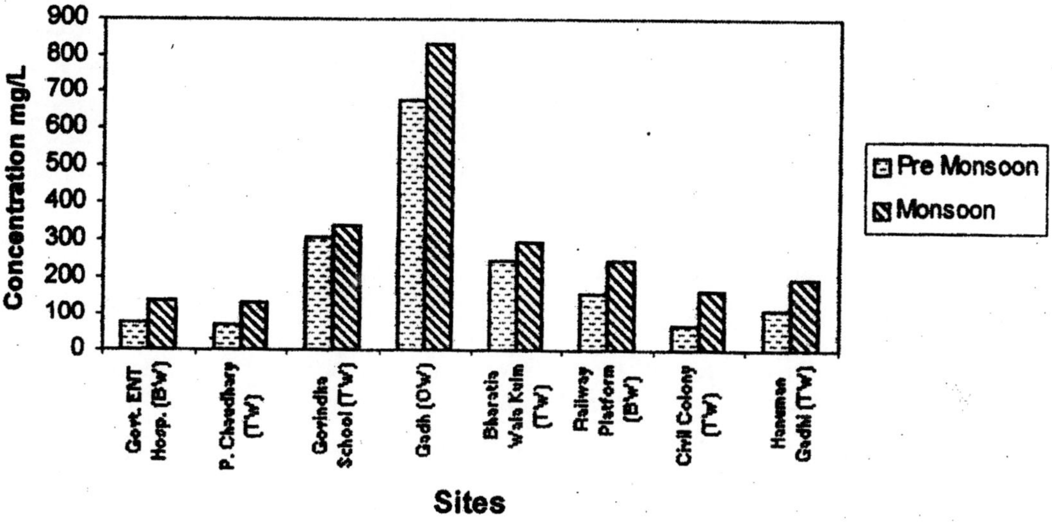

Figure 44.3: Nitrate in mg/L for the Water Samples from Churu District

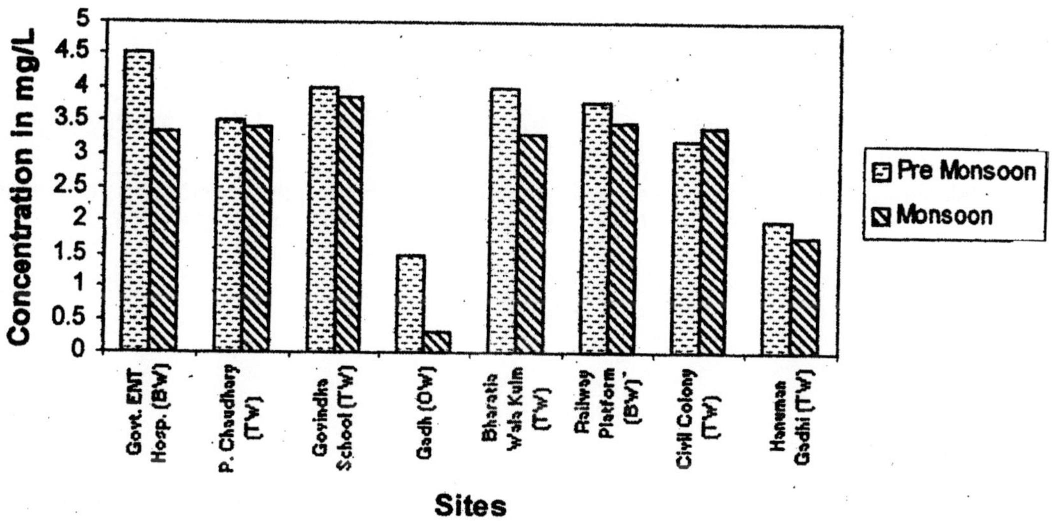

Figure 44.4: Fluoride in mg/L for the Water Samples from Churu District

Total Solids

In the present study, contents of total solids in groundwater range from 1755 mg/L to 2415 mg/L during monsoon and range from 2823 mg/L to 4000 mg/L during pre-monsoon.

Total Dissolved Solids

The monitored values of TDS of groundwater samples of Churu district were found higher than the permissible limit. This may be due to the presence of dissolved organics in water discharged by domestic effluents.

Conclusions

Excess Fluoride ingestion in children has resulted in dental and skeletal fluorosis. Fluorosis disease will increase in children and may be found in adults in near future of rural areas. Based on the results, it is concluded that due to the use of raw vermiculite, besides SiO_2 and Al_2O_3, Fe_2O_3, CaO, and MgO are also present in considerable amounts.

Nitrate concentration at sample site No. 4 is much higher than permissible limit prescribed by WHO. Thus the use of the water of such places should be avoided for drinking and other purposes without any pre-treatment. The authors have also suggested to the residents of these places to use technique for denitrification of drinking water.

References

Abbasi, S.A. and Vinithan, S., 1999. Water quality in and around an industrialized suburb of Pondicherry, *Indian J. Environ. Hlth.*, 39(4): 265–273.

Chouhisa, S.L. and Leela, C., 2001. Endemie fluorosis in Rajasthan. *Indian J. Environ. Hlth.*, 43: 177.

Doctor, P.B., Paiyani, C.V., Desai, N.M., Kulkarni, P.K., Ruparelia and Ghosh, S.K., 1998. Physico-chemical and microbial analysis of dye contaminated river water. *Indian J. Environ. Hlth.*, 40(1): 7–14.

Kudesia, V.P., 1985. *Water Pollution*. Pragati Prakashan, Meerut.

Madhavan, N. and Subramanyam, V., 2001. Fluoride concentration in river water of South Asia. *Current Science*, 80: 1312.

Majumdar, D. and Gupta, N., 2000. Nitrate pollution of groundwater and associated with human health disorders. *Indian J. Environ.*, 42: 28.

Tyagi, A., Bhotiak, K.S and Sharma, 2003. The study of temporal and spatial trends of water quality of river Kshipra using water quality index. *Indian Environ. Hlth.*, 45(1): 15–20.

Chapter 45

Population Fluctuation of Soil Mites in Relation to Some Important Abiotic Factors in the Pine Forest Ecosystem, Manipur, N.E. India

Kh. Latasana Devi and Th. Binoy Singh
Department of Life Sciences, Manipur University, Imphal – 795 003

ABSTRACT

Soil mites are one of the important group inhabiting in the soil ecosystem that plays important roles in the nutrient cycling as well as in the decomposition process and maintenance of soil fertility. The distribution pattern of these mites in soil is limited by various biotic and abiotic factors. The present investigation was conducted to determine the distributional pattern of soil mites in the two pine forest ecosystems at Gwaltabi forest range, Manipur. The study site lies between 24°52′ and 25°22′ North latitude and 94°2′ and 94°9′ East longitude. The soil samples were collected with the help of soil corer from three different soil depths *i.e.* 0–10, 10–20, 20–10 cm for a period of one year and the soil animals were extracted by modified Tullgren funnels. Maximum total population density of mites was recorded in July and minimum in May in both the study sites. Among the soil mites, oribatid exhibited the dominant group. The highest value of Sorensen's (Q/S) was observed during rainy season and followed by winter and summer seasons.

Keywords: *Soil mite, Oribatid, Tullgren Funnel, Community similarity, Population fluctuation, Abiotic factors.*

Introduction

Soil ecosystem supports a complex of animal communities of which microarthropods are of prime importance since they constitute the major component of soil mesofauna in all types of soils and these organisms have significant influences on soil microbial ecology, nutrient cycling and soil fertility.

Among the arthropods, mites are often the dominant group in the forest floors and trees (Behan Pelletier and Walter, 2000). Mites are the rival insects in the diversity of living habits and niches occupied (Lindquist *et al.*, 1979) and they are inhabitants of all kinds of soil, from extremely acid to alkaline and from nutrient poor to nutrient rich. Still, investigations of mite diversity have largely been limited to the litter and the top part of the hemiorganic horizon (Andre *et al.*, 2002). The distribution pattern of these mites in soil is limited by various biotic and abiotic factors. Recently, Noti *et al.* (2003) showed that different factors acting at different scales influenced mite diversity, from the landscape level to *in situ* soil properties such as total nitrogen. In addition, soil physical and chemical characteristic were measured as potential determinants of mite density and diversity (Ducarme *et al.*, 2004). Among the suborder of mite, Cryptostigmata exhibited the most dominant group and Astigmata exhibited the least dominant group during the study period.

Materials and Methods

The present study has been carried out in two sites. Site 1 is located at reserve forest ecosystem of Gwaltabi dominated by *Pinus kesia* about 27 km away from Imphal lying between 24°52′ and 25 22′ north latitude and 94°2′ and 94°9′ east longitude and at an altitude ranging from 830 m to 1520 m above M.S.L. Site 2 is located at Napet Palli area in the same forest range, which is an open disturbed pine forest ecosystem.

Each study area were divided into three sub plots *i.e.* 1, 2 and 3 having a size of 10 10 m each in both the study sites. A total of about 648 soil samples were drawn from October 2001 to September 2002 at three different soil layers *i.e.* 0–10, 10–20 and 20–30 cm from both the study sites with the help of soil cores having a size of 5.5 cm in diameter and 23.76 cm^2 in surface area. The soil mites were extracted by modified Tullgren Funnel extractor for 5–7 days at the temperature range of 35–40°C depending on the moisture content of the soil samples and they were collected in vials containing 80 per cent ethanol with few drops of glycerine. The soil mites were identified and counted under a binocular microscope. These mites were classified into four suborders *i.e.* Cryptostigmata, Mesostigmata, Prostigmata and Astigmata. Moisture content of the soil was determined by gravimetric method as given by Wilde *et al.* (1985). Soil temperatures were measured for each soil layers during the sampling period. The soil organic carbon was estimated by oxidation calorimetric method after modified Walkley and Black method. The Cryptostigmatid mites of the two study sites were analysed to study the community similarity using the following statistical formula.

Community Similarity

The similarity or dissimilarities of cryptostigmatid mite communities in the protected and degraded pine forests were worked out using the following index.

The coefficient of similarity (CCS) of Sorensen (1948) also known as Quotient of Similarity (Q/S) was calculated as

$$Q/S = \frac{2C}{S_1 + S_2} \times 100$$

where,

C: Number of species common to both the communities

S_1: Number of species in community 1 (Protected site)

S_2: Number of species in community 2 (Degraded site)

Results and Discussion

The population fluctuation pattern of total soil mites (acarina) exhibited two peaks in the month of July and October in Site I whereas in Site II one peak exhibited in the month of July. The minimum population density exhibited in the month of May in both the study sites (Figure 45.1).

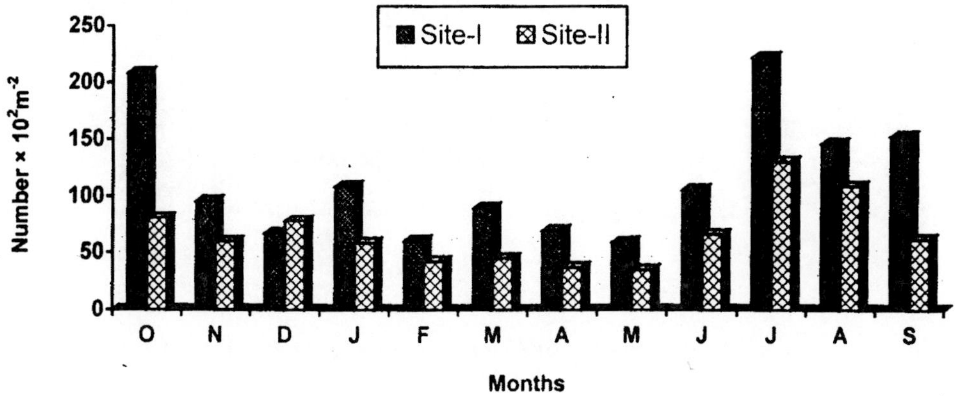

Figure 45.1: Monthly Fluctuation of Total Acarina Population Density in the Protected (Site 1) and Degraded (Site 2) Pine Forest Ecosystems (Number × 10^2m^{-2})

Fluctuation of the density of soil mites seems to involve several environmental factors (Figures 45.2 to 45.4). Of the various factors studied moisture content, organic carbon content, nitrogen content at the time of sampling showed significant positive correlation ($P < 0.05$) with the total soil mites population in both the study sites. Contrary to these, pH of the soil showed significant negative correlation with soil mites (Bhattacharya and Raychaudhuri, 1979). However, in the present investigation, the soil temperature showed insignificant correlation with soil mites population. Soil

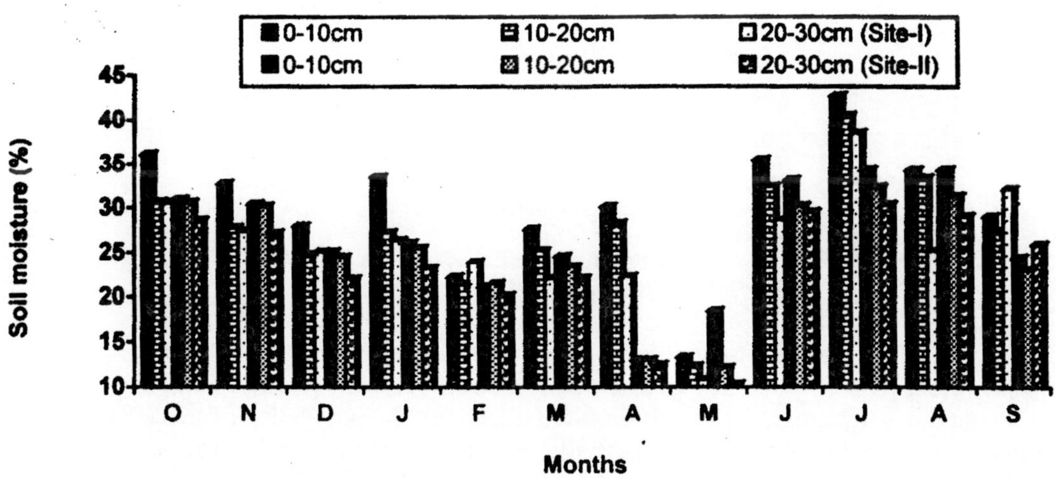

Figure 45.2: Monthly Variations of Soil Moisture (%) at the Three Soil Depths (cms) in Study Sites 1 and 2

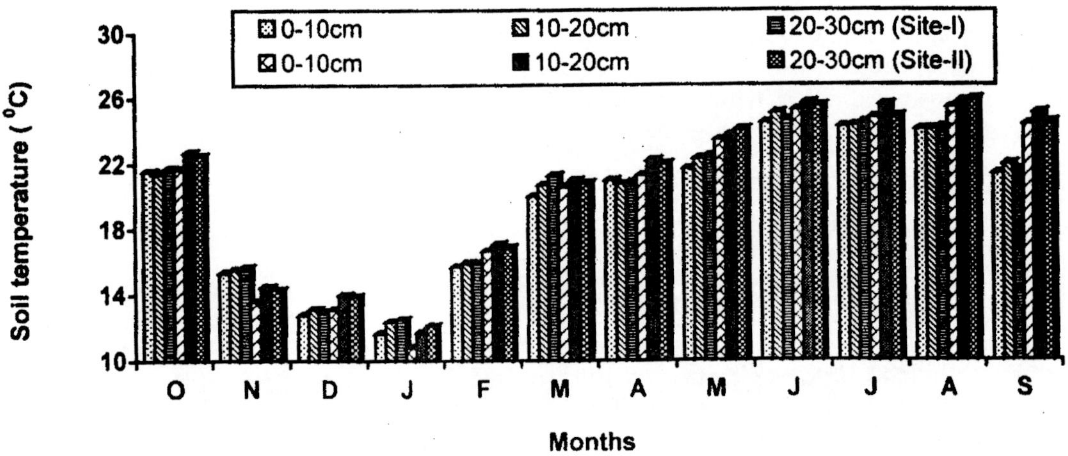

Figure 45.3: Monthly Variations of Soil Temperature (°C) at the Three Different Soil Depths (cms) in Study Sites 1 and 2

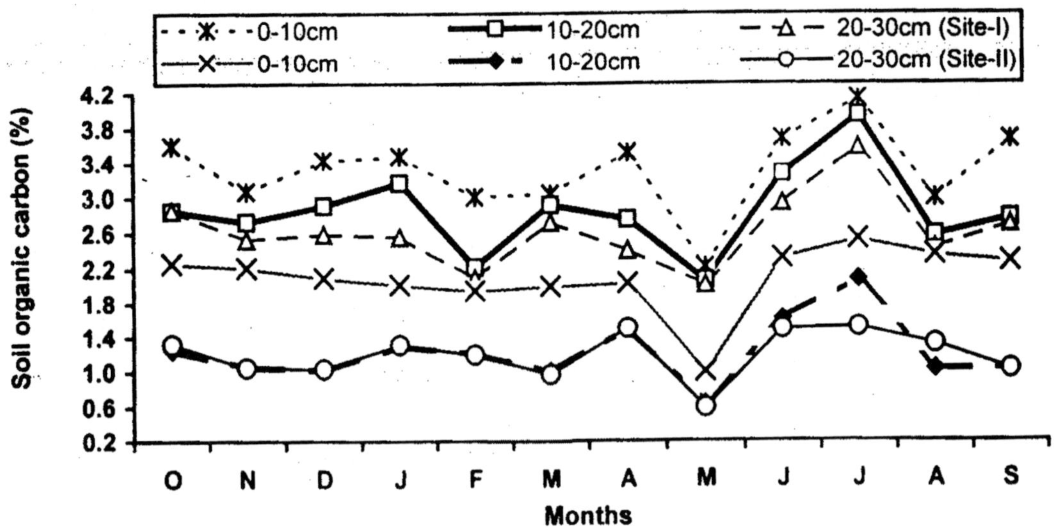

Figure 45.4: Monthly Variations of Soil Organic Carbon (%) at the Three Soil Depths (cms) in Study Sites 1 and 2

mites, mainly oribatids are less susceptible to temperature fluctuations perhaps because of the hard cuticular covering as has been suggested by Moursi and Hussain (1970). Seasonal population fluctuation trend of soil mites in two study sites were same *i.e.* the population decreased during the dry summer season and then increased gradually with the onset of the rainy season (Table 45.1). Maximum population density during the rainy season is due to the fact that all the favourable microclimatic conditions *viz.*, moisture content, organic carbon content were found to be maximum during this season which was revealed from the results whereas this favourable conditions were not available in the summer season. Similar findings of the present study was also recorded by Davis (1963).

Table 45.1: Seasonal Variation of Acarina Population Density in Site 1 (Protected Pine Forest) and Site 2 (Degraded Pine Forest Ecosystem) (Mean ± S.E.) × 10^2m^{-2}

Study Sites	*Seasons*	*Soil Layers (cm)*			*Total*
		0–10	*10–20*	*20–30*	
Site 1	Summer	26.19 ± 2.65	24.16 ± 0.74	22.76 ± 1.87	73.11 ± 5.26
	Rainy	64.92 ± 4.56	54.45 ± 4.42	48.36 ± 5.24	167.73 ± 14.22
	Winter	29.81 ± 2.54	27.71 ± 1.86	25.60 ± 2.77	83.12 ± 7.17
	Annual	43.54 ± 4.29	37.96 ± 3.46	34.37 ± 3.43	115.87 ± 11.18
Site 2	Summer	14.81 ± 0.72	11.53 ± 0.59	12.62 ± 0.47	38.96 ± 1.78
	Rainy	36.76 ± 4.30	25.81 ± 2.91	27.12 ± 2.19	89.69 ± 9.40
	Winter	21.04 ± 1.46	21.51 ± 2.27	16.95 ± 1.14	59.50 ± 4.87
	Annual	26.03 ± 2.67	20.81 ± 1.84	20.11 ± 1.60	66.95 ± 6.11

Among the soil mites, the Cryptostigmata (oribatid) were the most dominant group as well as the most diverse in species composition recorded during the present study in both sites and therefore more emphasis was given to this group. The dominant group, oribatid mite constituted 44-46 per cent whereas mesostigmata constituted 30–31 per cent, prostigmata 21–23 per cent and astigmata constituted only 2 per cent to the total soil mite population during the study period in both the study sites. Wallwork (1967) reported that the cryptostigmata occur in greatest numbers in coniferous forest soils where they may represent as much as 75 per cent to the total acarine fauna. The dominant character of Cryptostigmata as a group in most of the systems agrees with the reports of workers both from tropical and temperate regions (Usher, 1975; Luxton, 1982).

Cryptostigmata was represented by eight families belonging to eight genera in site 1 whereas in site 2 only seven families belonging to seven genera (Tables 45.2 and 45.3). In both the study sites *Oppia* sp. and *Scheloribates* sp. have been found as most abundant genera which was also reported by Choudhuri and Banerjee (1975). Abundance of these genera in different soil types may be ascribed to their tolerance to different ecological conditions.

Table 43.2: Seasonal Fluctuation of the Genera/Species of Oribatid Mites Belonging to their Families at the Depth of 0–30 cm in the Protected Pine Forest Ecosystem (Number × 10^2m^{-2})

Family	*Genus/Species*	*Seasons*			*Total*
		Summer	*Rainy*	*Winter*	
Chaunoproctidae	*Chaunoproctus* sp.	17	57	12	86
Scheloribatidae	*Scheloribates* sp.	18	82	31	131
Oppiidae	*Oppia* sp.	46	105	47	198
Haplozetidae	*Xylobates* sp.	17	29	15	61
Zetorchestidae	*Zetorchestes* sp.	–	34	–	34
Phthiracaridae	*Haplophorella* sp.	3	7	–	10
Galumnidae	*Galumna* sp.	–	8	4	12
Euphthiracaridae	*Rhysotritia* sp.	–	27	19	46

Table 45.3: Seasonal Fluctuation of the Genera/Species of Oribatid Mites Belonging to their Families at the Depth of 0–30 cm in the Degraded Pine Forest Ecosystem (Number × 10^2m^{-2})

Family	*Genus/Species*	*Seasons*			*Total*
		Summer	*Rainy*	*Winter*	
Chaunoproctidae	*Chaunoproctus* sp.	3	29	19	51
Scheloribatidae	*Scheloribates* sp.	18	47	28	93
Oppiidae	*Oppia* sp.	34	73	47	154
Haplozetidae	*Xylobates* sp.	–	8	5	13
Phthiracaridae	*Haplophorella* sp.	–	12	9	21
Euphthiracaridae	*Rhysotritia* sp.	7	12	–	19
Galumnidae	*Galumna* sp.	–	5	–	5

The present study revealed that the highest species diversity and abundance of oribatid mites were found in the protected than the degraded pine forest site. According to Badejo *et al.* (1999) the highest average density of oribatid mites were recorded in the forest reserve as in the present study. The highest value of Sorensen's Quotient (Q/S) was observed during rainy season and followed by winter and summer seasons (Table 45.4). This study indicated high degree of similarity in the rainy season and low similarity in the summer season.

Table 45.4: The Community Similarity of Total Oribatid Mites in Different Seasons for 0–30 cm Soil Layers

Seasons	*Sorensen's Coefficient (Q/S) %*	*Average Faunal Resemblance (%)*
Summer	66.67	67.5
Rainy	93.33	93.75
Winter	72.73	73.33

References

Andre, H.M., X. Ducarme and Ph. Lebrun, 2002. Soil biodiversity: Myth, reality or conning? *Oikos*, 96: 3–24.

Badejo, M.A., A.M. Makinde and N.M. Straalen, 1999. The Oribatid mite fauna of soil under different vegetation cover in Ile-Ife; Nigeria. *Tropical Ecology*, 40(2): 275–280.

Behan-Pelletier, V. and D.E. Walter, 2000. Biodiversity of oribatid mites (Acari: Oribatida) in tree canopies and litter. In: *Invertebrates as Webmasters in Ecosystems*, (Eds.) Coleman, D.C. and Hendrix, P.F. CAB International Wallingford, p. 187–202.

Bhattacharya, T. and D.N. Raychoudhuri, 1979. Monthly variation in the density of soil microarthropods in relation to some climatic and edaphic factors. *Entomon.*, 4(4): 313–318.

Choudhuri, D.K. and S. Banerjee, 1975. Qualitative and quantitative composition of Acari and Collembola in relation to soil organic matter microbes complex. *Oriental Insect*, 9(3): 313–316.

Davis, B.N.K., 1963. A study of microarthropod communities in mineral soils near corby Northants. *J. Animal Ecol.*, 32: 49–71.

Chapter 46

Limnology of Lentic Freshwater Systems in North Cachar Hills, Assam, India

***T. N. Majumdar*[1], *A. Gupta*[2] *and M. Daolagupu*[1]**

Department of Zoology, Haflong Government College, Haflong – 788 819, Assam

[1]Department of Ecology and Environmental Science, Assam University, Silchar – 788 011, Assam, India

E-mail: agecol@rediffmail.com

ABSTRACT

A study was carried out on the limnology of five lentic systems, at different altitudes in North Cachar Hills district of Assam. The influence of altitude on temperature is evident from the results obtained. Marked seasonal variations were observed in dissolved oxygen and free carbon dioxide content which also showed variations among the different lentic systems. Total alkalinity and conductivity of water were found to be low. Comparatively higher levels of total alkalinity, conductivity, phosphates and nitrates were recorded in the dry season. A strong correlation between primary productivity and phosphates, nitrates and total alkalinity could be observed. Statistical analysis using ANOVA and Duncan's Long range Test revealed significant difference among the different lentic systems in several physico-chemical properties such as dissolved oxygen, total alkalinity, phosphates, nitrates and conductivity of water.

Keywords: *Lentic system, Physico-chemical properties, North Cachar Hills, Assam.*

Introduction

The North Cachar Hills District is one of the two hill districts of Assam. It lies between *c* 25 to 26 North latitude and *c* 92 to 94 East longitude. The district consists of undulating hills, valleys and small patches of plain lands. The total geographical area of North Cachar Hills is approximately 4888 sq. km. On its eastern side are the states of Nagaland and Manipur, while Cachar district of Assam lies towards its south. On the western side is the state of Meghalaya, while Nowgaon and Karbianglong district of Assam are located towards its north (Figure 46.1). The main mountain range of the district is the Borail spreading from east to west in the southern part of the district.

Although extensive work has been done on the limnology of lentic systems in other parts of India (Singh and Ray, 1995; Tiwari, 1999; Sahat *et al.*, 2001; Mariappan and Vasudevan, 2002), few such investigations have been carried out in the lentic bodies of North-East India (Das *et al.*, 2002; Dutta Gupta *et al.*, 2004) with no published account on those in North Cachar Hills district. This article, therefore, attempts to generate base line information on the limnology of five lentic systems at different altitudes in this district. Besides physico-chemical variables, the vegetation near the systems along with aquatic macrophytes were also studied, since they have a pronounced effect on ecosystem processes (Carpenter and Lodge, 1986) and can be used to monitor the quality of water (Dewanji and Matai, 2000).

Materials and Methods

Study Area

For the present investigation five water bodies were selected at three different places in N.C. Hills District. The Fishery Tank, the Haflong Lake and the Upper Lake are situated in Haflong, (altitude

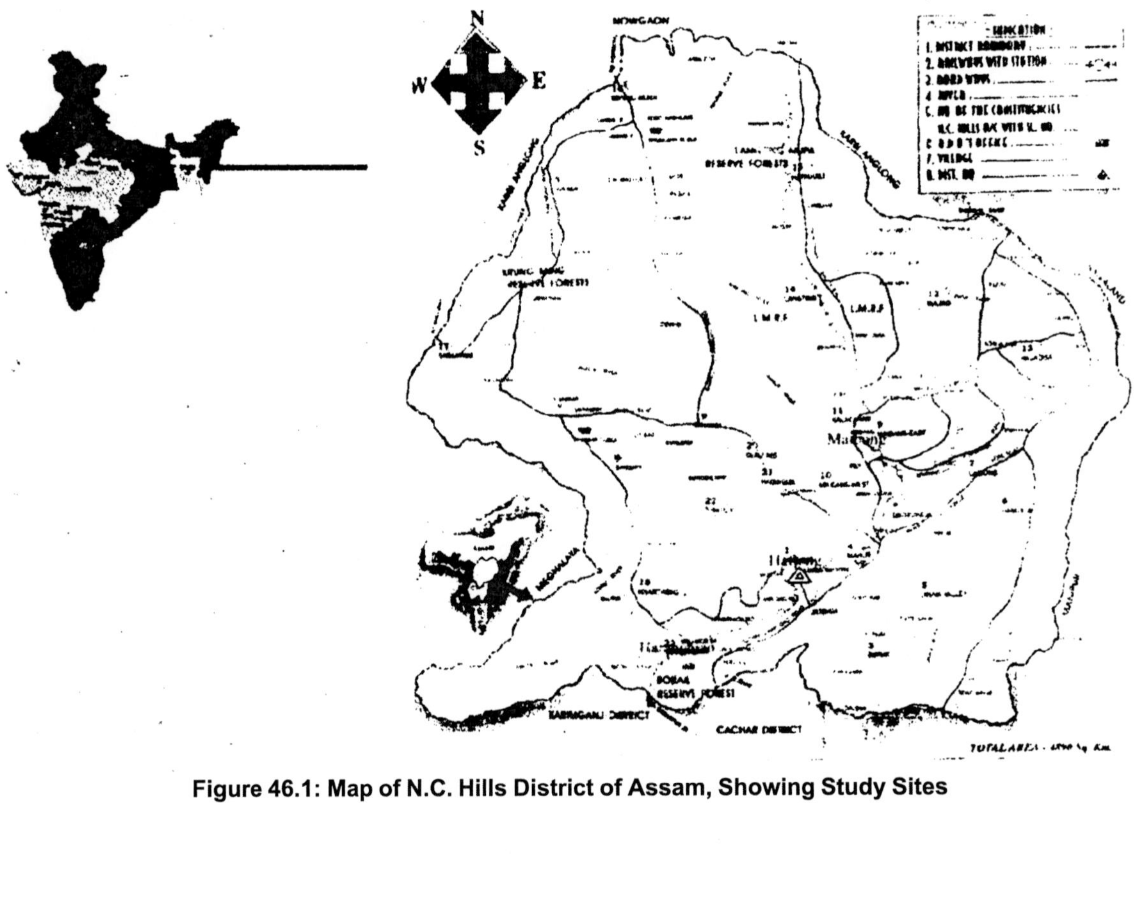

Figure 46.1: Map of N.C. Hills District of Assam, Showing Study Sites

c 935.10 msl.), the headquarter of N.C. Hills District, while a pond each in Maibong (altitude *c* 365,87 msl), a small town in the northern part of the district, and Harangajao (altitude *c* 240.79 msl) in the south, were investigated (Figure 46.1). The Fishery Tank is *c* 250 m long and *c* 50 m wide with an average depth of 2 metres. It is surrounded by dense human habitation, public buildings and roads. The Haflong Lake is an artificial lake *c* 1.0 km long and *c* 135 m wide with an average depth of 2 m. It is situated in a low lying area and receives sewage from a large part of Haflong town. The Upper Lake is located about 100 m above the level of the other two waterbodies in Haflong. It lies between two steep hills and is *c* 250 m long, *c* 30.0 m wide, and *c* 12.0 m deep. Human habitation around the lake is very thinly dispersed. The pond in Maibong, located near the railway station, has a length of *c* 80 m and width and depth of 40 m and 2 m, respectively, and is surrounded by paddy fields and human habitation. The pond in Harangajao is a small waterbody located amidst paddy fields near the Haflong-Silchar Road about 6.0 km. from Harangajao town. It is *c* 45 m long with an average width and depth of 18 m and 2.1 m, respectively.

Climate and Vegetation

Information on the, climate and vegetation of North Cachar Hills District was collected from different secondary sources (Champion and Seth, 1968; Parashar, 1994). Primary data on vegetation in and around the waterbodies were also collected.

Physico-chemical Properties of Water

Physico-chemical properties of water were studied in three lentic system at Haflong either monthly (Fishery Tank and Haflong Lake) or seasonally (Upper Lake) from August, 1999 to July, 2000. These systems along with the two ponds at Maibong and Harangajao were also investigated in three seasons (pre-monsoon, monsoon and winter) during 2001–2002. Air and water temperature were measured with a mercury bulb thermometer, pH and conductivity with a pH and conductivity meter (Merck-WTH, Germany), respectively, and total alkalinity, dissolved oxygen and free carbon dioxide by standard methods (Michael, 1984). Nitrate and phosphate were measured using a spectrophotometer (Thermo Spectronics) by phenol disulphonic acid digestion method and stannous chloride-ammonium molybdate method respectively. Primary productivity was estimated by the light and dark bottle method (Michael, 1984).

Results and Discussion

Climate and Vegetation

The climate in North Cachar Hills District is moderate with maximum temperature ranging between 24 C and 30 C and minimum between 14 C to 19 C. Annual rainfall ranges from 2200 mm to 3000 mm. Distribution of rainfall from August, 1999 to July 2000 and from March, 2001 to February, 2002 is depicted in Figures 46.2a and 46.2b, which show that there was no precipitation from December to February in the first year of study and in the second year, November, December and February were totally devoid of any precipitation. The highest precipitation was recorded in the month of August (475.00 and 569.00 mm respectively) in both the years. Total rainfall was higher in the second study year (2889.3 mm) than that in the first (2278.75 mm).

The hills of the Borail range have thick vegetation. The forest is basically moist Semi-Evergreen type with patches of east Himalayan Moist-Mixed Deciduous and pure Bamboo forest. Trees, shrubs and herbs around each water body and the aquatic macrophytes in them were studied in the present investigation (Table 46.1). Relatively fewer trees species were present in the catchment of Upper Lake

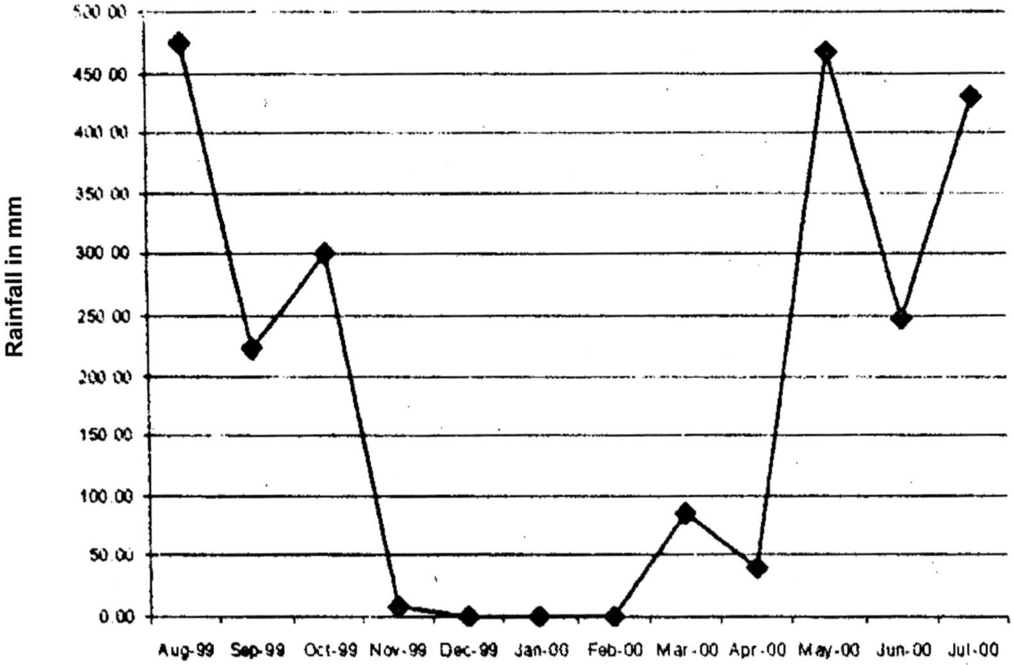

Figure 46.2a: Rainfall in N.C. Hills District from August 1999 to July 2000

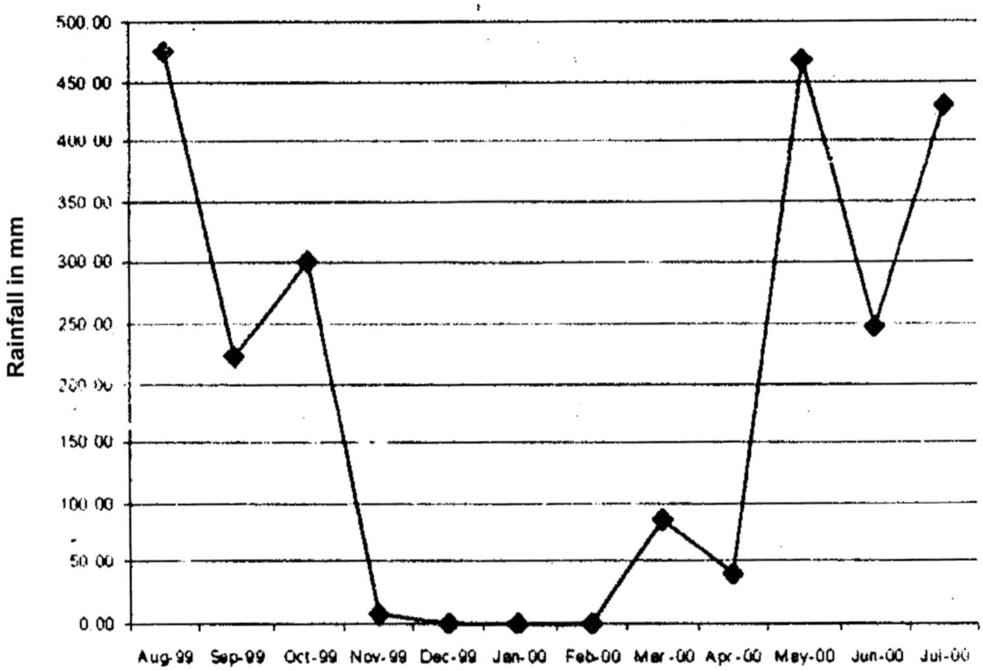

Figure 46.2b: Rainfall in N.C. Hills District from March 2001 to February, 2002

Table 46.1: Vegetation in and Around the Water Bodies of North Cachar Hills, Assam

Types of Vegetation	*Fishery Tank*	*Haflong Lake*	*Upper Lake*	*Maibong Pond*	*Harahgajao Pond*
Marginal					
Trees/Shrub	*Mangifera indica*	*Pinus* sp.	*B. tulda*	*Tectona grandis*	*T. grandis*
	Areea catechu	*Margo* sp.	*M. paradisiaca*	*Ficus carica*	*B. tulda*
	Dioscorea bulbifera	*Acacia* sp.		*Dioscorea bulbifera*	
	Anthocephalus cadamba	*Cassia* sp.		*Acacia nilotica*	
	Musa paradisiaca	*Delonix regia*		*Gmelina arborea*	
	Bambusa tulbla	*Eucalyptus* sp.			
		Ficus benghalensis			
Herbs/Small shrubs	*Lantana camara*	*Digitaria sanguinalis*	*Eupatorium odoratum*	*Chrysopogon aciculatus*	*Digitaria sanguinalis*
	Chrysopogon aciculatus	*C. aciculatus*	*Hedyotis lineata*	*Cynodon dactylon*	*Imperata cylindrica*
	Cynodon dactylon	*C. benghalensis*	*C. aciculatus*	*Digitaria sanguinalis*	*Chrysopogon aciculatus*
	Digitaria sanguinalis	*A. conyzoides*	*Achyranthes aspara*	*Commelina benghalensis*	*Ageratum conyzoides*
	Commelina benghalensis	*Parthenium* sp.	*C. benghatensis*	*Ageratum conyzoides*	*Lantana camara*
	Ageratum conyzoides	*A. spinosus*	*Bambusa subtilis*	*Lantana camara*	*Achyranthes aspara*
	Amaranthus spinosus	*Centella asiatica*	*Blachnum* sp.	*Eupatorium odoratum*	
	Amaranthus viridis		(Fern)		
	Blainvillea latifolia				

Contd....

Table 46.1–Contd...

Types of Vegetation	*Fishery Tank*	*Haflong Lake*	*Upper Lake*	*Maibong Pond*	*Harahgajao Pond*
Aquatic Macrophytes					
Marginal	*Axonopus compressus*	*Axonopus compressus*	*Axonopus compressus*	*Axonopus compressus*	*Axonopus compressus*
	Polygonum perfoliatum	*Sagittaria* sp.	*Alternanthera sessilis*	*Monochoria hastaefolia*	*Sagittaria* sp.
	Polygonum hydropiper	*Cardanthera triflora*	*Amaranthus* sp.	*Sagittaria* sp.	*Amaranthus* sp.
	Ipomea aquatica	*Limnophila heterophylla*	*Polygonum perfoliatum*	*Amaranthus* sp.	*Monochoria hastaefolia*
	Phylanthus nirurii	*Monochoria hastaefolia*	*Polygonum hydropiper*	*Limnophila heterophylla*	*Limnophila heterophylla*
	Amaranthus sp.	*Monochoria vaginalis*			*Ipomea aquatica*
		Amaranthus sp.			
Floating	*Eichhorrnia crassipes*	*Trapa natans*	*Trapa natans*	*Trapa natans*	*Azolla pinnata*
	Azolla pinnata	*Eichhornia crassipes*		*Pistia stratiotes*	*Utricularia* sp.
	Pistia stratiotes	*Pistia stratiotes*		*Azolla pinnata*	*Lemna* sp.
	Utricularia sp.	*Azolla pinnata*		*Lemna* sp.	*Hydrilla* sp.
		Utricularia sp.		*Utricularia* sp.	*Eichhornia*
		Lemna sp.		*Eichhornia crassipes*	*crassipes*
Submerged	NIL	*Hydrocharis* sp.	*Salvinia* sp.	*Salvinia* sp.	NIL
		Salvinia sp.		*Vallisneria* sp.	
		Vallisneria sp.		*Hydrilla* sp.	
		Hydrilla sp.		*Chara* sp.	
Emergent	NIL	*Nelumbo* sp.	NIL	*Nelumbo* sp.	NIL
		Nymphaea sp.		*Nymphaea* sp.	
		Oryza sp.		*Oryza* sp.	

and Harangajao Pond when compared with those around the other waterbodies studied. Emergent type of aquatic macrophyte was totally absent in Fishery Tank, Upper Lake and Harangajao Pond. Submerged aquatic macrophyte was absent in both Fishery Tank and Harangajao Pond whereas in Upper Lake only one species of submerged (*Salvinia* sp.) and one species of floating macrophyte (*Trapa natans.*) were found covering a very small area of the waterbody. Several species of marginal macrophytes were present in all the lentic systems, studied.

Physico-chemical Properties of Water

The physico-chemical properties of water in differet lentic systems studied are depicted in Tables 46.2 and 46.3. The influence of altitude on temperature is evidenced by the lowest temperature regime in Upper Lake, Haflong, and the highest in Harangajao. It is a well-known fact that no other single factor has so profound influence on aquatic ecosystem than temperature (Welch, 1952). Hydrogen ion concentration (pH) is another important environmental factor influencing the aquatic ecosystem (Welch, 1952). In the present investigation, the Fishery Tank at Haflong showed higher pH values (6.5 to 8.35) than the other systems during the first annual cycle (1999–2000). This probably reflects the effect of periodic liming and manuring of the tank by the Department of Fisheries during that period. In the absence of liming during 2001–2002, the pH regime became much lower (6.5–7.5). Lentic systems located at a lower altitude (< 100 msl) in both Brahmaputra and Barak valleys in Assam exhibited a somewhat higher pH range (Das and Baruah, 2002; Dutta Gupta *et al.*, 2004). In contrast, the pH values recorded in certain lentic systems of Shillong, Meghalaya, which are located at a higher altitude (1495–1900 msl) than those in the present study, were somewhat lower (Gupta *et al.*, 1992). Thus the lentic systems of North Cachar Hills, located at an intermediate altitude, exhibit a pH regime lower than that in the valleys, but higher than those at a higher altitude. Conductivity was found to range from 17.6–78.1 $\mu S\ cm^{-1}$, which is lower than those reported from the other regions of India (Habib *et al.*, 1997). The entry of nutrient-rich sewage into Haflong Lake and liming/ manuring during the first annual cycle in Fishery Tank is reflected in the relatively high conductivity in these two systems. The conductivity in the latter system declined in the second annual cycle when liming was discontinued. The range of dissolved Oxygen is wide in small waterbodies and may vary greatly from one waterbody to the other in the same area (Kaushik *et al.*, 1991). In the present study, dissolved oxygen was found to vary from 1.6 mg l^{-1} to 10.4 mg l^{-1} in the different study sites. Relatively higher dissolved oxygen values were recorded in Upper lake and Haranbrajao pond (6.2–10 mg l^{-1} and 5.2–10.4 mg l^{-1} respectively) when compared to those in Fishery tank, Haflong Lake and Maibong Pond (2.7–10 mg l^{-1}, 1.6–6.4 mg l^{-1} and 2.8–6.4 mg l^{-1} respectively) as a result of the entry of large amount of organic matter in the form of manure and sewage in the latter. Again, in Haflong Lake lower oxygen values were recorded in the first annual cycle (1.6–6.4 mg l^{-1}) than that in second annual cycle (4.8–6.4 mg l^{-1}). This might be due to the accidental drainage of water from this system in the first annual cycle when the water level was much lower. In most of the water bodies higher oxygen values were recorded during winter. Higher values of dissolved oxygen in colder months even in tropical systems were reported by some workers (Kauashik *et al.*, 1989; Prasad, 1990). Higher level of free carbon dioxide (6–13 mg l^{-1}) could be observed during post monsoon in the first annual cycle (1999–2000) in Fishery Tank, Haflong Lake and Upper Lake. In contrast, in second annual cycle (2001–2002), Fishery Tank, Haflong Lake and Harangajao Pond showed the highest level of free CO_2 (15.6 mg l^{-1}, 11 mg l^{-1}, and 8 mg l^{-1} respectively) during monsoon whereas it was higher during pre-monsoon in Upper Lake and Maibong Pond (5.2 mg l^{-1} and 9 mg l^{-1} respectively). The rain absorbs small amount of CO_2 from the atmosphere and delivers it to the water bodies (Unni, 1972). In the second study year rainfall was considerably heavier in the monsoon months. This might be the cause of higher CO_2 during Monsoon.

Table 46.2: Physico-chemical Properties of Water in the Lentic Systems of North Cachar Hills, Assam, in the First Annual Cycle (1999–2000)

Month	Water Body	Air Temp. °C	Water Temp. °C	pH	Conductivity μScm^{-1}	Dissolved Oxygen mgl^{-1}	Free Carbon dioxide mgl^{-1}	Total Alkalinity mgl^{-1}	Phosphates μmgl^{-1}	Nitrates mgl^{-1} μgml^{-1}	Gross Primary Productivity mgC/m³/24 hrs (Range)	Respiration mgC/m³/24 hrs (Range)	Net Primary Productivity mgC/m³/24 hrs (Range)
Aug'99	Fishery Tank	27	28	7.5	73.9	5.8	8	68	26.6	0.94	1576.5±523.53 (975.9–1951.9)	925.9±114.66 (825.8–900.9)	650.6±458.72 (150.1–1051)
	Haflong Lake	27	28	7.5	52	4.2	10	64	15.3	1.2	1764.9 (1576.5–1951.9)	1351.3	412.9 (225.23–600.58)
	Upper Lake	24	26	7.24	20.9	8.6	3	48	5.1	1.8	2402.3 (1726.66–3077.45)	2026.9 (1501.4–2552.4)	375.36 (225.2–525.5)
Sep'99	Fishery Tank	26	28.5	6.75		4.8	12.4	64					
	Haflong Lake	28	29	6.9		3.2	10.5	64					
	Upper Lake												
Oct'99	Fishery Tank	24	25.5	6.5		3.2	13	104					
	Haflong Lake	25.5	26	6.7		3.2	14	68					
Nov'99	Fishery Tank	21	19.5	7.15	78.1	5.6	6	60	127	4.6	1701.63±312.55 (1351.3–1951.9)	925.10±144.66 (825.80–1051.0)	775.73±263.64 (525.5–1051)
	Haflong Lake	20	19	6.8	64.2	4.0	8	64	12.7	4.6	1351.3 (1051–1651.6)	488 (150.14±825.8)	863.33 (150.14–1776.5)
	Upper Lake	19	18	6.5	30.1	6.8	6	30	18.1	5.9	1426.4 (1351.3–1501.4)	825.8 (750.7–900.9)	750.7
Dec'99	Fishery Tank	17	17	6.55		5.6	6	80					
	Haflong Lake	17	16	6.6		3.8	6.8	92					
	Upper Lake												
Jan'00	Fishery Tank	16	14.5	6.78	73.9	3.2	12	92	30	1.6	625.60±188.94 (450.41±825.8)	425.4±263.69 (150.1–675.7)	200.2±86.69 (150.1–300.3)
	Haflong Lake	16	14	6.5	59	3.6	6.8	94	10	2.8	750.7 (450.4–1051)	375.36 (300.3–450.4)	375.36 (150.14–600.58)
	Upper Lake	19	17.5	6.95	24.1	7	2.8	36	12.5	6.0	1126.1	825.8 (750.7–900.9)	300.3 (325.2–375.36)

Contd...

Table 46.2–Contd...

Month	*Water Body*	*Air Temp. °C*	*Water Temp. °C*	*pH*	*Conductivity µScm^{-1}*	*Dissolved Oxygen mgl^{-1}*	*Free Carbon dioxide mgl^{-1}*	*Total Alkalinity mgl^{-1}*	*Phosphates µmgl^{-1}*	*Nitrates mgl^{-1} µgml^{-1}*	*Gross Primary Productivity mgC/m^3/24 hrs (Range)*	*Respiration mgC/m^3/24 hrs (Range)*	*Net Primary Productivity mgC/m^3/24 hrs (Range)*
Feb'00	Fishery Tank	20	19	6.95		2.8	13	64					
	Haflong Lake	20	19	7.0		1.6	9	80					
	Upper Lake												
Mar'00	Fishery Tank	23	22.5	7.6		3.6	11.5	60					
	Haflong Lake	22	23	7.1		5.0	8.4	68					
	Upper Lake												
Apr'00	Fishery Tank	24	21	8.2		5.8	6.4	66					
	Haflong Lake	24	20	7.7		6.4	9.5	64					
	Upper Lake												
May'00	Fishery Tank	24.5	25	7.8	64.2	2.7	13	68	17	0.48	1676.63±991.23 (600.6–2552.2)	1151.1±740.63 (300.3–1501.4)	525.53±456.7 (225.2–1051.1)
	Haflong Lake	25	26	7.4	48	4.2	10	66	9.5	0.38	2552.4 (1651.6–3453.3)	1764.8 (1051–2477.4)	825.8 (600.58–975.9)
	Upper Lake	22	21	7.01	22.9	6.2	5.8	40	5.0	2.5	825.8 (450.43–1201.15)	337.8 (75.1–600.6)	487 (375.36–600)
Jun'00	Fishery Tank	25	22	8.2		4.6	12.4	68					
	Haflong Lake	24.5	21	7.3		4.8	11	66					
	Upper Lake												
Jul'00	Fishery Tank	28	29	8.35		4.8	12.4	66					
	Haflong Lake	28.5	29	7.6		4.6	11	68					
	Upper Lake												

Table 46.3: Physico-chemical Properties of Water in the Lentic Systems of North Cachar Hills, Assam, in the Second Annual Cycle (2001–2002)

Season	*Waterbody*	*Air Temp. °C*	*Water Temp. °C*	*pH*	*Conductivity $\mu\ Scm^{-1}$*	*Dissolved Oxygen (mgl^{-1})*	*Free Carbon Dioxide (mgl^{-1})*	*Total Alkalinity (mgl^{-1})*	*Phosphates $\mu g\ ml^{-1}$*	*Nitrates mgl^{-1}*
Premonsoon (May'01)	Fishery Tank	24	25.2	7.5	23.6	6	8	84	1.08	5.8
	Haflong Lake	23	24	7.6	55.8	5.6	8.4	80	21.2	45.9
	Upper Lake	20	20.5	7.2	19.2	8	5.2	20	–	2.8
	Maibong Pond	26	28	6.52	55.4	2.8	9	56	2.4	2.2
	Harangajao Pond	24	22.5	7	20.2	6.4	5.2	56	1.02	0.76
Monsoon (Aug'01)	Fishery Tank	26	29	6.5	20.2	5.6	15.6	76	0.24	6.7
	Haflong Lake	28	29	7.3	63	4.8	20	68	24.6	23
	Upper Lake	24	26	6.45	18.9	9.6	4.8	20	–	2.05
	Mailbong Pond	22	23	7	59	5.6	7	46	11.68	19.7
	Harangajao Pond	30	32	6.05	18.2	5.2	8	48	1.33	0.88
Winter (Jan'02)	Fishery Tank	20	18	7.15	28.6	6	6	84	4.62	4.85
	Haflong Lake	16	14.5	6.8	74.6	6.4	8.4	94	21.3	45.9
	Upper Lake	18	17	7	20.1	10	3.4	26	–	2.8
	Maibong Pond	20.5	19	6.2	61	6.4	5.6	68	5.35	4.6
	Harangajao Pond	20	18	6.85	17.6	10.4	2.4	20	0.72	0.68

Table 46.4: Comparison Among Physico-chemical Properties of Water in Different Waterbodies Using One Way ANOVA and Duncan's Long Range Test

Period	*Physico-Chemical Parameters*	*F Value*	*df*	*Level of Significance*	*Duncan's LRT*
First Annual Cycle	Dissolved Oxygen	9.8	9,2	P ≤ 0.01	UL > HL, FT; HL ≅ FT
	Free Carbon Dioxide	0.13	9,2		
	Total alkalinity	5.016	9,2	P ≤ 0.05	FT, HL < UL
	pH	0.88	9,2		
	Air Temp.	0.09	9,2		
	Water Temp.	0.05	9,2		
	Conductivity of Water	69.63	9,2	P ≤ 0.001	FT > HL > UL
	Phosphates	0.67	9,2		
	Nitrates	1.34	9,2		
	Gross Primary Productivity	30.07	9,2	P ≤ 0.001	
	Respiration	0.076	9,2		
	Net Primary Productivity	0.36	9,2		
Second Annual Cycle	Dissolved Oxygen	3.398	10,4		
	Free Carbon Dioxide	2.15	10,4		
	Total Alkalinity	14.36	10,4	P ≤ 0.001	FT, HL > MP, HL, UL; MP > UL, FT ≅ HL; MP ≅ HP; HP ≅ UL
	Phosphates	41.30	10,4	P ≤ 0.001	HL > MP, FT, HP, UL; MP > HP, UL
	Nitrates	13.42	10,4	P < 0.001	HL > MP, FT, UL, HP
	pH	1.17	10,4		
	Conductivity	64.35	10,4	P ≤ 0.001	HL > MP > FT, UL, HP
	Air Temp.	0.36	10,4		

Higher total alkalinity could be observed in Fishery Tank (60 mg l^{-1}–104 mg l^{-1}) and Haflong Lake (64 mg l^{-1}–94 mg l^{-1}) followed by Maibong (46–68 mg l^{-1}) and Harangajao Ponds (20–56 mg l^{-1}), while the lowest was in Upper Lake (20–48 mg l^{-1}). Alikunhi (1957) stated that in highly productive waters, alkalinity ought to be more than 100 mg l^{-1}. Alkalinity greater than 100 mg l^{-1} was recorded only once (October 1999 in Fishery Tank) during the present study, although concentrations of 80–94 mg l^{-1} were also recorded in Fishery Tank and Haflong Lake in the post monsoon and winter seasons, while in all the other systems, alkalinity values were consistently much lower. Phosphate levels were higher in Fishery Tank during first annual cycle (26.6–127 µg l^{-1}), which again may be due to manuring and management. Nitrates were found to fluctuate widely between 0.38 mg l^{-1} and 45.9 mg l^{-1}. The least concentration of nitrate was recorded from Harangajao Pond. During the second annual cycle, the range of nitrate content was the highest (23.0–45.9 mg l^{-1}) in Haflong Lake among all the waterbodies studied. This might be due to the sewage disposal into the lake. However, the fluctuations were very narrow during the first annual cycle, when there was a stress in the lake ecosystem due to the accidental draining out of its water. Conductivity, total alkalinity, phosphates and nitrates were relatively high in all the systems during dry seasons which might be due to evaporation of water (Furch and Junk, 1993). In the present study, higher phosphate concentrations were recorded in Haflong Lake, Maibong Pond and Harangajao Pohd during monsoon. Both Maibong and Harangajao Ponds being situated in the vicinity of paddy fields received nutrient enriched surface runoff during monsoon, while the bulk disposal of sewage into Haflong Lake during monsoon might be the reason behind the observed high level of phosphate in these waterbodies during this season. Fluctuations in both gross primary productivity and net primary productivity could be observed in all the waterbodies. The highest GPP was observed in Haflong Lake (2552.4 mgC/m^3/24 hrs) in pre-monsoon, although highest NPP was recorded in the same system during post-monsoon (863.33 mgC/m^3/24 hrs). A strong correlation between primary productivity and phosphates, nitrates and total alkalinity contents of the lentic systems was apparent.

Statistical Analysis

Statistical analysis of data for physico-chemical properties of water in different waterbodies with ANOVA and Duncan's Long Range Test is shown in Table 46.4. The table revealed a significant difference among the lentic systems only in dissolved oxygen, total alkalinity and conductivity during the first annual cycle and in total alkalinity, phosphates, nitrates and conductivity during the second annual cycle. During the first annual cycle, Upper Lake showed significantly higher dissolved oxygen content than Haflong Lake and Fishery Tank, while both Fishery Tank and Haflong Lake showed significantly higher total alkalinity than Upper Lake. Fishery Tank had higher conductivity of water than Hatlong Lake, which in its turn had higher conductivity of water than Upper Lake. During the second annual cycle, significantly higher total alkalinity was observed in both Fishery Tank and Haflong Lake than that in the other systems. Halfong Lake also showed higher concentrations of phosphates, nitrates and conductivity compared to the others.

Acknowledgements

TNM is thankful to Mr. GK. Das, Principal, Haflong Government College, for his help and cooperation. Help rendered by Sri Arun Chandra Das in collection of samples is also acknowledged.

References

Alikunhi, K. H., 1957. Fish culture in India. *Farm Bulletin*, ICAR, New Delhi, 20: 1–6.

Carpenter, S.R. and Lodge, D.M., 1986. Effects of submerged macrophytes on ecosystem processes. *Aquat. Bot.* 26: 341–370.

Champion, H.G. and Seth, S.K., (1968. *Forest Types of India*. Forest Research Institute, Dehradun.

Das, A C., Baruah, B.K., Baruah, D. and Sengupta, S., 2002. Study on wetlands of Guwahati city-1. Water quality of ponds and beets. *Poll. Res.*, 21(4): 511–513.

Dewanji, A. and Matai, S., 2000. Aquatic weeds as indicators of water quality. In: *Environmental Stress: Indication Mitigation and Eco-Conservation*, (Eds.) Yunus, M., Sing, N. and Dekok, L.J. Kluwer Academic Publishers, Boston, p. 251–258.

Dutta Gupta, S., Gupta, S. and. Gupta, A., 2004. Euglenoid blooms in the flood plain wetlands of Barak Valley, Assam, Northeastern India. *J. Environ. Biol.*, 25(3): 369–373.

Furch, K. and Junk, W.J., 1993. Seasonal nutrient dynamics in an Amazonian Lake Arch. *Hydrobiol.*, 128: 277–295.

Gupta, S., Michael, R.G. and Gupta, A., 1992. A preliminary investigation on physico-chemical factors, periphyton and invertebrate communities in a protected water works of Shillong, Meghalaya. In: *Proc. Nat. Acad. Sci.*, India, 62(B)III: 333–337.

Habib, O.A, Tripett, R. and Murphy, K.J., 1997. Seasonal changes in phytoplankton community structure in relation to physico-chemical factors in Lack Lomond, Scotland. *Hydrobiologia,* 350: 63–79.

Kaushik, S., Agarkar, M.S. and Saxena, D.N., 1991. Water quality and periodicity of phytoplantonic algae in Chambal tank, Gwalior, Madhya Pradesh. *Bionature,* 11: 87–94.

Kaushik, S., Sharma, S. and Saksena, M.N., 1989. Physico-chemical factors and the aquatic insect density of a pond receiving cotton mill effluent at Gwalior. *Indian J. Ecol.,* 16(1): 64–67.

Mariappan, P., and Vasudevan, T., 2002. Correlation coefficients of some physico-chemical parameters of drinking water ponds in eastern part of Sivagangai District, Tamil Nadu. *Poll. Res.,* 21(4): 403–407.

Michael, P., 1984. *Ecological Methods for Field and Laboratory Investigations.* Tata McGraw-Hill Publishing Company, Limited, New Delhi.

Parashar, D. (Ed.), 1994. *N.C. Hills: The Enchanting Land and the Bird Mystery of Jatinga.* North Cachar Hills Autonomous District Council, Haflong in Association with Konark Publishers. Pvt. Ltd., New Delhi.

Prasad, D.Y., 1990. Primary productivity and energy flow in Upper Lake, Bhopal (India). *Indian J. Environ. Health,* 32(2): 132–199.

Sahat, T., Manna, N.K., Som Majumdar, S. and Bhattacharjya, I.N., 2001. Primary productivity of the Subhas Sarobar Lake in East Calcutta in relation to some selected physicochemical parameters. *Poll. Res.,* 20(1): 47–52.

Singh, J.P. and Roy, S.P., 1995. Limnobiotic investigation of Kawar Lake, Begusarai, Bihar. *Env. Eco.,* 13(2): 330–335.

Tiwari, D.R., 1999. Physicochemical studies of the Upper Lake Water, Bhopal, Madhya Pradesh, India. *Poll. Res.,* 18(3): 323–326.

Unni, K.S., 1972. An ecological study of the macrophytic vegetation of Doodhadhari lake, Raipur (M.P.). 3. Chemical factors. *Hydrobiologia,* 40: 25–36.

Welch, P.S., 1952. *Limnology.* Mc-Graw Hill Book Company, New York.

Chapter 47

Antibacterial Activity of Selected Veterinary Herbs Against Cattle Diarrhoea Causing Pathogens

K. Nandakumar[1], M. Sivashanmugam[1], G. Perumal[2], N. Thangaraj[1] and S. Sargunavalli[1]*

[1]Department of Botany, [2]Department of Microbiology
Kandaswami Kandars College, Velur–Namakkal – 638 182, Tamil Nadu, India

ABSTRACT

The present study was exclusively aimed to study the dysentery causing organisms namely *Escherichia coli, Salmonella typhi, S. typhi para A, S. typhi para B, Shigenella dysenteriae* and *Vibrio cholerae* and their antibacterial activities. The present study was, the Traditional Veterinary Medicinal plants and their uses were screened. The Antimicrobial activity of selective veterinary herbs against diarrhoea causing pathogen in cattles was performed. Selected plants namely *Saccharum arundinaceum, Cissus quadrangularis, Citrullus colocynthis, Zehneria scabra, Tamarindus indica, Bambusa arundinacea* were employed to study the antibacterial activity.

Keywords: *Antibacterial, Diarrhoea, Cattle, Pathogens, Veterinary herbs.*

Introduction

Cattle wealth in India plays a significant role in economic status of our nation. In India cattle are being used for almost all agricultural purposes such as ploughing, lifting water from wells and the

* Corresponding Author.

transport of products from the field to markets. Cattle provide us milk and milk product. Cattle also provide meat. The sheep and goats provide about 75 per cent of the meat.

Diet induced diarrhoea is not infrequent in animals. This is a most common problem on newly born animals. Calf, fold, lamb, piglet, pup and kitten mostly suffer from such diarrhoea at 2 to 4 weeks of their age. Ingestion of excess milk, inferior quality milk, milk powder, milk replacer or excess starchy food is the causes of such diarrhoea. At this time the stomach or intensive accumulates undigested casein protein. Bacteria like *E. coli* and *Salmonella* sp. Proliferate on the undigested substrate of the intestine and colonize there causing enteric colibacillosis or salmonellosis. Affected animal remain dehydrated and pass soft light yellow coloured foul smelling faeces. Diarrhoea is considered as the characteristic symptom of disturbances and it is a global health problem (Kavitha *et al.*, 2003).

The diarrhoeagenic organisms are versatile in their pathogenesis and well equipped with phenotypic virulent fractures leading to the different forms of diarrhoea (Daswani *et al.*, 2001). Infectious diseases account for approximately one-half of all deaths in tropical countries. In industrialized nations, despite the progress made in the understanding of microbiology and their control, incidents of epidemics due to drug resistant microorganisms and the emergence of hitherto unknown disease-causing microbes, pose enormous public health concerns. Historically, plants have provided a good source of anti-infective agents; emetine, quinine, and berberine remain highly effective instruments in the fight against microbial infections. Phytomedicines derived from plants have shown great promise in the treatment of intractable infectious diseases including opportunistic AIDS infections.

Plant compounds are routinely classified as "antimicrobial" on the basis of susceptibility tests that produce MICs in the range from 100 to 1,000 μg/ml, orders of magnitude weaker than those of typical antibiotics produced by bacteria and fungi (MICs, 0.01 to 10 μg/ml). A compound that is synthesized in response to pathogen invasion and is required to protect the plant from a pathogen but that shows little activity in an in vitro susceptibility test is not necessarily an antimicrobial. Such a substance might have a regulatory function, indirectly increasing the level of resistance of the plant. This analysis suggests that we lack a solid rationale for making a functional assignment for the vast majority of plant compounds that have been classified as antimicrobials.

Satya *et al.* (2003) stated that antimicrobial activity of alcoholic extract of *Memecylon umbellatum,* showed maximum effect against *Staphylococcus aureus* and it also showed antibacterial activity against gram-negative bacteria. Prema (2004) investigate the acetone, chloroform, ethanol and diethyl ether fractions of eleven medicinal plants including *Eucalyptus globulus,* were tested against five bacterial pathogens namely, *S. typhi, P. aeruginosa, S. aureus, Shigella flexneri* and *K. pneumoniae.* The maximum zone of inhibition was recorded in *E. globulus* against *S. aureus* and it also showed inhibition action against all the five pathogen.

Historically, plants have provided a source of inspiration for novel drug compounds, as plant derived medicines have made large contributions to human health and well-being. Their role is two fold in the development of new drugs: (1) they may become the base for the development of a medicine, a natural blueprint for the development of new drugs, or; (2) a phytomedicine to be used for the treatment of disease. There are numerous illustrations of plant derived drugs. The roots, leaves and stem of these plants are widely used in Ayurvedic system of medicine for many health disorders like Bronchitis, Jaundice, Scurvy etc. (Onkarappa, *et al.*, 2003). Puratchikody *et al.* (2001) reported that the effect of anti-bacterial activity of *Cyperus rotundus.* Rhizomes were extracted with petroleum ether, benzene, chloroform, acetone and ethanol against the bacterial strains of *S. aureus, E. coli, B. subtilis, P. aeruginosa* and *P. vulgaris.*

Materials and Methods

The present study was exclusively aimed to study the dysentery causing organisms namely *Escherichia coli, Salmonella typhi, S. typhi para A, S. typhi para B, Shigenella dysenteriae* and *Vibrio cholerae* and their antibacterial activities by using medicinal plants. Selective six plants namely *Saccharum arundinaceum, Cissus quadrangularis, Citrullus colocynthis, Zehneria scabra, Tamarindus indica* and *Bambusa arundinacea* were employed to study the antibacterial activity.

Plant Material Collection

Fresh, healthy aerial parts of selected medicinal plants were collected from wild population of Kolli hills, Namakkal district. The collected plant materials were weighed 20 gm each.

Extract Preparation

The methodology was followed as per Onkarappa *et al.* (2003) with slight modifications. Each 20 gm of plant materials were washed with running water for 5–6 times and then it was surface sterilized with 0.1 per cent $HgCl_2$ solution and finally washed in distilled water for three times. Then they were crushed in mortar and pestle by adding 100 ml of 50 per cent methanol. The crushed extract was allowed to over night for extraction. The extracts were filtered and centrifuged at 5000 rpm for 5 minutes. The supernatant was collected and made up to different concentration like 25, 50, 75 and 100 per cent by adding 50 per cent methanol.

Screening for Antibacterial Activity

Culture plates for bacteria are prepared by pouring 15–20 ml of Mueller Hinton Agar medium into sterile Petri plates. The test bacteria were inoculated into nutrient broth and incubated at 37 C for 24 hours. After the incubation period the culture tubes were compared with the Turbidity (opacity) standard.

In vitro antibacterial activity of plant extracts was tested by well-in-Agar method. The inoculum suspension was spread uniformly over the agar medium using sterile cotton swabs to get uniform lawn of bacteria. Using a flamed cork borer well of 0.5 cm diameter was made in the media at a distance of 1–2 cm from the periphery and one well in centre of the plates. These plates were labeled and 0.2 ml of each plant extracts was aseptically filled into the well. Then the plates were incubated for 24 hours at 37 C. The affectivity of these extracts was recorded by measuring the diameter of inhibition zone at the end of 24–48 hours incubation.

Results and Discussion

The study the dysentery causing organisms namely *Escherichia coli, Salmonella typhi, S. typhi para A, S. typhi para B, Shigenella dysenteriae* and *Vibrio cholerae* and their antibacterial activities by using medicinal plants present in the Table 47.1. Selective plants namely *Saccharum arundinaceum, Cissus quadrangularis, Citrullus colocynthis, Zehneria scabra, Tamarindus indica, Bambusa arundinacea* were employed the antibacterial activity.

Antibacterial Activity Test

The results obtained by using the extract of plant species studied against the *Escherichia coli, Salmonella typhi, S. typhi A, S. para typhi B, Shigella dysenteriae* and *Vibrio cholerae.*

Antibacterial Activity on *E. coli*

Among the selected plant extracts studied, the methanolic extract of *Saccharum arundinaceum* showed maximum inhibition zone against *E. coli* (8 mm) (The minimum inhibition zone (2 mm) was observed in *Tamarindus indica, Citrullus colocynthis, Bambusa arundinacea* and *Cissus quadrangularis.* The antibacterial activity was moderate in mature tubers of *Zehneria scabra* (3 mm).

Table 47.1: Antimicrobial Activity of Selective Veterinary Herbs Against Diarrhoea Causing Pathogen

Sl.No.	Name of the Plants	Zone of Inhibition (mm in diameter)					
		E. coli	S. typhi	S. para typhi A	Shigella	S. para typhi B	V. cholerae
1.	*Tamarindus indica*	2	–	–	10	–	–
2.	*Citrullus colocynthis*	2	–	–	14	3	–
3.	*Bambusa arundinacea*	2	–	–	–	–	–
4.	*Cissus quadrangularis*	2	6	–	–	–	–
5.	*Zehneria scabra*	3	–	–	2	–	–
6.	*Saccharum arundinaceum*	8	–	–	–	–	–

Antibacterial Activity on *Salmonella typhi*

The methanolic extract of *Cissus quadrangularis* showed antibacterial activity against *Salmonella typhi* (6 mm). No inhibition zone was observed in the ethanolic extracts of *Tamarindus indica, Citrullus colocynthis, Bambusa arundinacea, Zehneria scabra* and *Saccharum arundinaceum.*

Antibacterial Activity on *Salmonella para typhi A*

It was interesting noted that, no inhibition zone was observed against *Salmonella para typhi A* by the ethanolic extract of all the study materials.

Antibacterial Activity on *Salmonella para typhi B*

The methanolic extract of *Citrullus colocynthis* exhibited a moderate antibacterial activity against *S. para typhi B* (3 mm). The antibacterial activity was not observed in the other species studied.

Antibacterial Activity on *Shigella dysenteriae*

The methanolic extract of *Citrullus colocynthis* showed maximum inhibition zone on *Shigella dysenteriae* (14 mm). The minimum antibacterial activity was found in the ethanolic extract of *Zehneria scabra* (2 mm). The antibacterial activity was moderate in *Tamarindus indica* (10 mm) and no inhibition zone was found in *Bambusa arundinacea, Cissus quadrangularis* and *Saccharum arundinaceum.*

Antibacterial Activity on *Vibrio cholerae*

The methanolic extract of all the six plant species studied shows no inhibition zone against *Vibrio cholerae.*

A vast majority of small-molecule plant antimicrobials are agents with weak or narrow-spectrum activities, while bacteria, yeast, and fungi produce antibiotics that both are often effective and have broad spectra of activity. The nature of this disparity is puzzling. One radical possibility is that plant "antimicrobials" actually have other functions in the plant and their low level of antimicrobial activity is accidental and largely irrelevant.

Antibacterial activity of volatile components from various solvent extracts of *Anisomeles indica, A. malabarica, Blumea lacera* and *Melia azadirachta* were tested against *E. coli, P. aeruginosa, Serratia marscens* and *Staphylococcus aureus* by tube dilution method. The result showed acetone and methanol extracts of all plants exhibited strong antibacterial activity (Ramasamy and Charles Manoharan, 2004).

An even more extreme opinion, not uncommon and stemming from the same puzzle, is that plants happen to make many secondary metabolites for no good reason, and some of them will inevitably have antimicrobial properties, very much like random substances from a library of synthetic compounds.

Jay Prakash and Hosmani (2003) screened for the antibacterial activity of methanolic extracts of *Nostoc muscorum* against *E. coli* and *S. aureus* with different concentration of extracts and the result showed the extract had great potential as antibacterial agent against both bacteria when compared with standard antibiotics.

Wide range of result variation was observed in the methanolic extract of different parts of the study materials against the disease causing bacteria. Among the six plants used *Tamarindus indica, Citrullus colocynthis, Cissus quadrangularis* and *Zehneria scabra* showed an effective positive results against the bacteria such as *Shigella, Salmonella typhi* and *S. typhi para B*. But the pathogen *Vibrio cholerae* and *S. typhi para A* were totally resistant against the methanolic extracts of all the six plant materials used. The antibacterial activity results variation is due to the concentration of plant extracts and the solvent which is used for extract preparation.

References

Daswani, P.G., Birrdi, T.J., and Antia, N.H., 2001. Study of action of Cyperus *rotundus* root decoction on the adherence and enterotoxin production of Dirrhoeagenic *E. coli. Indian J. of Pharmacology,* 33: 116–117.

Jay Prakash, N. and Hosmani, S.P., 2003. Antibacterial activity of *Nostoc muscorum* against *E. coli* and *Staphylococcus aureus. J. Ecobiol.,* 15(6): 477–479.

Kavitha, D., Shila, P.N. and Niranjali Devaraj, S., 2003. Antibacterial effect of conessine, an alkaloid from *Holarrhena antidysenterica* against entropathologenic *E. coli. Biomedicine,* 23(1&2): 66–77.

Onkarappa, R. Shobha, K.S. and Chaya, K., 2003. Screening of antibacterial activity of medicinally important plant extracts. *Biomedicine,* 23(1&2): 9–120.

Prema, P., 2004. Antibacterial activity of selected medicinal plants. *J. Ecobiol.,* 16(5): 333–337.

Puratchikody, A., Jaswanth, A., Nagalakshmi, A., Angumeenal, P.K. and Ruckmani, K., 2001. Antibacterial activity of *Cyperus rotundus* Linn. *Indian Journal of Pharmaceutical Science,* p. 326–327.

Ramasamy, S. and Charles Manoharan, A., 2004. Antibacterial effect of volatile components of selected medicinal plants against human pathogens. *Asian J. of Microbiol. Biotech. Env. Sc.,* 6(2): 209–210.

Satya, S. Recta Vijayarani, K. Srividhya, R. Gangatharan, N., Francis Xavier, M., Arunprasad, S. and Puratchi Kody, A., 2003. Antimicrobial screening of Leaves of *Memecylon umbellatum. Ancient Science of Life,* 23(2): 120–122.

Chapter 48

Preparation of Value Added Products (Fish Sausage) from Low Cost Deep-Sea Fish Bull's Eye (Priacanthus hamrur)

Udaykumar Nidoni, Jaya Naik and Raju, C.V.
All-India Coordinated Research project on Post Harvest Technology, College of Agricultural Engineering, Raichur, University of Agricultural Sciences, Dharwad, Karnataka, India

ABSTRACT

The Investigation was carried out to know the shelf life of Development of Value Added products, which is prepared from the low cost fish Bull's eye (*Priacanthus hamurur*), for this product. An attempt was made to prepare a kneaded, broiled Japanese fish paste product (sausage) using low cost technologies. The technology of the product preparation was standardized to adopt by common fisher folk and develop small scale or cottage business, which would ultimately bring financial benefits. The prepared product was stored at different storage temperatures and the quality changes during storage and its acceptability were studied. It was found that, Sausage Stored at 0–5 C was acceptable up to 12 days, and at –20°C the product was found to be good up to 120 days. Statically Significant results were obtained ($P < 0.05$).

Keywords: *Fish processing, Water washed, Minced meat.*

Introduction

Bulls'eye (*Pricanthus hamurur*) commonly known as Disco fish in Mangalore. It is a deep sea underutilized fish species found all along the West coast of India. Fish in general considered being an important food item from the nutritional point of view (James, 1984; Sikorski *et al.*, 1990). It provides about 14 per cent of the worlds needs for animal proteins and 4.5 per cent of the total protein requirement (Whittal, 1984). Fish supplies a good balance of proteins, Vitamins and Minerals and relatively low calories (Sikorski *et al.*, 1990). In addition, certain species of sea foods are excellent sources of n-3 polyunsaturated fatty acid, which have therapeutic Value (Kinsella, 1987). The most important break trough in the utilization of low value by-catch has been the production of fish mince and mince based products, where by considerable value addition is achieved (Ravindran, 2004). Fish mince or mince based fish is the flesh separated in a communicated form free from scales, Skin, bones and fins of fish.

Fish mince in recent years has become increasingly important for the seafood industry, both as an intermediate in surumi production and as the main raw material for breaded fish sticks and various types of formed fish sticks. One of the important uses of fish mince is preparation of surumi; Surumi is the myofibrilar protein concentrate produced by repeated washing of fish mince in order to remove water, nitrogenous matter and flavour compounds. In India onshore plants dominate current Sausage production. The main species that are used in the production of fish Sausage are bull's eye fish (*Priacanthus hamurur*), Croaker (*Jhonius* spp.) and Pink perch (*Nemptrius japonicus*). Other species that are used for fish Sausage production are Croaker and Pink perch. Fish Sausage is exported to Japan, Korea, Taiwan, Singapore, Hong Kong, China, Australia and European Union from India. The frozen fish Sausage stored at 0–5 C was acceptable upto 12 days, and at –20 C the product was found to be good upto 90 days. Statistically significant results were obtained ($P < 0.05$).

Materials and Methods

Fresh fish (*P. hamurur*) was procured from the Mangalore Fish Landing Centre and brought to the processing hall in iced condition. Fishes were washed and dressed to remove the head, entrails, and fins and again washed to remove blood and peritoneal membrane. After dressing fishes were fed into meat picking machine and meat was separated. The picked meat was then fed into the meat mincer. Then the minced meat was mixed with additives (Table 48.1) in Silent cutter was taken for the preparation of fish sausage.

The paste weighing about 20 g was filled into Natural casing of one side sealed by using Sausage stuffer, then sealed to other side of the Natural casing by using aluminium wire, and wash with soap water, then fresh water and process heat at 88–90 C for 15 min. and then cooling can be done to store at refrigerated temperature.

During processing > 200 g of paste were collected for proximate bio-chemical, Microbiological analysis of raw paste. The bio chemical, Microbiological and organoleptical changes of the product stored at refrigerated temperature was analyzed once in three days and the product stored at cold storage was analyzed once in fort fight upto 90 days. For overall acceptability 10 expert panelists have been made a taste and the additives and flow chart Process have been engaged.

Results and Discussion

The raw material characteristics were studied and presented in Table 48.2. The standardizations such as starch based filler, colour using phosphate, monosodium glutamate, sugar, sodium chloride, chilly powder, coriander, ginger and garlic were used for standardization using different proportion based on the over all acceptance scores. The selection type and level of starch is determined based on factors such as its effect on gel strength, freeze thaw stability and cost (Lee *et al.*, 1992) with a marked gel strengthening effect include corn and wheat starch (Kim and Lee, 1987) sausage prepared with nine per cent corn floor got higher organoleptic scores. Byadgi chilly powder was used to get attractive colour to the product. Sausage containing 0.4 per cent chilly powder scored higher over all acceptability scores. Moulded paste was boiled in hot water for different time intervals. It was observed that the paste boiled for sixty minutes scored higher over all acceptability.

Table 48.1: Recipe Used for the Preparation of Fish Sausage

Sl.No.	*Materials*	*Percentage Ingredients*
1.	Minced meat	70
2.	Sodium chloride	2.5
3.	Sugar	1.5
4.	Poly phosphate	0.20
5.	Monosodium glutamate	0.2
6.	Colour solution	0.13
7.	Spice Mixture	
	(*a*) Chilly powder	0.40
	(*b*) Pepper powder	0.20
	(*c*) Coriander	0.30
	(*d*) Ginger	0.05
	(*e*) Garlic	0.05
8.	Preservative	
	(*a*) Potassium sarbate	0.2
9.	Corn starch	9.00
10.	Crushed ice	10.00
11.	Fat (oil)	5.00

Changes in physical, biochemical and microbiological characteristics of sausage stored at refrigerated temperature were given in Table 48.3, and the frozen storage temperature of sausage was given in Table 48.4 the increase in expressible water in both temperature were noticed, this may be due to decomposition of starch by the activity of bacteria and enzymes resulting in conversion of bound water to free water, that results in decrease in gel strength. There was a progressive decrease in folding test grades both temperatures of stored products. Grade A for fish sausage is highly acceptable and D is not acceptable. There was a good agreement between folding test grades and texture score of organoleptic analysis of stored sausage.

The decrease in pH was noticed in the product stored at refrigerated temperature, as increase in pH was noticed in products stored at cold storage. The changes in pH during storage of sausage may due to microbial activity in the product during storage. The increase in titratable acidity is due to the production of acids from sugars by microorganisms and may be due to the formation of free fatty acids and some amino acids due to hydrolysis of lipids and proteins respectively. Both TMA-N and TVB-N is increased steadily up to ninety days in case of product stored at cold storage temperature and increased steadily up to twenty-one days in case of products stored at refrigerated temperature. The increase in TMA-N and TVB-N may be due to microbial activity in the products. No rancid odour was observed until the storage, PV in both temperatures increased steadily but well within the limits of acceptability throughout the storage periods. Freshly prepared Sausage paste had a TPC of 8.01 10^{-7} cfu/gm that reduced to 7.6 10^{-6} cfu/gm, which may be due to the boiling process.

Figure 48.1: Flow Chart for Fish Sausage

Raw Material (Fresh –0°C)

↓

Dressing-removal of Scales, head and Viscera

↓

Washing in ice cold water

↓

Meat Picking (Separation of meat from bone and skin using meat separator)

↓

Mincing (To reduce the size of meat using meat mincer)

↓

Grinding with ingredients in silent cutter

↓

Ground fish paste is fed into the hand stuffer

↓

Stuffing into the HDPE casing (one end sealed previously)

↓

Ringing using aluminum wire

↓

Washing first in soap water, than in fresh water

↓

Heat processing at 88–90°C for 60 minutes

↓

Cooling at 10–15°C for 15 minutes to attain a temperature of 20–25°C

↓

Wiping and drying under fan, storage at refrigerated temperature

Table 48.2: Raw Material Characteristics of Bull's Eye (*P. hamurur*)

(A)	**Physical Characteristics**	
	Total length (cm)	18.50
	Standard length (cm)	15.00
	Total weight (g)	83.50
	Dressed yield (%)	65.00
	Picked meat yield (%)	39.80
	Minced meat yield (%)	31.50
(B)	**Biochemical Characteristics**	
	Moisture (%)	75.53
	Protein (%)	17.87
	Total lipids (%)	01.40
	Total ash (%)	01.20
	pH	06.50
	Total volatile base nitrogen (Mg %)	07.10
	Trimethyle amine nitrogen (Mg %)	02.77
	Peroxide value (Millimoles of oxygen/kg fat)	14.87
	Free fatty acid (% of oleic acid)	02.37
(C)	**Microbiological Characteristics**	
	Total plate count (cfu/gm)	3.2 × 1
	Yeast and mold count (cfu/gm)	–
(D)	**Organoleptic Characteristics**	
	Appearance	08.70
	Colour	08.20
	Odour	08.20
	Texture	08.10
	Overall acceptability	08.30

Mean of hedonic scores as assessed by 10 panelists.

In the product stored under cold storage temperature TPC steadily increased up to 21 days storage, when the product was considered spoiled. The sensory score gradually decrease during storage from 8.2–6.3 in 12 days of storage at 0–5 C and from 8.1–5.7 during frozen storage in 90 days of storage.

As the storage period increase there was a gradual reduction in mean panel scores of all the attributes and this reduction was significant at 5 per cent level. However, between the attributes there was no significant difference, which means that there was uniform reduction in scores of all attributes during storage. Srinivas *et al.* (1999) have also made similar observations when fish sausage was stored in cold storage.

During frozen storage white discoloration was noticed after 120 days but not before 90 days, which was spread throughout the storage period. The emergence of fast foods in India has provided

Table 48.3: Changes in Physical, Biochemical and Microbiological Characteristics of Sausage Stored at Refrigerated Temperatures (–5°C)

Parameters	*Days*						
	0	*6*	*9*	*12*	*15*	*18*	*21*
Expressible water (%)	8.52	7.73	8.08	8.57	8.88	9.14	9.44
Folding test (Grades)	A	A	B	C	C	D	D
pH	6.52	6.50	6.44	6.41	6.33	6.27	6.40
Total titratable acidity (%)	0.108	0.187	0.245	0.248	0.300	0.341	0.382
TMA-N (%)	1.65	1.60	2.20	3.62	4.07	5.82	8.48
TVB-N (%)	6.26	8.23	10.42	12.71	15.03	20.36	22.92
FFA (% of oleic acid)	2.81	3.26	5.10	6.19	8.15	9.26	10.24
PV (Millimoles of oxygen/kg fat)	8.19	9.58	11.57	12.52	15.47	20.03	21.17
TPC (cfu/gm)	7.6×10^6	3.0×10^{-6}	8.11×10^{-6}	1.01×10^8	5.54×10^8	1.2×10^9	9.0×10^9
Overall acceptability	8.19	7.61	7.10	5.42	–	–	–

Table 48.4: Changes in Physical, Biochemical and Microbiological Characteristics of Sausage Stored at Refrigerated Temperatures (–20°C)

Parameters	*Days*						
	0	*15*	*30*	*45*	*60*	*90*	*120*
Expressible water (%)	7.62	8.29	9.11	10.38	11.53	12.59	14.58
Folding test (Grades)	A	A	A	B	B	C	D
pH	5.72	5.75	5.77	5.81	5.84	5.88	6.00
Total titratable acidity (%)	0.98	0.116	0.129	0.143	0.155	0.170	0.245
TMA-N (%)	1.572	3.331	3.142	4.402	5.415	6.81	8.850
TVB-N (%)	5.475	7.020	9.49	10.28	11.72	13.84	16.78
FFA (% of oleic acid)	2.81	2.82	3.83	4.55	5.31	5.98	8.05
PV (Millimoles of oxygen/kg fat)	8.29	10.59	12.17	12.77	13.55	15.89	18.73
TPC (cfu/gm)	7.6×10^6	1.02×10^6	7.6×10^5	2.30×10^4	8.13×10^4	3.45×10^3	1.9×10^3
Overall acceptability	8.26	8.07	7.86	7.06	6.88	6.79	5.81

immense opportunity for marketing innovative products such as Sausage, a value added product from underutilized and low priced fish. Bull's eye fish a deep-sea variety can effectively be utilized for the preparation of Japanese styled fish product.

References

Kim, J.M. and Lee, C.M., 1987. Effect of Starch on textural properties of Surumi gel. *J. of Food Sci.*, 52: 722.

Lee, C.M., Wu, M.C. and Okada, M., 1992. Ingredients and formulation technology for Surumi based products. In: *Proceeding of the International Symposium in Engineered Seafood Including Surumi*, National Fisheries Institute, Washington, D.C., pp. 168.

Srinivas, T.V., Bhandari, M.H and Raju, C.V., 1999. Standardisation of colour of sausage using chilly, turmeric and by smoking process. *Mysore J. Agric. Sci.*, 33: 253–260.

James, D., 1984. *The future of fish in nutrition. Info Fish Marketing Digest*, 4: 41–44.

Kinsella, J.E., 1987. *Seafoods and Fish Science in Human Health and Diseases.* Marcel Dekker Inc., New York.

Ravindran, K., 2004, utilization Fisheries resources of the Indian exclusive economic zone. In: *Large Marine Ecosystems: Exploration and Exploitation for Sustainable Development and Conservation on Fish Stocks,* (Ed.) V.S. Somavamshi. Fishery survey of India, Mumbai, pp. 252–262.

Sikorski, Z.F., Kolakowasti, A. and Pan, B.S., 1990. The nutritive composition of the major groups of Marine foods organisms In: *Sea food: Resources, Nutritional Composition and Preservation,* (Ed.) Sikoriski, Z.E. CRC Press, Boca Raton, Florida, pp 29–54.

Whittle, K.J., 1984. A vital role for aquatic resources in feeding a hungry world. *Info Fish Marketing Digest*, 5: 20–24.

Chapter 49

Studies in the Hornworts of Sitamata Wildlife Sanctuary, Rajasthan, India

B.L. Chaudhary and C.S. Dulawat*

Bryology Laboratory, Department of Botany, College of Science, Mohanlal Sukhadia University, Udaipur – 313 001, Rajasthan, India

ABSTRACT

Six hornworts have been collected from Sitamata Wildlife Sanctuary Rajasthan. These are *Anthoceros erectus* Kash., *A. subtilis* Steph., *Phaeoceros laevis* subsp. *carolinianus* (Michx.) Prosk., *Phaeoceros himalayensis* (Kash.) Prosk., *Notothylas indica* Kash., *Notothylus levieri* Schiffn. ex Steph.

Keywords: *Bryophyta, Hornworts, Sitamata Wildlife Sanctuary, Rajasthan.*

Introduction

Rajasthan is located in the northwestern part of the subcontinent. It is the largest state in India with an area of 3,24,274 sq. km and situated between latitude 23 12′N and longitudes 69 31′E and 78 17′E at a height of 578.76 m above mean sea level.

The Sitamata Wildlife Sanctuary lies in the civil districts of Chittorgarh and Udaipur in southwest region of Rajasthan state. It is situated between 24 04′N and 24 23′N North latitude and 74 24′E and 74 40′E East longitude. It covers 422.95 km^2 of area. Geographically there are three major systems namely the Aravallis, the Vindhyas and the Malwa Plateau. The Sanctuary has an average rainfall of

* Corresponding Author: E-mail: csdulawat@yahoo.com.

756 mm with a maximum up to 986 mm and minimum 510 mm. Further it has a highly varied physiography, from plateau lands to hilly tracts. Due to high rainfall, average around 756 min per annum and humidity, it is one of the richest spot of angiospermic vegetation in general and bryophytes in the state. The study area of the forest is of dry deciduous type, which consists of mixed tree growth with *Tectona grandis L. f. nom cons.* as a predominant species.

No reports are available regarding to the occurrence of hornworts from this area. However, Bapna (1969), Bapna and Chaudhary (1989) and Chaudhary and Deora, (2001), studied the bryophytes of Rajasthan, therefore author undertook this study.

Materials and Methods

The present study was conducted to identify the bryophyte flora of the sanctuary. Area was visited several times during different seasons of the year (2004–2005), especially rainy season. When bryophytes flourish well due to high humidity and optimum temperature. Field notes were taken at the time of collection to observed habit, habitat and localities during period under study and brought to the laboratory in the sealed bags. Plants were air dried, pressed and stored in well-labelled standard size packets.

For the description of genera, publication by Bharadwaj (1965), Udar (1976), Udar and Singh (1978) and standard books by Asthana and Srivastava (1991), Schuster (1969), and Singh (2002) were consulted. For morphological studies, material was dissected and observed under Olympus Binocular Microscope. Camera Lucida sketches were made under a binocular microscope. The specimens were identified and submitted in the Bryology Laboratory, Department of Botany, University College of Science, Mohan Lal Sukhadia University, Udaipur.

Results and Discussion

The following Hornworts were identified:

Anthocerotales; Anthocerotaceae

1. *Anthoceros subtilis* Steph. (Plate 49.1, Figure 1. A–H, Specimen no. CSD/SM/24)

Plants green, prostrate, fleshy in small patches, rarely forming isolated rosettes. Thallus with variable morphoforms, 4–5 mm long, 2–3 mm broad, irregularly lobed with dissected margin, cavernous. Highest stomatal frequency (20–32 stomata/sq. mm) among all the indian species of *Anthoceros.* Rhizoids simple. Surface cells polygonal, each having a simple, large chloroplast with central pyrenoid region. Internally, thallus has large mucilaginous cavities in the median region and scattered *Nostoc* colonies. Thallus usually with 4–8 capsule but the number may vary, capsules 15–30 mm long. Capsule wall 4–5 layered, epidermal layer of capsule wall with numerous stomata. Spores brown, 30–50 μm in diameter with spiriulate-baculate projections, often apically bifurcated. Elaters light brown, thin-walled, short, sometimes stout, 112–168 mm long, complete elaters usually 4-celled, elater cells usually with dark thin bands.

Field Notes

Plants grow on moist soil near Damdama nadi, Kalakhet and Mamadev.

Distribution

Maharashtra (Purandhar, Lonavala), Karnataka (Manglore), Gujarat and Rajasthan.

Plate 49.1

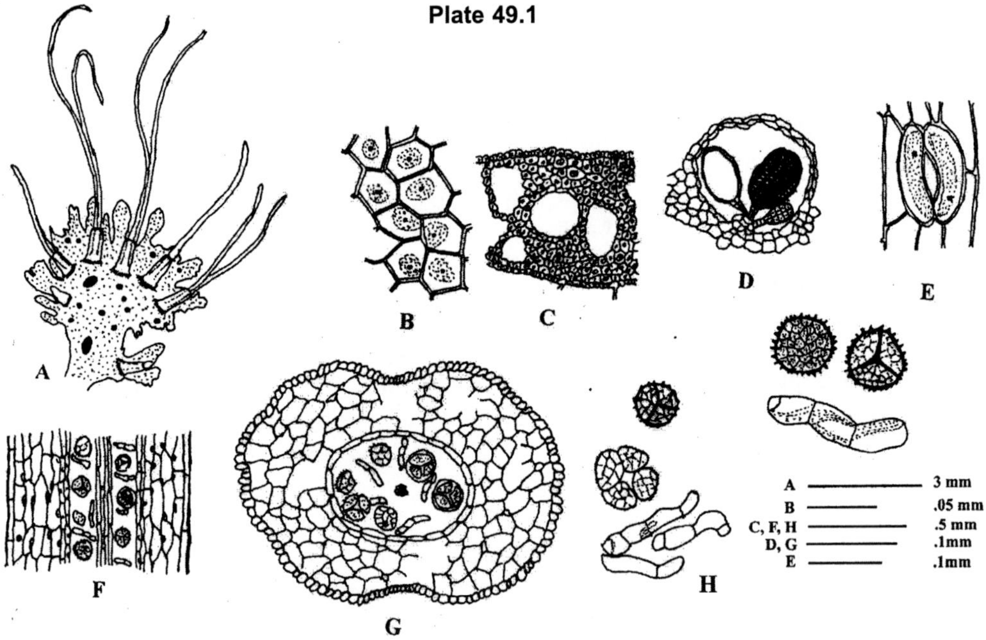

Figure 1 (A–H): *Anthoceros subtilis* Steph

(A) Thallus with sporogonium; (B) Surface cellsl; (C) Cross section of thallus; (D) Antheridia with cross section of thallus; (E) Stomata from capsule; (F) L.S. of sporogonium; (G) T.S. of sporogonium; (H) Spores and elaters.

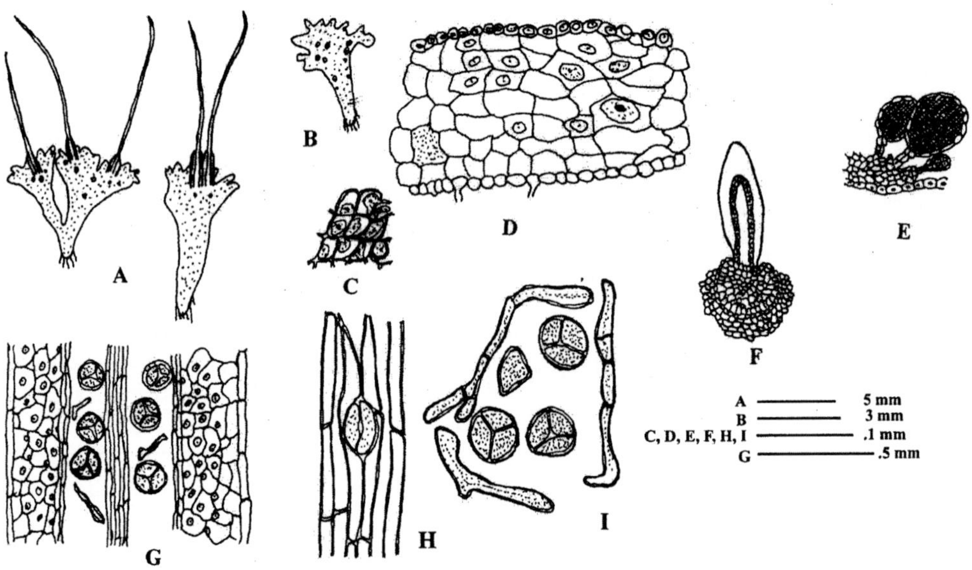

Figure 2 (A–I): *Anthoceros erectus* Kash

(A) Plant with sporogonium; (B) Male plant showing androecia; (C) Surface cells; (D) Cross section of thallus; (E) Anthredia magnified; (F) L.S. of young sporogonium showing foot and sporogenous tissue; (G) L.S. of sporophytes portion; (H) Stomata from capsule; (I) Spores and elaters

2. *Anthoceros erectus* Kash. (Plate 49.1, Figure 2, A–I, Specimen no. CSD/SM/21)

Plants light green, erect, fleshy, cavernous and in dense tufts or isolated patches. Male and female plants intermixed. Thallus with a short cylindrical stalk expended at distal and which is deeply lobed with rounded or somewhat toothed margins. In male thalli, expended portion often cup-shaped but fen like or funnel shaped in female plants. Rhizoids smooth and confined to basal stalk. Surface cell more or less rectangular or polygonal with a large chloroplast. Interior of thallus with large mucilage cavities often separated by one or two cell thick partitions; but tissue compact around the foot of sporogonium. Nostoc colonies of varying size lie deeply embedded in the tissue of thallus and irregularly distributed. Capsules 10–30 mm long and 0.4 mm thick, involucere cavernous. Spores rounded, black, reticulate 40–60 µ in diameter, tri-radiate, reticulate with broad ridge on outer face elaters simple or branched.

Field Notes

It is very common often growing in moist soil, rocks on slopes near water channel of Bhagibawari and Ambaretti.

Distribution

Mussoorie, Kullu, Nainital, Manali, Punjab, Shimla, Rajasthan and Gujarat.

Phaeocerotaceae

3. *Phaeoceros laevis* (L.) Prosk. (Plate 49.2, Figure 3. A–I, Specimen no. CSD/SM/27)

Plants monoecious, thalli lobed with wavy margin, usually broad at apex and narrowing towards base. Epidermal layer of capsule-wall stomatiferous, each stomata with two reniform guard cells surrounded by 6 more longer than broad epidermal cells having uniformly thickened radial and end-walls. Spores yellowish-green, 33–36 µm in diameter with a prominent equatorial crossitudo, sporoderm minutely papillate-granulose, proximal face marked with distinct, thin, tri-radiate mark. Elaters light yellow, 156–257 µm long, thin-walled short, complete elater 4 celled, often with irregular, dark thin bands, sometimes branched.

Field Notes

Plants grow on moist soil associated with Bryum species in Pungatalab, Charni and Anuppura.

Distribution

Western Himalayas (Shimla, Pauri, Massoorie), Eastern Himalayas (Ukhrul, Mongpo, Nathula road, Darjeeling, Cherrapunji, Gangtok, Shergaon), South India (Ooty, Mukurty, Kodaikanal, Shembaganur, Devicolam, Vagavurrai, Mysore) and Gujarat.

4. *Phaeoceros himalayensis* (Kash.) Prosk. (Plate 49.2, Figure 4. A–F, Specimen no. CSD/SM/32)

Plant monoecious. Thalli fan shaped, dark green, blackish when dry, prostrate, overlaping in large patches. *Nostoc* colonies commonly growing in dense mats. Rhizoids smooth-walled but in some cases contain granular substance. Sterile plants may bear tubers or margins, tubers also formed on female plants at the end of growing season along the lateral margins. Surface cells are mere or less hexagonal and bear a large chloroplast. Capsules cylindrical, spores pale to yellowish-green, 33–42 µm in diameter with a prominent equatorial-crossitudo, sporoderm marked with hump like projection often elongated or without rounded and broad. Elaters light brown, 123–212 µm, thin walled and sometime branched.

Plate 49.2

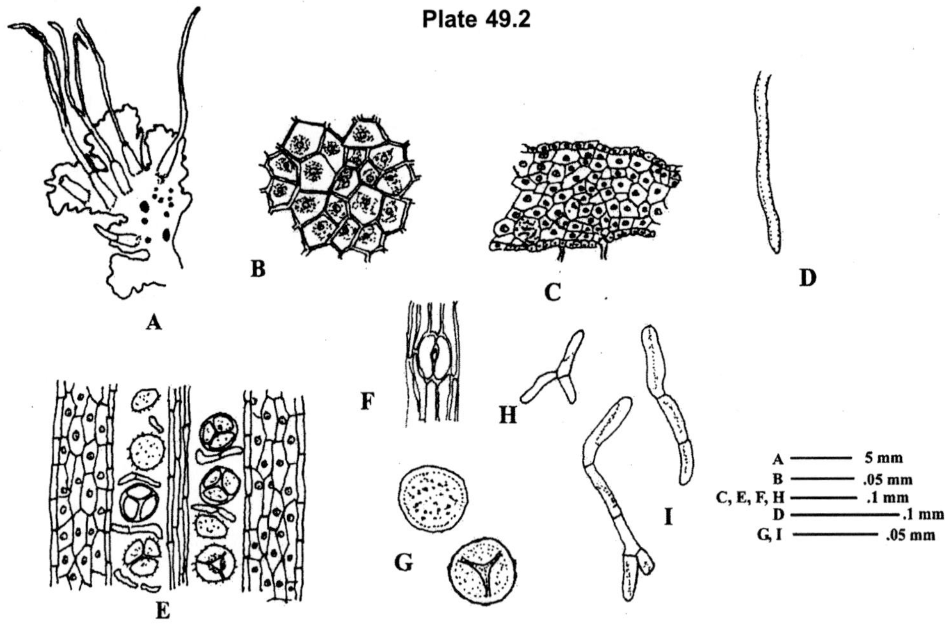

Figure 3 (A–I): *Phaeoceros laevis* (L.) Prosk.

(A) Thallus with sporogonium; (B) Surface cells; (C) Cross section of thallus; (D) Rhizoid; (E) L.S. of sporogonium; (F) Stomata from capsule; (G) Spores; (H–I) Elaters

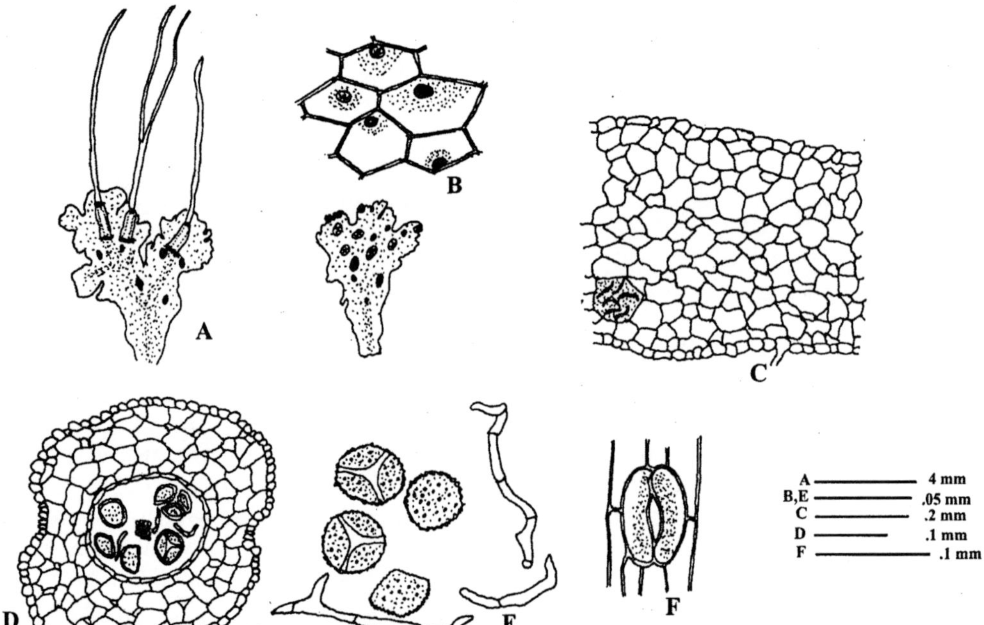

Figure 4 (A–F): *Phaeoceros himalayensis* (Kash.) Prosk.

(A) Female and Male Thallus; (B) Surface cells; (C) Cross section of thallus; (D) Cross section of sporogonium; (E) Spores and elaters; (F) Stomata from capsule

Field Notes

The plants are common on shaded moist soil, along creek bank and streams, growing gregariously near Ambaretti, Bhenwa and Pungatalab area of sanctuary.

Distribution

Sagar, Bhopal, Mahabaleshwer, Mussoorie, Kullu, Nainital, Manali, Punjab, Shimla, Rajasthan and Gujarat.

Notothylaceae

5. *Notothylas indica* Kash. (Plate 49.3, Figure 5. A–F, Specimen no. CSD/SM/39)

Plants light green, prostrate, forming rosettes, dichotomously branched. Thalli compact, 6–12 cells thick in the middle, 1–2 cells at the margins, many lobed chloroplast with central pyrenoid complex, chloroplast attached with inner cell walls by narrow strands, nostoc colonies bluish green, irregularly distributed in the thallus. Sporogonia exserted, 1–2 in each involucre, 2.5–4 mm long, having bulbous foot, a small basal meristem and oblong cylindrical, brown, bivalved capsule. Sutures 2 celled, capsule wall 3–5 cell layer thick, columella well developed, persistent, surface cells of columella spirally thickened. Spores deep brown 48–64 µm, with prominent equatorial crassitudo and conspicuous, pseudoelaters light brown, 1–2 celled, with spiral bands.

Field Notes

Plants growing on moist shady places in small patches, often associated with mosses and with *Cythodium* and canals the plains of Panchagura, Punga talab and Amba.

Distribution

Dehradun, Mussoorie, Kullu, Nainital, Manali, Punjab, Shimla, Kumaon, Gauhati, Rajasthan and Gujarat.

6. *Notothylus levieri* Schiffn. ex Steph. (Plate 49.3, Figure 6. A–F, Specimen no. CSD/SM/36)

Plants monoecious. Plant light green, thin, delicate, overlapping in small patches, forming rosettes. Thallus much lobed at the periphery, lobes irregular with toothed margin 3–4 mm in width. Capsule exserted, usually in single or in pairs and completely enclosed in a thin involucre, suture usually 4–8 cells thick. Rhizoids simple. *Nostoc* colonies of varying size are frequent. Columella totally is lacking from the capsule. Spores dark brown 32–43 µm, with minutely tuberculated sporoderm denticulate flange. Pseudoelaters 1–3 celled with spiral thickning bands.

Field Notes

Plants growing on moist shady slopes in small patches, often associated with mosses and with *Cythodium* and canals in the plains of Punga talab, Kalakhet and Pal.

Distribution

Dehradun, Mussoorie, Lucknow, Allahabad, Pachmarhi, Mumbai, Nagpur, Tikamgarh, and Rajasthan.

Acknowledgement

Authors are thankful to Dr. Virendra Nath and Dr. A.K. Asthana (NBRI Lucknow) for valuable suggestions and identification.

Plate 49.3

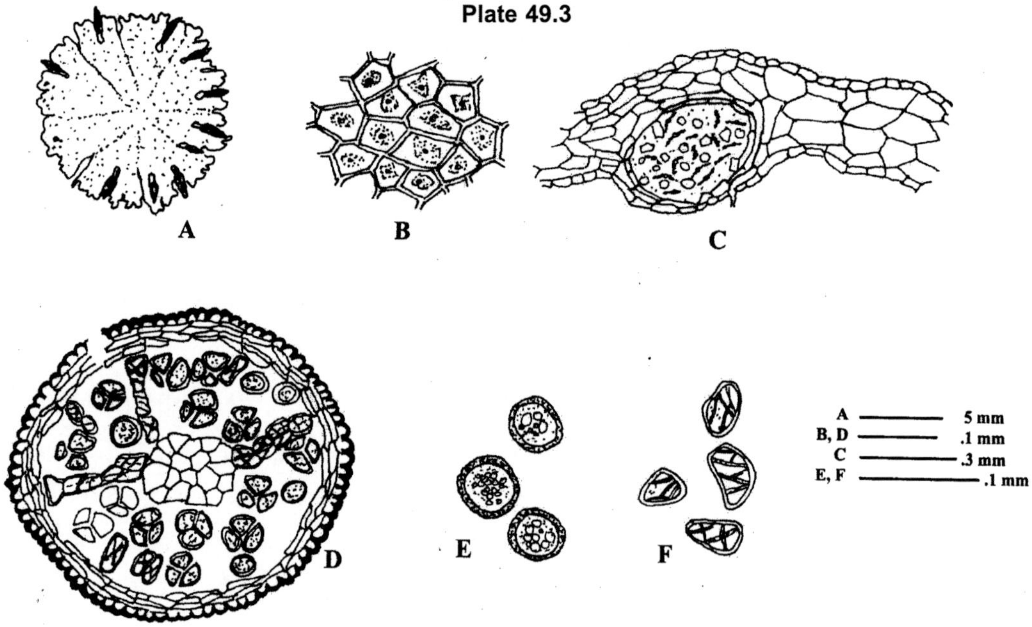

Figure 5 (A–F): *Notothylas indica* Kash

(A) Thallus; (B) Surface cells; (C) Longitudinal section of thallus portion with *nostoc* colony; (D) Cross section of sporogonium; (F) Spores; (F) Elaters

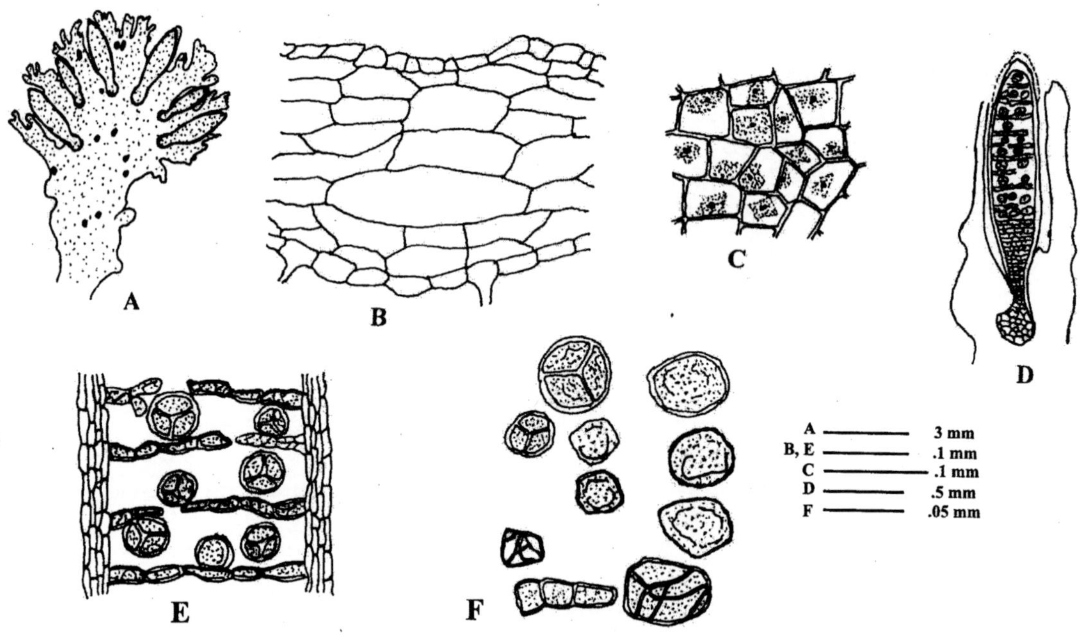

Figure 6 (A–F): *Notothylus levieri* Schiffn. ex Steph.

(A) Thallus; (B) Longitudinal section of thallus; (C) Surface cells; (D) L.S. of sporogonium; (D) L.S. of sporogonium, a portion change; (F) Spores and Elaters

References

Asthana, A.K. and Srivastava, S.C., 1991. Indian Hornworts: A taxonomic study. *J. Cramer inder Gebriider Borntraeger Verlagsbuehhandlung*, Berlin, 42: 1–158.

Asthana, A.K. and Nath, V., 1993. A new *pheocerose* from Western Himalayas, *P. Udarii* sp.nov.

Bapna, K.R., 1969. Morphological and ecophysiological studies of bryophytes of Rajasthan with special reference to hepatics. *Ph.D. Thesis*, University of Udaipur.

Bapna, K.R., 1977. On the occurrence of *Anthoceros longii* Staph. in Rajasthan and it's validity. *J. Indian Bot. Soc.*, 56: 258–261.

Bapna, K.R. and Kachroo, P., 2000. *Hepaticology in India* Vols. 1 & 2. Himanshu publication, New Delhi, India.

Bharadwaj, D.C., 1958. Studies in Indian Anthocerotaceae II. *J. Indian Bot. Soc.*, 37: 75–92.

Bharadwaj, D.C., 1965. Studies in India Anthocerotaceae VI. Some aspects of morphology of *Pheocerose Prosk. Phytomorphology*, 15: 140–150.

Bharadwaj, D.C., 1960. Studies in Indian Anthocerotaceae III. The morphology of *Anthocerose erectus* Kash. and some other species. *J. Indian Bot. Soc.*, 39: 568–502.

Mehra, P.N. and Handoo, O.N., 1953. Morphology of *Anthoceros erectus* and *Anthoceros himalayensis* and phylogeny of *Anthocerotales*. *Bot. Gaz.*, 114 (4): 371–382.

Chapter 50

Forest Types of Neeru Valley, Bhadarwah, Jammu and Kashmir

Harish Chander Dutt and Shashi Kant
Department of Botany, University of Jammu, Jammu

ABSTRACT

Two of the four forest Ranges of Bhadarwah Forest Division *viz.*, Neeru and Kellar are situated altogether in Neeru valley. Total area of all the four forest ranges is about 110038 hectares, out of which 18086 hectares or Neeru range and 14239 hectares of Kellar range is computed from topographical map sheet numbers 43-O/12, O/16, P/9, P/13, 52-C/4, 52-D/i at 1 : 50,000 scale respectively. The total area of these two ranges is about 64024 hectares out of which 32325 hectares is computed as Forests from topographical map sheet 43/O, 43/P and 43-O/12, O/16, P/9, P/13, 52-C/4, 52-D/i at 1 : 25,00000 and 1 : 50,000 respectively.

In present communication these forests are classified in seven sub-types *viz.*, 10/C1a with *Olea ferruginea* Royle., *Terma politoria* Planch. and *Quercus baloot* Griff.; 12/C1a, with *Q. dilatata* Lindl., *Q. incana* Roxb., *Rhododendron arboreum* Sm. and *R. campanulatum* D.Don.; 12/C1b, with *Q. dilalata, Q. incana, Abies pindrow* Spach. and *Pinus wallichiana* A.B. Jackson; 12/C1c, with *Cedrus deodara* Roxb. and *Aesculus indica* Hiern.; 12/C1d, with *Abies pindrow* Spach., *Cedrus deodara, Picea smithiana* (Wall) Boiss and *Pinus wallichiana;* 12/C1f, with pure stands of *Pinus wallichiana;* 12/C2a, with pure stands of *Q. incana* as dominant tree species and three types 13/C1, with *Pinus roxburghiii* Sargent., *Q. dilatata, Fraxinus* species, *Zizyphus mauritiana* Lam., *Cedrus deodara, Alnus nitida* Endl. and *Spiraea sorbifolia* Linn. and *Ficus palmata* Forsk., 15/C1, with *Betula utilis* D. Don., *R. arboreum, R. campanulatum* and 16/E1, with *Juniperus* species as dominant tree/shrub species or five groups in accordance with the classification of Champion and Seth (1964).

Keywords: *Temperate forest, Coniferous forest, Neeru valley, Bhaderwah.*

Introduction

For most of recorded history forests have been recognized not only as a source of personal succor, even wealth, but also as a communal resource, a source of water, game, land, fuel and timber for all. India is one of the richest countries in the world in both inorganic and organic natural resources. Mother India is connected by land with the, Chinese and Asiatic regions of the world which exercise a powerful influence on the natural characteristics of the physical landscape, climate and plants, animals and human populations. Both geologically and geographically these influence have contributed a great deal in determining what may be called the Indian and the non-Indian or the exotic elements in Indian Biology.

Study Area

The study area Neeru watershed is stretched between latitude 33 08′24″–32°52′26″N and longitude 75 48′18″–75°32′41″E is tributary of Chenab. Neeru stream is duel in origin and originated from Kaplash Kund (4341 masl) and Ashapatti Glacier (3300 masl) and open into Chenab at Pul Doda (821 masl). The area of its catchment is 64,024 hectares and is surrounded by Kishtwar and Dada tehsils of district Doda in north and north-west, by Chamba district of Himachal Pradesh in east and Kathua and Udhampur district of J&K in the south-cast respectively. The study area is explored by various workers *viz.*, Raina and Kant (1993) reported 488 plant species from Bhadarwah (Neeru valley), among macrofungi about 80 species are collected from Bhadarwah forests by Rampal (1988). Kumar (1987) recorded 129 economic important plant species from the area in his Ph.D. thesis. But a complete forest map of the area is not published for the time, because it needs a detailed and extensive exploration. Thus, the present study was advised to explore the area and construct a detailed forest map of the area. The study area Neeru watershed, bounded on all sides of Himalayan ranges, is dominated by coniferous trees *viz., Abies pindrow, Cedrus deodara, Picea smithiana, Pinus roxburghiii* Sargent., *Pinus wallichiana.*

In some parts below 1200 masl trees species like *Olea ferruginea, Trema politoria, Zizyphus mauritiana, Quercus* species, *Punica granatum* Linn., *Ficus palmata,* dominates the flora of the area. These species are very much common on hillocks of right bank of Neeru drainage, while *Pinus roxburghii* is one more addition on the slopes on left bank. Above 1200 masl to 2000 masl. *Alnus nitida* a broad leaved tree dominates the vegetation along both the banks of the Neeru Nallah. *Pinus wallichiana* and *Cedrus deodara* extend over an elevation of 1,500 masl to 3,000 masl in forest area. At higher elevations (above 2,500 mals) the stands of *Abies pindrow* and *Picea smithiana* are stretched on the mountain peaks, while *Taxus wallichiana* Zucc. occurs above 2,000 masl elevation and is very poorly represented in the area.

Quercus incana Roxb., *Q. dilatata, Q. semecarpifolia* J.E. Smith, *Alnus nitida, Prunus cornuta* Wall., *Fraxinus excelsior* Linn., *Aesculus indica, Rhus succedanea* Linn., *Juglans regia* Linn., *Ficus palmate, Populus ciliata* Wall. and *Acer* species are important broad leaved species in India.

In shrub vegetation, *Berberis lycium* Royle., *Daphne oleoides* Schreib., *Elaeagnus umbellata* Thund., *Cotoneaster bacillaris* Wall., *Prinsepia utilis* Royle., *Ficus palmate, Parrotiopsis jacquemontiana* Dene., *Rabdosia rugosus* Wall., *Rubus ellipticus* Smith., *Spiraea canescens* D.Don., *Rosa macrophylla* Lindl., *Indigofera gerardiana* Wall., *Lonicera quinquelocularis* Hardw. are dominating species in the area. At low elevation (below 1200 masl) *Punica granatum* is one more addition to the flora.

In the present study the forest types of Neeru valley are C1assified in accordance with the classification of Champion and Seth (1964).

Observations

The forests of Neeru-watershed have been divided into seven sub-types of five groups in accordance with the classification made by Champion and Seth (1964). These sub-types and types are described below:

Group 10: Sub-Tropical Dry Evergreen Forests

Type 10/C1: Sub-Tropical Dry Evergreen Forest

Sub-Type 10/C1a: Olea cuspidata Scrub Forest

This types of forest extend from Pul-Doda to Paranoo on the slops of right bank of Neeru Nallah. The forest is characterized by *Olea ferruginea* syn. *Olea cuspidata* scrubs followed by few stands of *Quercus baloot* on the right bank of Neeru stream, while towards the top of right bank slopes *Olea ferruginea* is followed by *Trema politoria.* The slopes of left bank of Neeru stream between Pul-Doda and Paranoo are enriched with *Olea ferruginea* followed by *Q. baloot, Zizyphus mauritiana* and *Punica granatum.* Beside these species patches of trees of *Pinus roxburghii* and *Trema politoria* are found on these slopes. It is also observed that *Olea cuspidata* scrub forest, on left banks doesn't extend above Nali Dangri, Bigotha, Bigotha, Grondi, Bhubba, Bata and Masiri villages. On the slopes towards the mouth of Neeru stream, patches of *Dalbergia sissoo* Roxb. *Nerium indicum* Mill., *Calotropis procera* Br. and *Opuntia vulgaris* are common, while *Zizyphus mauritiana* and *Punica granatum* are distributed towards the Paranoo. A broad leaved tree species of *Alnus nitida* shows a transition towards temperate region. The area covering sub-type 10/C1a is directly exposed to the sunrays thus low moisture contents in soil is a characteristic of the slopes of this zone. The soil of these slopes is loose, gray in colour with neutral to alkaline pH.

Group 12: Himalayan Moist Temperate Forests

Type 12/C1: Lower Western Himalayan Temperate Forests

Sub-Type 12/C1a: Ban Oak Forest

Sub-type 12/C1a *i.e.* Ban Oak forest is restricted to pockets at high elevations in Neeru Valley. The dominant tree species of this zone are *Quercus dilatata, Quercus leucotrichophora,* A. Camus syn. *Q. incana* while *Rhododendron arboreum, Rhododendron campanulatum, Juniperus recurva* Ham. are represented as associated species. The forest sub-type extends from Gul-Danda Dhar to Ramtund including some parts of Chuncholu slope. Seoj-Dhar slope facing Ramtund peak doesn't fall in this sub-types but stands of *Q. leucotrichophora* are distributed towards the top of this slope also. Generally moist soil is the characteristic feature of this zone while some places are devoid of shrubby vegetation due to rocky surface of earth. When we move towards Sundrikot and Kirmiri Gali, sub-type 12/C1a forest is stretched on the lower belt of Sunderikot and Kirmiri Gali. It is observed that Ban Oak Forest *i.e.* sub-type 12/C1a with *Quercus* species in lower stratification and *Rhododendron* species in upper stratification are stretched in the area.

Sub-Type 12/C1b: Moru Oak Forest

Moru Oak forest *i.e.* 12/C1b is represented by the stands of *Quercus dilatata, Abies pindrow* and *Pinsu wallichiana* in some grooved pockets in between the hills of left bank of Neeru stream. However this sub-type is totally absent on the light bank of Neeru stream. Thus, it is clear that the forest sub-type 12/C1b is restricted to some upper pockets in between the hills of left bank of Neeru stream. Between Dhar Lachran to Mushdeo-ro-Nal the stands of *Quercus dilatata, Abies pindrow, Pinus wallichiana* represent sub-type 12/C1b in deep grooved valleys, while *Abies pindrow* flourish well towards the periphery of these grooves where sunlight reach easily. Due to heavy precipitation in the form of snowfall in winters and rainfall in summers the soil remain moist throughout the year. Moderate acidic pH between 5–6.5 is the characteristic feature of the soil of this forest type.

Sub-Type 12/C1c: Moist Deodar Forest

Dominating sub-type 12/C1c of the area is represented by pure *Cedrus deodara* stand with few countable trees of *Aesculus indica.* On hills and peaks of right bank of Neeru stream sub-type 12/C1a is stretched between Subar Dhar and Thanala. This stretched area includes Lanchan Dhar in east and south facing slope of Thubba in west, width gets narrowed between Kansaru and Duggi slopes. The forest sub-type being extended upto Thanala forms a pure continuous belt of *Cedrus deodara.* Beside some grooves, 12/C1c forest sub-type is also distributed on mountains and sub-valleys of left bank of Neeru steram. Forest sub-type 12/C1c extend from slopes of Nalthi to Dhar Lachran in continuous stand while sub-type 12/C1b is suspended in the grooved pockets in this (12/C1c) forest sub-type. It is also observed that sub-type 12/C1c also extend in the form of narrow stretch between Dhar Lachran and Bigotha in Kellar range. This narrow stretch includes Dranga, Gutasa, Darodoo, Galian, Niota, Panj Gram, Chanti, Upper Bhalla, Traown and Bigotha.

Although the area of this forest sub-type remain moist throughout the year due to frequent rains but evidence of forest fires are common in this forest sub-type due to the availability of abundant resin. Acidic soils is a characteristic feature of this region.

Sub-Type 12/C1d: Western Mixed Coniferous Forest

The sub-type 12/C1d in the area is represented by *Abies pindrow, Cedrus deodara, Picea smithiana* and *Pinus wallichiana.* Just above sub-type 12/C1c, on high elevation of left bank of Neeru stream, sub-type 12/C1d is stretched from upper portion of Dharaphar to Ramtund in the form of narrow continuous stand. On right bank of Neeru stream, this forest type is restricted to the upper peaks of Nagni Mandir and peaks above Duggi-Nal. Some patches have been located in between Padri Gali and Katari-di-Gali. Mixed forest stands of *Abies pindrow, Cedrus deodara,* and *Picea smithiana* are also stereteched between Ashapati tributary and Kaplash tributary of Neeru-watershed. Few plants of *Betula utilis* on rocky surface and *Taxus wallichiana* in the grooves are also located in between these two tributaries. Zamanari-dhar is known for Himalayan Yew *i.e. Taxus wallichiana.* The soils of the forest type remain wet due to frequent rainfall in summer while winters are characterized by heavy snowfall.

Sub-Type 12/C1f: Low Level Blue Pine Forest

Sub-Type 12/C1f forest sub-type is represented by pure *Pinus wallichiana* stand in Bhadarwah-Himalayan Moist temperate forests. These stands are patched in the lower belt of sub-type 12/C1c on the peaks of left bank of Neeru stream. The soil of the area remain acidic and humus rich.

Type 12/C2: Upper West Himalayan Temperate Forest

Sub-Type 12/C2a: Kharsu Oak Forest

Sub-type 12/C2a *i.e.* Kharsu Oak forest is represented by pure stands of *Quercus semecarpifolia* and few *Trema politoria* on the hillock slops of right and left banks near Paranoo. The soils of the area remain dry and gray in colour.

Group 13: Himalayan Dry Temperate Forests

Type 13/C1: Dry Broad Leaved and Coniferous Forest

Climatically intermediate between sub-tropical and temperate zone, type 13/C1 lie from Paranoo to Bhalla, while upper peaks of Kellar forest range doesn't fall in this type. On the slopes of right bank between Paranoo to Bhalla *Pinus roxburghii, Quercus dilatata, Fraxinus excelsior* are distributed. *Zizyphus mauritiana, Cedrus deodara* and *Pinus wallichiana* represent the 13/C1 type above Bhalla and in vicinity of Gurakha (slopes on right bank). Along both the banks of Neeru, the population of *Alnus nitida* and *Ficus palmata,* two broad leaved species are distributed along both the banks of Neeru stream and its tributaries from Bhalla to Bheja. Slopes of left bank between Paranoo and Bhalla are characterized by *Pinus wallichiana, Pinus roxburghii, Quercus dilatata* at low elevation while *Cedrus deodara* dominate the vegetation at higher elevation. Snowfall is the prime requisite for this forest type. Among shrubs, *Berberis lycium, Rosa macrophylla, Rubus ellipticus, Daphne oleoides, Viburnum grandiflorum* Wall. dominate the shrubby vegetation. *Ficus palmata* grow individually near the water streams.

Group 15: Moist Alpine Scrub

The forest of this group form the vegetation of very high elevation, along the snow line.

Type 15/C1: Birch/*Rhododendron* Scrub Forest

This forest type mainly is represented by dwarfed stunted trees with grassy patches scattered in between, and is restricted to the slopes or Ashapatti only. The flora of this zone include *Betula utilis, Rhododendron arboreum, R. campanulatum, Corydalis* species and *Lonicera quinquelocularis* etc. The trunks are short and extremely branched and attain a girth about 50 cms. The area is characterised by very high snowfall.

Group 16: Dry Alpine Scrub

Type 16/E1: Dwarf Juniper Scrub

Dwarf Juniper scrub forest type is restricted in distribution to the upper portion of the Neeru Valley particularly Ramtund area towards Kaplash Lake. Adjacent to the open meadows the species like *Juniperus communis* Linn., *Juniperus recurva, Caragana* species are distribution patches. The species grow well in the rocky area, which become dry even due to the small sunshine. Although frequent rainfall in summer is characteristic feature of the area, but area remain dry due to rocky surface. Winters are known for heavy snowfall in the area.

Conclusions

The range of elevation varies from 821 masl to 4500 masl. The area embraces climate from sub-tropical to alpine, cut up in numerous valleys big and small with different aspects and slopes, presents varied environment for vegetation to grow. The dominant coniferous species composing the forests, in order of their economic importance are *Cedrus deodara, Pinus wallichiana, Abies pindrow* and *Pinus roxburghii.* Negligible sprinkling of *Picea smithiana* and *Taxus wallichiana* is also present. Incidence of shrubs is very high in the upper forests of Fir, the most common shrub met with is *Viburnum grandiflorum, Parrotiopsis jacquemontiana* restricted to cooler area. *Berberis lycium, Plectranthus rugosus, Principia utilis, Daphne oleoides* and *Sarcococca saligna* Muell. is locally abundant in deodar forests. But this green wealth is under a great threat of destruction due to heavy grazing, cutting and lopping for torchwood and fodder, girding for sheet-bark roofing, parasitic infection, wood damaging fungi, insects like Scolytidae bark borers, wild animals and climate injuries.

Acknowledgements

Author is thankful to G.B.P.I.H.E.D., Almora for providing the funds for the project Cultivation and propagation of Traditional Herbs of Jammu. The present work is the part of the project.

References

Champion and Seth, 1964. *Forest Types of India.* Government of India.

Kumar, A., 1987. Physiosociological and productivity studies in Bhadarwah forests, Jammu (J&K). *Ph.D. Thesis,* University of Jammu, Jammu.

Raina, A.K. and Kant, S., 1993. Life forms and biological spectra of Bhadarwah forests, Jammu. *The University Review,* University of Jammu, Jammu.

Rampal, R.K., 1988. Ecological studies of the macrofungi of Bhadarwah forests, Jammu (J&K). *PhD. Thesis,* University of Jammu, Jammu.

Chapter 51

Studies on the Impact of Phytopesticide, Nimbecidine on Histological Changes in the Testis of Adult Male, Insect *Spheárodema rusticum* (Heteroptera : Belostomatidae)

C. Elanchezhiyan, V. Shoba and Mrs. Selvisabhanayakam
Department of Zoology, Annamalai University Annamalainagar – 608 002, Tamil Nadu, India

ABSTRACT

Insects are highly evolved groups since Ordovician period. They have wide adaptation to different habitats. Hence the insects are the chief biological organisms not only causing severe damage to crops on land but also causes severe injury to economically important game fishes such as Catla, Rohu etc. For successful aquaculture, the control of predator insects are taken into consideration, works in relation to the control of predator insects by using phytopesticide like nimbecidine are rare. Therefore, it has been programmed in the present study to find out the effect of nimbecidine on the histological changes in the testis of *Sphaerodema rusticum.* The sublethal concentration was found out about 0.13 per cent for nimbecidine. The insects were exposed to nimbecidine for 7 days, 14 days and 21 days have revealed certain histopathological changes in the testis. The testis consisting apical cells, gonial cells, primary and secondary spermatocytes and sperm bundles in the control insects. In treated insects, several histopathological changes have been observed such as atrophy of apical cells, swollen spermatogonial cells, pycnotic and necrotic testis follicle, sperm bundles were disorganized and looks like whorl than 7 days and 14 days exposure. It is suggested that the phytopesticide causing several histopathological damages in the testis of *S. rusticum* and inturn causes severe reproductive potentiality of *S. rusticum,* a predator insect in aquaculture.

Keywords: *Nimbecidine, Sphaerodema rusticum, Testis.*

Introduction

In India, neem extracts are used as insect repellent to protect the agricultural crops, household and stored grains. The Neem Seed Kernel Suspension (NSKS) at 2 per cent level function as a dependable repellent in *Spodoptera littura* (Joshi *et al.*, 1975). The neem products are used in agricultural field to reduce terrestrial pollution and thus prevent food poisoning. The neem based pesticides are safe, non-toxic, non-polluted and avoid the accumulation of residual effects on agricultural products (Hedge, 1993). Azadirachtin enriched neem extract is used in the management of human and animal pests, management of stored products pests, management of garden and horticultural pests and management of crop pests. However, interest in botanical pesticides received during recent years because of some of the draw backs of synthetic insecticides, including lack of selectivity, impact on the environment and the emergence and spread the pest resistance. At present time, isolation, identification and development of natural products are the focus of numerous research programmes around the globe. Extracts of the Indian neem–tree, *Azadirachtica indica* are gaining prominence, the foremost biotoxical insecticides of this decade (Prabhakar *et al.*, 1986 and Stone, 1992).

Several studies have been made on histological changes in the testis due to pesticide treatment in few orders such as Coleoptera (Alallah and Newson, 1965); Isoptera (Rajinder *et al.*, 1965, Hamstead, 1966); Diptera, (Kennedy, 1947; Beard 1965; Benschoter, 1967; Choudhry and Tripathi, 1975; Deepak and Choudhry, 1978) and Hemiptera (Redfren *et al.*, 1980).

In recent years, there has been a spurt in research concerning the reproduction of invertebrates, as they exhibit complexity in their reproductive physiology. The reproductive system of insect is a complicated one. Testicular development and spermatogenesis have been studied in many lepidopteran insects (Numata and Hidake, 1980 and Lai-Fook, 1982).

With this in view, several studies have been undertaken to understand the histological changes in the testicular tissues of insects due to treatment with heavy metal, pesticide and other chemicals such as endrin (Irrinmount, 1970); alpholate (Gangrade and Sand, 1970); penfulron (Redfren *et al.*, 1980); phosphamidon (Balakrishnan, 1990 and Nirmala Devi 1990); nimbeciline (Thiruvasagam, 1994); *Pongamia glabra* leaf extract (Ramanathan, 1995); heavy metal mercuric chloride (Palanisamy, 1997); monochrotophos (Ravichandran, 1996) and endosulfan (Sumathi, 2002).

However, the action of nimbecidine, which is a insecticide compound with relatively high toxicity has not been well documented with reference to testicular tissues in the adult male insect, *Sphaerodema rusticum*, a predator insect in aquaculture.

Materials and Methods

Sphaerodema rusticum is found in freshwater pond and streams, clinging to waterweeds. The insects were daily fed with mosquito larvae, aquatic plants and pieces of earthworms. The troughs were cleaned properly every alternative days include changing the water.

Histopathological Preparations

The control and treated insects were kept separately. After 48 hours they were dissected under binocular microscope by using insect Ringer solution (Ephrassi and Beadle, 1936). The Ringer was subsequently removed and the tissue was fixed in Bouin's fluid for 24 h, later the tissue was processed by adopting standard histological techniques. (Gurr, 1958).

Results and Discussion

A transverse section of the testis composed of a large number of cysts compactly held together. Each follicle has a thin wall consists of large apical cells, which were mainly meant for providing nourishment for the germ cells called trophocytes. The spermatogonia was invariably found closely associated with the germinal epithelium. The secondary spermatocytes were comparatively more in number than the primary spermatocytes where, reduction division taken place and given rise to spermatids and further, it developed into sperms. The sperms were elongated and have hair like structure arranged in the form of bundles (Figures 51.1 and 51.2).

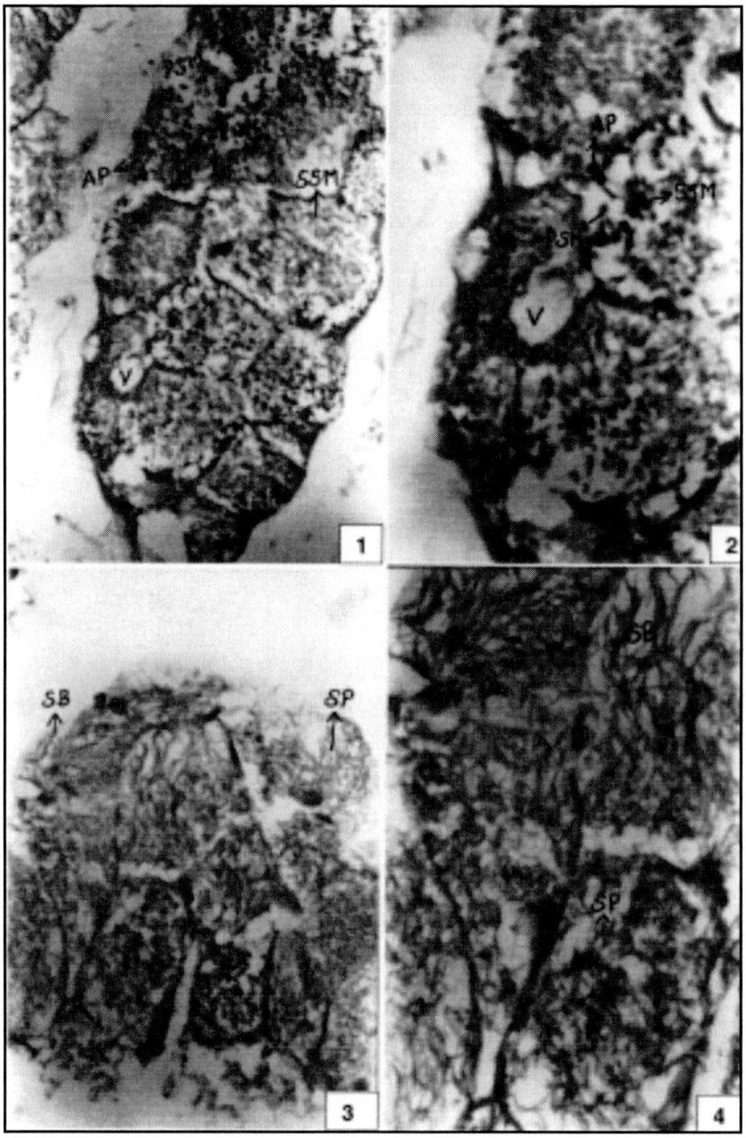

Figures 51.1–51.4

The nimbecidine treated insects exhibited remarkable histological changes in the testis such as the occurrence of pycnotic and necrotic peritoneal membrane with disintegrated nutritive cells indicate, the less amount of nutrition for the developing germ cells and spermatids. The spermatids seemed to be less in number and clumped. The primary and secondary spermatocytes appeared to be increased in their size and volume (Figures 51.3 and 51.4).

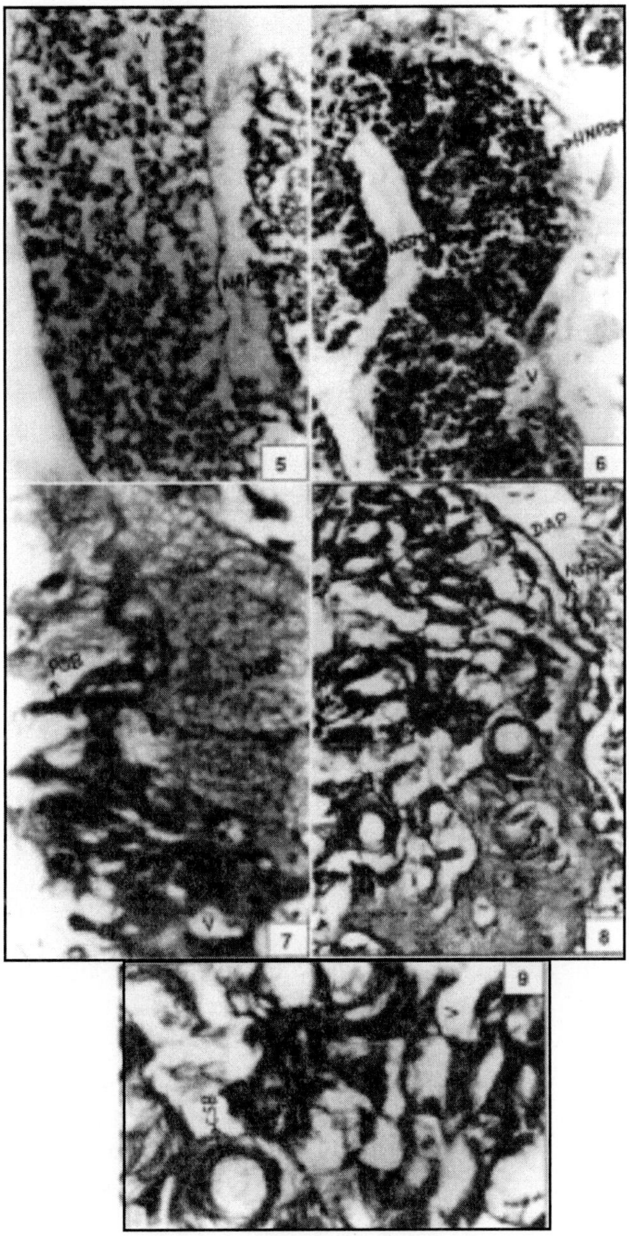

Figures 51.5–51.9

The chronic exposure of nimbecidine to the testis of *Sphaerodema rusticum* showed certain histopathological changes such as the occurrence of apical cells were highly disorganized and disintegrated. The primary and secondary spermatocytes were highly pycnotic and necrotic. The sperm bundles were found to be disintegrated, disorganized and highly pycnotic indicate that the sperms were not active which impaired the reproductive potentiality of the insect (Figure 51.5). The follicle of the treated insects contained more vacuoles. The sperm bundles were coiled, pycnotic and necrotic. The spermatids were broken into pieces. It indicates that the sperms were not considered as an effective male gametes which reduced the reproductive potential than the control (Figure 51.6.) Highly obliterated testis with necrotic and pycnotic spermatocytes. The sperms were highly nectrotic, coiled and conglomerated (Figures 51.7–51.9) indicates that sperms were not viable seems to be sterile than compared to the control.

In the present study, it has been observed that there are several changes in the morphology of the treated testis such as disintegration of primary and secondary spermatocytes, spermatids and chumping of sperms compared to the control. These findings are in parallel with the works on *Odontopus varicornis* (Jayakumar, 1988); *Pheropsophus lissoderous* (Balakrishnan, 1990); *Periplaneta americana* (Ramanathan, 1995); *Sphaerodema rusticum* (Rathinam, 1997), *Oxya nitidula* (Prasanna, 1998) and *Sphaerodema rusticum* (Rajathi, 2004), respectively.

It has been observed that the changes in the testis of treated insects for 7, 14 and 21 days were more obvious than the insects exposed for 48 h. The sperm bundles were coiled, pycnotic and necrotic sperms affect the motility and ultimately caused the sterility of insects. Thus, it is evident from the present observations that chronic exposure of the phytopesticide, nimbecidine which act as antifertility agent brought about certain reproductive impairment in this test insect, *Sphaerodema rusticum.*

Acknowledgement

The authors are very grateful to the authorities of Annamalai University and to the HOD of Zoology for providing all necessary facilities to complete this work.

References

Atallah, Y.H. and L.D. Newson, 1965. Ecological and Nutritional studies on *Coleonigilla Maculata* De Gew (coleoptera) III. The effect of DDT, toxaphere and endrin on the reproductive and survival potential. *J. Econ Ent.*, 59: 1181–1187.

Balakrishnan, S., 1990. Studies on the effect of Roger on the histology and histochemistry of the testis, fat body and accessory reproductive glands of the adult male *pherosophus lissoderus. M.Phil. Thesis*, Annamalai University.

Beard, R.L., 1965. Ovarian suppression by DDT and resistance in the Housefly, *Musca domestica. Entamon. Exp. Appl.*, 8: 193–204.

Benschoter, C.A., 1967. Effect of dietary biotin on reproduction of the Housefly. *J. Econ. Ent.*, 60.

Choudhry, H.S. and C.P.M. Tripathi, 1975. Effect of thiourea on the Nucleocytoplasmic relations of Nurse cells and developing occyte in sarcophagi ruficornis of all India insect. In. *Proceedings Chemostoriliant Research Workers Conference*. U.N.D.P. ICAR Project on Plant Production U.A.S. Bangalore, p. 78–81.

Deepak, V. and H.S. Chaudhry, 1978. Histopathological effects of thiourea on ovarion development of *Calliphora erythrocephala* (Diptera : Calliphoridae). *Appl. Ento. Zool.*, 13: 34–37.

Epherussi, B. and G.N. Beadle, 1936. A technique of transplantation of Drosophila. *Amer. Nat.*, 70: 218–225.

Gangrade, G.A. and N.C. Sant, 1970. Effect of apholate on reproductive organs of almond Moth *Cadra caultella* (Walker) (Lepidoptera : Phycitidae). *Indian J. Ent.*, 32(2): 134–139.

Gurr, E., 1958. *Methods of Analytical Histology and Histochemistry.* Leonard Hill (Brooks) Ltd., London.

Hamstead, E.O., 1966. Reduction in reproductive capacity of European Red Mite by Niagra, 9203. *J. Econ. Ent.*, 59(2): 481.

Hedge, N.G., 1993. Neem production and development constraints at grass root level. *World Neem Conference*, p. 1–7.

Irvinmount, D., 1970. Chronic effects of endrin on blunt-nose minnouws, (pipe pholasnotatus) and *Quippies octiculatus*). *Ann. Fish. Soc.*, 96: 185.

Jayakumar, C., 1988. Effect of dimethoate (Roger) on the accessory reproductive gland, testis, fat body, haemolymph neurosecretory cells and corpus allatum in *Odontopus varicornis* (Dist) (Hemiptera : Pyrrhocoridae). *Ph.D. Thesis*, Annamalai University.

Joshi, B.G., Ramaprasad, G. and Sitharamaiah, S., 1975. Effect of neem seed kernel suspension on *Telenomus remus* an egg parasite of *spodoptera littura*. *Phytoparasitica*, 10: 16–63.

Kennedy, J.S., 1947. The excitant and repellant effects on mosquitoes of sublethal contacts with DDT. *Bull. Ext. Res.*, 37: 593–607.

Lai-Fook, K., 1982. The role of male accessory glands in insects reproduction. *A. Rev. Ent.*, 21: 199–221.

Nirmala Devi, 1990. Effect of phosphamidon on the histology and Histochemistry of the testis and accessory gland of the adult male *catacanthus incarnates* (Spin) (Hemiptera : Pentatomidae). *M.Phil. Thesis*, Annamalai University.

Numata, H. and Hidaka, T., 1980. Development of male sex cells in the swallow tail *papilio xuthus* L. (Lepidoptera : Papilonidae) in relation to pupal diapause. *Appl. Ent.*, 200(15): 851–859.

Palanisamy, K., 1997. Studies on the effect of heavy metal (Mercuric chloride) on histopathological and biochemical changes in the selected tissue of the adult male Laecoterphes rubber (Linn) (Heteroptera : Nepidae). *M.Phil. Thesis*, Annamalai University.

Prabhakar, N.D.C., Condrict, A.N. Kishaba and D.E. Negar Dirk, 1986. *J. Beon. Entomol.*, 79: 39–41.

Prasanna, V., 1998. Effect of nimbeciline on the histopathological changes in the testis of the adult male paddy pest, *oxya nitidula* (walker) (Orthoptera : Acrididae). *M.Sc. Dissertation*, Annamalai University.

Rajathi, V., 2004. Impact of heavy metal mercury in the adult male reproductive physiology of *sphaeromderma rusticum* (Heteroptera : Blastomatidae). *Ph.D. Thesis*, Annamalai University.

Rajinder, S., Saini and L.K. Cutkomp, 1965. The effects of DDT and sublethal dose of bicofol on reproduction of the two spotted spider mite. *J. Econ. Ent.*, 49: 249–254.

Ramanathan, B., 1995. Studies on the effect of leaf extract of *Pongamia glabra* (vent) (Fabaceae) on histological and biochemical changes in testis and fat body of the adult male *Periplaneta americana* (Linn) (Orthoptera : Blattidae). *M.Phil. Thesis*, Annamalai University.

Rathinam, 1997. Impact of Nimbeciline on the histological changes in the testis of *Sphaerodema rusticum* (Duf.) (Hemiptera : Belastomatidae). *M.Sc. Dissertation*, Annamalai University.

Ravichandran, S., 1996. Toxicological studies on the ovary, fat body and haemolymph of the adult female *Laceotrephes rubber* (Linn) (Heteroptera : Nepidae). *Ph.D. Thesis*, Annamalai University.

Redfren, R.E., A.B. Demilo and A.B. Borkovec, 1980. Large milk weed bug: Effect of penfulron and its analogoues on reproduction. *J. Econ. Entomol.*, 73: 682–683.

Stone, C., 1992. *Science*, Washington, 255: 1070–1071.

Sumathi, 2002. Studies on the impact of endosulfan on certain selected tissues of the adult male insect *Gryllotapa africana* (Palisot de Beavrols) (Orthoptera : Gryllotalpidae) in relation to reproduction. *Ph.D. Thesis*, Annamalai University.

Thiruvasagam, M., 1994. Studies on the effect of nimbeciline on the histology, histochemistry and biochemistry of male accessory reproductive gland, fat body and haemolymph in the adult male *Aspongopus janus* (Fabr.) (Hemiptera : Pentatomidae). *M.Phil. Thesis*, Annamalai University.

Chapter 52

Bhitarkanika Mangrove Forest: A Sensitive Fragile Ecosystem

Lakshman Nayak* and Durga Prasad Behera

P.G. Department of Marine Sciences, Berhampur University, Bhanja Bihar – 760 007, Orissa

ABSTRACT

Out of 110 species of mangrove identified in the globe, 63 species are found in Bhitarkanika. The mangrove supports the marine life of 60 genera and 64 species of microfauna, 34 genera and 55 species of meiofauna, 19 species of amphibians, 42 species of reptiles including the rare and endangered sea turtle, more than 200 species of birds, 60 species of mammals and 6 species of marine mammals. 300 species of fishes and 229 species of crustacea are also inhabiting this sanctuary. The wild animals such as leopard, cat, fishing cat, jungle cat, hyena, wild boar, spotted dear, sambar, porcupine, pythons, dolphins, king cobra, saltwater crocodiles, water monitor lizard, terrapins, common mangoose, Indian field mouse, Indian mole rat are living in this sanctuary. The water temperature of the sanctuary varied from 25 C to 32 C. The air temperature varied from 28 C to 36 C. The pH value varied from 6.98 to 8.7. Salinity fluctuated from 9 ppm to 33.8 ppm. The dissolved oxygen varied from 2.325 ml/l to 7.250 ml/l. The carbonate alkalinity value varied from 25 ppm to 60 ppm and bicarbonate alkalinity varied from 65 ppm to 120 ppm.

Keywords: *Bhitarkanika, Sensitive, Fragile, Ecosystem.*

Introduction

Mangroves are the tidal forests of coastal wetlands, existing in the intertidal zones of sheltered shores, estuaries, tidal creeks, backwaters, lagoons, marshes and mud-flats of the tropical and sub-tropical regions of the world. The Bhitarkanika mangrove forest is the second largest mangrove in

* Corresponding Author: E-mail: laxmankabita@yahoo.co.in.

India. Mangroves of Bramhani and Baitarani delta of Kendrapara district have been declared as Bhitarkanika Wildlife Sanctuary in April 1975 covering an area of 672 sq.km. The core area of the sanctuary has been declared as Bhitarkanika National Park in September 1998 covering an area of 145 sq.km. Bordering the Bhitarkanika Sanctuary/National Park, an area of 1435 sq.km. has been declared as Gahirmatha Marine Wildlife Sanctuary in September 1997. Bhitarkanika is endowed with very complex sub-ecosystem of freshwater, marine and terrestrial, which are intricately mixed with each other. The maintenance of such an ecosystem depends on regular influence of freshwater from the adjoining land and tidal inflow from the sea. The horizontal and vertical zonation of the mangrove plant communities influenced by freshwater influx, tidal inflow, degree of inundation, seasonal rainfall and salinity gradient etc. have its influence on the wildlife and their occurrence. Mangrove swamps are one of the richest and most productive area and form the base of the food chain in the sea and coastal waters. The mangroves, wetlands and coral reefs, which are fragile ecosystems face threat due to various anthropogenic activities. Its over exploitation will certainly hamper the existence of many communities of plants and animals in this fragile ecosystem. The mangrove ecosystem play a very vital role in contributing to the food web in general and detritus food chain in particular supporting rich estuarine and adjacent marine fisheries (Chadha and Kar, 1999). The litter fall of mangrove trees amount to a large quantity of nutrients which are circulated both within the mangrove forest and in the surrounding marine ecosystem. The leaf material is utilised by bacteria and fungi and forms the base of an aquatic food web (Odum and Heald, 1975; Snedker, 1978). The mangroves play a very significant role in protecting the hinterland against cyclones and the ingress of seawater during tidal surge. Mangroves stabilise coastal landmass against sea erosion. They are repositories of immense biological diversity and are also the nursery and breeding ground of several marine life forms like species of shrimps, crabs, fishes and molluscs. Mangroves sustain the ecological security of the coastal zones. They also provide livelihood to thousands of fishermen and other villagers who resides in and around these areas. Mangrove forest acts as a buffer against any natural disasters. Depending on the recommendations of National Committees on Wetlands, Mangroves and Coral Reefs, 24 wetlands, 33 mangroves and four coral reefs in the country have been identified for conservation and management by the Ministry for Conservation and Management (Mandal and Rastogi, 2004).

Several works have been done on the distribution of mangroves in India (Sidhu, 1963; Nayak, 1993 and Pande, 2002). Some works have been done on the floral characteristics of Bhitarkanika by Banerjee and Das (1972), Mishra and Panigrahy (1987), Pattnaik and Choudhury (1989) and Banerjee and Rao (1990). Little work has been carried out on the hydrographical features of Mahanadi estuary by Upadhyay (1988) and Raju *et at.* (2000). No work has been done on the hydrological characteristics of Bhitarkanika mangrove forest. Therefore, the present work is an attempt to study the hydrological parameters of Bhitarkanika with flowering and fruit-bearing seasons of different mangrove trees.

Present Status of Flora and Fauna of Mangrove Forest in Bhitarkanika

At present, nearly 110 species of mangrove plants occur in different parts of the world (Ramdhai and Sundararaj, 1981). In India the area covered by mangrove is around 4871 km^2 (Pande, 2002). There are 63 species of mangrove and 8 species of their associates in Bhitarkanika (Kar, 1999) (Table 52.1). These species are most common and found in the wetland and play an important role in building land mass, arresting soil erosion, controlling flood damage and savage of earthquake and tsunami.

Table 52.1: Mangrove and their Associates in Bhitarkanika

Sl.No.	*Name of the Species*	*Common Name*	*Habit*	*Family*
1.	*Acanthus illicifolius*	Harakancha	Herb	Acanthaceae
2.	*Acanthus solubilis*	Harakancha	Herb	Acantaceae
3.	*Acrostichum aureum*	Kharkhari	Fern	Polypodiaceae
4.	*Aegialities rotundifilia*	Banarua	Tree	Plumbaginaceae
5.	*Aegiceras comiculatum*	Kharsi	Tree	Plumbaginaceae
6.	*Aegiceras Mujus*	Kharsi	Tree	Plumbaginaceae
7.	*Amoora cucullata*	Oonara	Tree	Meliaceae
8.	*Avicennia alba*	Dhala bani	Tree	Avicenniaceae
9.	*Avicennia marina*	Singala bani	Tree	Avicenniaceae
10.	*Avicennia officinalis*	Bani	Tree	Avicenniaceae
11.	*Brownlowia tersa*	Lati sundari	Herb	Tiliaceae
12.	*Brugueiera cylindrical*	Bandari	Tree	Rhizophoraceae
13.	*Bruguieira gymnorrhiza*	Bandari	Tree	Rhizophoraceae
14.	*Bruguieira parviflora*	Kaliachua	Tree	Rhizophoraceae
15.	*Bruguieira sexangula*	Bandari	Tree	Rhizophoraceae
16.	*Caesalpinia cristata*	Nentei	Creeper	Caesalpiniaceae
17.	*Caesalpinia nuga*	Nentei	Creeper	Caesalpiniaceae
18.	*Cerbera odallun*	Pani amba	Tree	Apocynaceae
19.	*Ceriops decandra*	Garani	Tree	Rhizophoraceae
20.	*Ceriop tagal*	Garani	Tree	Rhizophoraceae
21.	*Clerodendron inerme*	Chiani	Creeper	Verbanaceae
22.	*Crinum asiaticum*	Pani Kenduli	Herb	Amarylidaceae
23.	*Crinum defixum*	Pani Kenduli	Herb	Amarylidaceae
24.	*Cynometra ramiflora*	Singada	Tree	Leguminosae
25.	*Cynometra iripa*	Singada	Tree	Leguminosae
26.	*Cyperus compactus*	Tianshi	Grass	Cyperaceae
27.	*Cyperus corrymbosis*	Keutia ghasa	Grass	Cyperaceae
28.	*Dalbergia spinosa*	Goera Kanta	Creeper	Papilionaceae
29.	*Derris hetrophylla*	Katria nai	Creeper	Leguminosae
30.	*Derris scandens*	Katria nai	Creeper	Leguminosae
31.	*Dolichandrone spathacea*	Gosinga	Tree	Bignoniaceae
32.	*Excoecaria agallocha*	Guan	Tree	Euphorbiaceae
33.	*Fimbristylis ferruginea*	Luni ghasa	Grass	Cyperaceae
34.	*Finlaysonia obovata*	Lati rai	Creeper	Peripocaceae
35.	*Flagellaria indica*	Bahumruga	Creeper	Flagellariceae

Contd...

Table 52.1–Contd...

Sl.No.	Name of the Species	Common Name	Habit	Family
36.	*Heritier fomes*	Sundari	Tree	Sterculiaceae
37.	*Heritiera kanikensis*	Kania Sundari	Tree	Sterculiaceae
38.	*Hibiscus tiliaceus*	Bania	Herb	Malvaceae
39.	*Intsia bijuga*	Masitha	Tree	Legurninosae
40.	*Kandelia candel*	Sindhuguan	Tree	Rhizophoraceae
41.	*Lumnitzera racemosa*	Churanda	Tree	Combretaceae
42.	*Lumnitzera littorea*	Churanda	Tree	Combretaceae
43.	*Merope angulata*	Bana lembu	Herb	Rutaceae
44.	*Myriostachya wightina*	Nalia ghasa	Grass	Poaceae
45.	*Nypa frticans*	Nypa	Tree	Palmae
46.	*Pandanus fascicularis*	Luni kia	Herb	Pandanaceae
47.	*Phoenix paludosa*	Hentala	Tree	Palmae
48.	*Phragmites karka*	Nala	Grass	Grarnineae
49.	*Iprteresoa cparctata*	Dhani dhana	Grass	Poaceae
50.	*Rizophora apiculata*	Rai	Tree	Rhizophoraceae
51.	*Rizophora mucronata*	Rai	Tree	Rhizophoraceae
52.	*Rizophora stylosa*	Rai	Tree	Rhizophoraceae
53.	*Salacia chinesia*	Batara lata	Creeper	Hippocrateaceae
54.	*Saliconia brachita*	–	Herb	Chenopodiaceae
55.	*Saladora persica*	Miriga	Herb	Salvadoraceae
56.	*Sapium indicum*	Batula	Shrub	Euphorbiaceae
57.	*Sarcolobus carinatus*	Riagidi	Shrub	Asciepiadaceae
58.	*Sesuvium proffulacastrum*	–	Herb	Aizoaceae
59.	*Sonneratia apetala*	Keruan	Tree	Sonneratieceae
60.	*Sonneratia caseolaris*	Orua	Tree	Sonneratieceae
61.	*Sonneratia griffithii*	Orua/Chakada	Tree	Sonneratieceae
62.	*Suaeda maritime*	Giria saga	Herb	Chenopodiaceae
63.	*Suaeda monoecia*	Giria saga	Herb	Chenopodiaceae
64.	*Suaeda nudiflora*	Giria saga	Herb	Chenopodiaceae
65.	*Tamarix ericoides*	Jagula	Herb	Tamaricaceae
66.	*Tamarix indica*	Jagula	Herb	Tamaricaceae
67.	*Tamarix troupii*	Jagula	Herb	Tamaricaceae
68.	*Thespesia populnea*	Habili	Tree	Malvaceae
69.	*Xylocarpus grantum*	Shshumar	Tree	Malvaceae
70.	*Xyclocarpus mekongensis*	Pitakorua	Tree	Malvaceae
71.	*Xyclocarpus molluccensis*	Pitamari	Tree	Malvaceae

Source: Kar, 1999.

Mangroves support rich faunal diversity (Rao, 1987), There are 11 major faunal taxa of which nematodes are the dominant in the Bhitarkanika mangrove sediments (Sarma and Wilsanand, 1994). Bhitarkanika mangrove forest provides an ideal niche for both aquatic as well as terrestrial organisms of diverse categories. The Olive Ridley sea turtle (*Lepidochelys olivacea*) generally migrate to Gahirmatha beach for nesting. Salt water crocodile (*Crocodylus porosus*) is another important reptile found in the rivers and creeks of Bhitarkanika. Other aquatic animals are found in Bhitarkanika mangrove forest. These are water monitor, lizard, mudskippers, limulus, mud crabs, dolphins and a variety of fishes. Migratory birds like open billed storks, white ibis, grey pelicans, benheaded goose, bramhiny ducks, etc. visit during winter season on large numbers for feeding and breeding. The resident birds like night herons, cattle egrets, cormorants, grey herons, kingfishers, water fowls, common sandpiper, collard dove are also found in this sanctuary. The terrestrial wild animals like python, king cobra, varanus, spotted deer, wild boar, chital, sambar, jackal, hyena, porcupine, fishing cat, mongoose, three striped palm squirrel, Indian field mouse, house rat, bat and marine dolphin, etc. are inhabitants of Bhitarkanika.

Materials and Methods

Bhitarkanika Sanctuary is one of the spectacular sanctuaries lies between 20°4′ and 20°8′ N latitude and 86 45′ and 87 5′E longitude covering an area of about 650 sq km of which about 150 sq km are under mangrove forest (Figure 52.1). The study was carried out during January 2004 to December 2004. The monthly water samples were collected from three different stations. The samples on hydrological parameters such as temperature, pH, salinity, dissolved oxygen and alkalinity were collected regularly. The air and water temperature were recorded using a thermometer (±0.1°C). The water samples were brought to the laboratory and analysed for pH, salinity, dissolved oxygen and alkalinity following the standard method (APHA, 1995). During the water sampling the flowering and fruit bearing periods of different species of mangrove trees were observed and recorded. Station 1 is chosen at Nalitapatia, which is located near to Dhamara river. Station 2 is Dangmal, located in between Nalitapatia and Bhitarkanika estuary. Station 3 is chosen at Bhitarkanika estuary, which is towards the sea, and the hydrological parameters are highly fluctuative.

Results and Discussion

The hydrological data collected from three stations of Bhitarkanika were represented in Tables 52.2 to 52.4. The temperature was highest being 36 C during May and lowest temperature was being 28.10 C during January at Station 1 *i.e.,* Nalitapatia. The surface water temperature was highest being 32 C in the month of May and the lowest temperature was observed to be 25.3 C in the month of December. In Station 2 (Dangamal), the highest water temperature was observed to be 32 C during April and the lowest water temperature was observed during the month of December being 25.10 C. The highest atmospheric temperature was observed being 36 C during April and lowest atmospheric 'temperature was observed to be 28.5 C during January. In Station 3 (Bhitarkanika estuary), the highest atmospheric temperature was observed during the month of May being 35 C and the lowest atmospheric temperature was observed during the month of January being 27 C. The highest and lowest water temperature in this station was observed being 32 C and 25 C in the month of May and December respectively. There has been gradual decrease in the temperature from Station-l to Station-3 towards the down stream along the river system. The finding are in general agreement with that of Upadhyay (1988) and Raju *et al.* (2000).

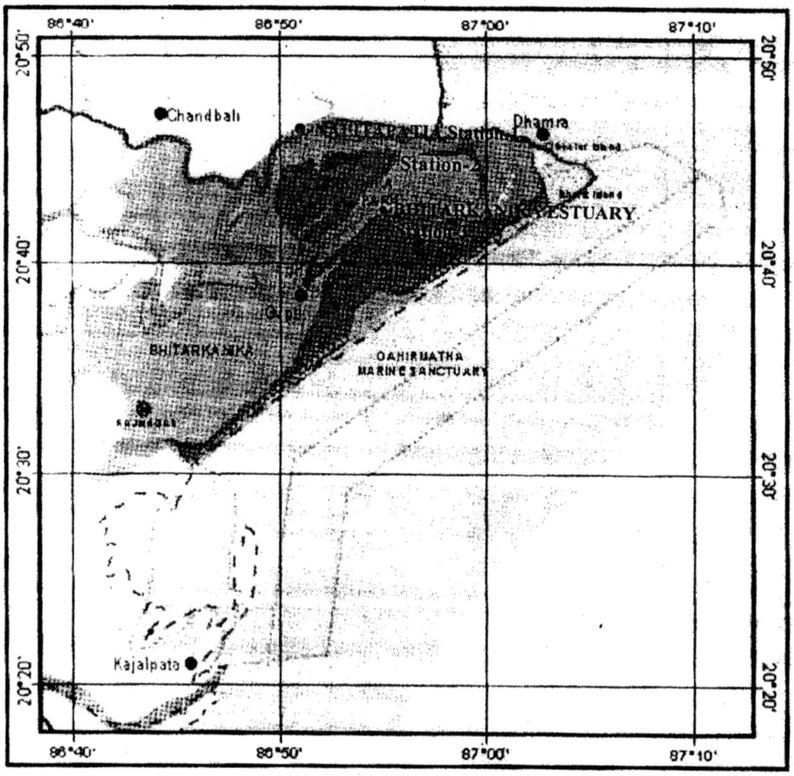

Figure 52.1: Map Showing Various Sampling Stations in Bhitarkanika

Table 52.2: Monthly Average Value of Hydrological Parameters of Bhitarkanika During January 2004 to December 2004 at Station 1

Month	*Temperature (°C)*		*pH*	*Salinity (ppm)*	*Dissolved Oxygen (ml/l)*	*Alkalinity in ppm*	
	Air	*Water*				*CO_3*	*HCO_3*
January	28.10	26.0	8.20	31.40	6.355	40	100
February	30.50	27.0	7.92	29.30	6.532	25	090
March	32.80	28.0	8.50	30.80	6.500	30	085
April	35.72	29.2	7.05	23.00	7.012	45	090
May	36.00	32.0	7.02	33.80	5.268	35	075
June	34.80	29.7	7.00	15.00	5.360	45	105
July	33.70	29.2	8.00	10.00	6.330	40	110
August	32.50	29.0	7.60	19.00	2.580	25	075
September	32.10	28.7	7.20	17.00	6.130	35	080
October	31.80	28.2	7.05	28.00	7.160	40	100
November	31.00	27.0	7.80	26.00	7.210	30	090
December	30.50	25.3	8.70	30.00	7.250	40	085

Table 52.3: Monthly Average Value of Hydrological Parameters of Bhitarkanika During January 2004 to December 2004 at Station 2

Month	*Temperature (°C)*		*pH*	*Salinity (ppm)*	*Dissolved Oxygen (ml/l)*	*Alkalinity in ppm*	
	Air	*Water*				*CO_3*	*HCO_3*
January	28.50	26.00	8.30	31.00	5.525	35	105
February	29.70	28.00	8.10	28.00	5.120	30	080
March	32.00	29.00	8.20	31.00	4.140	25	070
April	36.00	32.00	8.17	32.00	5.585	35	085
May	35.80	31.00	8.20	18.00	5.023	30	080
June	34.20	30.00	8.12	16.00	3.792	30	070
July	33.00	29.25	8.00	09.00	3.315	45	090
August	32.50	27.00	8.20	23.00	2.325	25	070
September	32.20	29.00	8.30	27.00	2.790	30	085
October	31.80	28.50	8.31	25.00	4.682	40	060
November	31.25	27.00	8.40	28.00	5.104	30	090
December	30.00	25.10	8.30	29.00	6.788	35	110

Table 52.4: Monthly Average Value of Hydrological Parameters of Bhitarkanika During January 2004 to December 2004 at Station 3

Month	*Temperature (°C)*		*pH*	*Salinity (ppm)*	*Dissolved Oxygen (ml/l)*	*Alkalinity in ppm*	
	Air	*Water*				*CO_3*	*HCO_3*
January	27.00	25.25	8.20	28.00	6.296	45	110
February	29.50	27.00	8.30	29.00	6.564	35	105
March	31.80	29.00	8.02	30.00	5.255	25	080
April	34.50	31.00	8.13	32.0	5.012	35	075
May	35.00	32.00	8.15	28.00	5.181	38	085
June	33.00	29.00	6.98	28.00	5.190	40	090
July	32.50	29.50	8.00	10.00	5.182	35	075
August	32.20	29.60	7.16	12.00	5.162	30	085
September	32.00	28.07	7.26	16.00	5.590	25	075
October	31.50	28.00	7.80	20.00	6.280	45	085
November	31.00	27.00	8.40	22.00	7.091	50	110
December	30.00	25.00	8.50	25.00	7.210	60	120

The highest pH value was observed in the month of December being 8.70 and lowest pH value was observed during the month of June being 7.00 at Station 1 (Nalitapatia). In Station 2 (Dangmal), the highest pH was observed during the month of November being 8.4 and the lowest pH value was observed during the month of July being 8.00. Similarly at Station 3 (Bhitarkanika estuary) the highest pH was observed during the month of December being 8.5 and the lowest pH value was recorded

Table 52.5: The Flowering and Fruit Bearing Season of Different Mangrove Species and their Associates of Bhitarkanika During 2004

Sl.No.	*Species*	*January*	*February*	*March*	*April*	*May*	*June*	*July*	*August*	*September*	*October*	*November*	*December*
1.	Sundari				Flower	Fruit							
2.	Bada Sundari					Flower	Fruit						
3.	Lati Sundari				Flower	Fruit							
4.	Kala Bani					Flower	Fruit						
5.	Dhala Bani					Flower	Fruit						
6.	Narkaa Bani					Flower	Fruit						
7.	Singali Bani					Flower	Fruit						
8.	Pitamari						Flower	Fruit					
9.	Pitakur						Flower	Fruit					
10.	Sisumar				Flower	Fruit							
11.	Guan						Flower	Fruit					
12.	Hental			Flower	Fruit								
13.	Habali												
14.	Jagula								Flower	Fruit			
15.	Kangara	Flower	Fruit										
16.	Garani					Flower	Fruit						
17.	Chiani	Flower	Fruit										
18.	Banrua										Flower	Fruit	
19.	Kerua				Flower	Fruit							
20.	Pani Amba				Flower	Fruit							
21.	Uara		Flower	Fruit									
22.	Bana Lamba									Flower	Fruit		
23.	Orua		Flower	Fruit									
24.	Charunda										Flower	Fruit	
25.	Harkanch					Flower	Fruit						
26.	Nentei						Flower	Fruit					
27.	Ragidi						Flower	Fruit					

Contd...

Table 52.5–Contd...

Sl.No.	*Species*	*January*	*February*	*March*	*April*	*May*	*June*	*July*	*August*	*September*	*October*	*November*	*December*
28.	Katukula		Flower	Fruit									
29.	Satra			Flower	Fruit								
30.	Masitna						Flower	Fruit					
31.	Sanga						Flower	Fruit					
32.	Pangulalata			Flower	Fruit								
33.	Sinduka				Flower	Fruit							
34.	Rai				Flower	Fruit							
35.	Kharsi				Flower	Fruit							
36.	Bania			Flower	Fruit								
37.	Bhairata											Flower	Fruit
38.	Dhala Katia							Flower	Fruit				
39.	Kala Katia						Flower	Fruit					
40.	Gohira			Flower	Fruit								
41.	Madhur Aunchar			Flower	Fruit								
42.	Pita Aunchar					Flower	Fruit						
43.	Masu										Flower	Fruit	
44.	Kayla										Flower	Fruit	
45.	Nargani					Flower	Fruit						
46.	Ajanjari									Flower	Fruit		
47.	Kasimla		Flower	Fruit									
48.	Sahada		Flower	Fruit									
49.	Kochla								Flower	Fruit			
50.	Jambo		Flower	Fruit									

Contd...

Table 52.5–Contd...

Sl.No.	Species	January	February	March	April	May	June	July	August	September	October	November	December
51.	Simili						Flower	Fruit					
52.	Hengula					Flower	Fruit						
53.	Sanga Banga				Flower	Fruit							
54.	Kankoli	Flower	Fruit										
55.	Ankokoli	Flower	Fruit										
56.	Kendu		Flower	Fruit									
57.	Aora									Flower	Fruit		
58.	Mehendi						Flower	Fruit					
59.	Chan Chana										Flower	Fruit	
60.	Tunda Poda										Flower	Fruit	
61.	Malanga	Flower	Fruit										
62.	Gila						Flower	Fruit					
63.	Kanta								Flower	Fruit			
64.	Kat Champa						Flower	Fruit					
65.	Panairi									Flower	Fruit		
66.	Katarkuria		Flower	Fruit									
67.	Lunulunia					Flower	Fruit						
68.	Thorania						Flower	Fruit					
69.	Dimiri			Flower	Fruit								
70.	Pimpuli	Flower	Fruit										
71.	Jajhanga											Flower	Fruit
72.	Fulcati									Flower		Fruit	
73.	Singada				Flower	Fruit							
74.	Habli		Flower	Fruit									

during the month of June being 6.98. The higher value of pH may be due to influence of seawater penetration. The lower value of pH was quite low during monsoon period, which may be due to the influence of fresh water influx and decomposition of organic matter carried out by the floodwater into the riverine system (Raju *et al.*, 2000 and Shriadah, 2000).

Seasonal and regional fluctuation in salinity was observed at all the Stations (Tables 52.2 to 52.4). The salinity values increased both in post-monsoon and pre-monsoon months. The highest value of salinity was observed in the month of May being 33.8 ppm and lowest in the month of July being 10 ppm at Station 1. In Station 2 the lowest salinity value was observed during the month of July being 9 ppm and highest in the month of April being 32 ppm. Similarly the higher and lower values of salinity were observed in the month of April and July respectively being 32 ppm and 10 ppm at Station 3. The salinity of the estuary did not go usually above 33.8 ppm or below 9 ppm although during the monsoon season there is freshwater influx into the estuary. The high salinity values may be due to high evaporation and low intensity of river run off into the estuarine system. In the present study the annual variation of salinity was 9 ppm to 33.8 ppm. Salinity values were maximum during peak summer month at all the stations. However, drastic reduction of salinity was noticed during the monsoon period, which is in conformity with the observation of Achuthankutty (1987) from the estuary of Goa.

The dissolved oxygen concentration showed a wide range of variation throughout the year (Tables 52.2 to 52.4). In the month of December the dissolved oxygen concentration was very high being 7.25 ml/lit and minimum being 2.58 ml/lit in the month of August at Station 1. The highest and lowest value was observed during December and August being 6.788 ml/lit and 2.325 ml/lit at Station 2. Likewise the maximum and minimum values were observed during December and August respectively at Station 3 being 7.21 ml/lit and 5.162 ml/lit. Comparatively high dissolved oxygen values were associated with high pH values along the estuarine system during post-monsoon period, which suggest the abundant growth of phytoplankton and related zooplankton. Low values of dissolved oxygen concentration correspond to monsoon period, which may be due to bio-degradation of organic matter in the water column (Raju *et al.*, 2000).

Alkalinity generally determines the magnitude of dial fluctuation of pH of water. Water with low alkalinity (less than 15 mg/l) results in wide fluctuation in pH value from 7.5 at dawn to 10 or even highest in the afternoon. Alkalinity value was generally higher at all the stations. The highest value was observed in the month of December and lowest value was observed in the month of April. The bicarbonate alkalinity varied from station to station. Bicarbonate alkalinity was more in the month of June and July, which may be due to high influx of fresh water and high pH in estuarine system of Bhitarkanika. This observation is in conformity with the observation of Nayak and Behera (2004) from Chilika lagoon.

The mangrove plants and their associates of Bhitarkanika bearing flowers and fruits in different seasons of the year were observed during the study period. The observations were represented in Table 52.5. Flowering and fruit bearing in plants is associated with photoperiodism. As temperature increases, growth of upper tepal tissue speeds up for a short time, while growth of lower tissue remains about constant causing flower opening (Salisbury and Ross, 2001).

References

APHA, 1995. *Standard Methods for the Examination of Water and Wastewater*. American Public Health Association. AWWA and WPCF, Washington D.C.

Achutankutty, C.T., 1987. Commercially important penaeid shrimp larvae in the estuaries of Goa. *Mahasagar Bull. of Nat. Inst. Oceanog.*, 20(4): 217–224.

Banerjee, L.K. and Das, G.C., 1972. New distributional records from Orissa coast. *Bulletin of Botanical Survey of India*, 14(1–4): 184–186.

Banerjee, L.K. and Rao, T.A., 1990. *Mangroves of Orissa Coast and their Ecology*. Bishen Singh Mahendra Pal Singh, Dehera Dun, p. 1–114.

Chadha, S. and Kar, C.S. 1999. *Bhitarkanika Myth and Reality*. Natraj Publisher, Dehra Dun, p. 55–170.

Kar, C.S., 1999. The unique mangrove ecosystem of Bhitarkanika. *Oriforest*, 5: 20–34.

Misra, S.C. and Panigrahi, G., 1987. Studies on the mangrove flora of Orissa with particular reference to the Rhizophoraceae. R.Br. *Journal of Economic and Taxonomic Botany*, 11(1): 121–132.

Mandal, N. and Rastogi, M., 2004. Living wealth: Conservation is the key–the preservation of biodiversity. *Science Reporter*, June: 24–27.

Nayak, S.T., 1993. Remote sensing application in the management of wetland ecosystem with special emphasis on management of mangrove ecosystem and coastal protection, Visakhapatnam. India, pp. 21.

Nayak, L. and Behera, D.P., 2004. Seasonal variation of some physio-chemical parameters of Chilika lagoon (east coast of India) after opening of new mouth, near Sipakuda. *Indian Journal of Marine Sciences*, 33(2): 206–208.

Pande, S.K., 2002. Mangroves: The Coastguard. *Science Reporter*, June: 40–44.

Patnaik, S.N. and Choudhury, B.P., 1989. Present status and future development of mangrove vegetation in Bhitarkanika. In *INDO-US Workshop on Wetlands, Mangroves and Biosphere Reserve*, pp. 89–95. Government of India, Ministry of Environment and Forests, New Delhi.

Ramdhai, V. and Sundararaj, V. 1981. Mangrove vegetation: A valuable fish feed. *Science Reporter*. December: 677–678.

Rao, A.N. 1987. Mangrove ecosystem of Asia and Pacific. In: *Mangroves of Asia and the Pacific: Status and Management*, (Ed.) R.M. Umali. Technical Report of the UNDP/UNESCO, Research and Training Pilot Programme of Mangrove Ecosystem in Asia and the Pacific. Quezon City, Metro Manila, p. 1–48.

Raju, V.A., Nayak, L. and Das, R.C., 2000. Seasonal abundance of post larvae of penaeid prawns in the Mahanadi estuary, Orissa. *Asian J. Zool. Science*, 9: 15–23.

Sarma, A.L.N. and Wilsanand, V., 1994. Littoral meiofauna of Bhitarkanika mangroves of River Mahanadi system, east coast of India. *Indian Journal of Marine Sciences*, 24(4): 221–224.

Shriadah, A.M., 2000. Chemistry of the mangrove waters and sediments along the Arabian Gulf Shoreline of the United Arab Emirates. *Indian Journal of Marine Sciences*, 29: 224–229.

Sidhu, S.S., 1963. Studies on the mangroves of India. In: *Proceedings of Indian Academy of Sciences*, 33(8): 129–136.

Salisbury, B.F. and Ross, W.C., 2001. *Plant Physiology*, 3rd Edition. CBS Publishers and Distributors, New Delhi, p. 405–417.

Upadhyay, S., 1988. Physico-chemical characteristics of the Mahanadi estuarine ecosystem, East coast of India. *Indian Journal of Marine Sciences*, 17: 19–23.

Chapter 53

Biological Pollution Abatement of Petroleum Refinery Wastewater Using *Pseudomonas* sp.

Sk Masud Hossain

Dept. of Chemical Engineering, Mohamed Sathak Engineering College,
Kilakarai – 623 806, Tamil Nadu
E-mail: skmhossain@yahoo.co.in, Fax: +91 4567 245344.

ABSTRACT

Many point and non point sources of wastewaters contribute to the refinery's wastewater loading. This is known that the majority of inflects in most petroleum refinery products are biodegradable. Biodegradation using bacteria *Pseudomonas* sp. is an effective pollution abatement solution for oil refinery wastewater. Batch aerobic studies of the bioprocess showed that the bacterium reduces the pollution load maximally in 120 h with pH of 7.5 and temperature 40 C. The maximum oil and grease reduction is 73.60 per cent with 78.45 per cent phenolic components reduction. Maximum sulphide reduction is 72.30 per cent. The BOD reduction is maximum 86.70 per cent with maximum total suspended solids reduction is 83.65 per cent. The bacteria reduce most of the pollution parameters.

Introduction

Most of the environmental problems that confront mankind today are connected to the use of energy in one-way or another. The concern over environmental degradation brought about by increased consumption of fuels such as oil has been growing. Petroleum refining operations generate the predominant share of hazardous waste produced in the petroleum industry. A major part of this must be treated to reduce or remove contaminants so that the effluent meets discharge requirements specified in the Clean Water Act. Major sources include processing operations and from only dischargers associated with heat exchangers and cooling towers. Other sources of wastewater include storage and handling operations, storm water run off, and spills of oils and oily waters.

The frequency and intensity of oil refinery pollution of rivers, streams, beaches and ocean has been increasing during the last century. This pollution is an almost inevitable consequence of the dependence of a rapidly growing population on oil-based technology. Damage caused by this pollution may range from spoiling of beaches, coastal resorts and riverbanks, destruction of fish and other aquatic flora and fauna, to risk of fire hazards in rivers, harbors and other enclosed areas and explosion in the sewers. The sulphur content of petroleum from different sources exists in the form of elemental sulphur, sulphate, sulphide and thiosulphate. Sulphides, mercaptans and phenols are tonic and taste and odor producing substances and destroy fish and other agnostic organisms in a receiving watercourse. These may also seriously effect the public water supply located down-stream. It is therefore necessary to reduce the pollution from the effluent to a level at which they would be won-tonic and will not create a water pollution problem. The noxious effects of oil pollution on marine communities are well chronicled, as these are the damaging impacts on marine and coastal habitats. All potential petroleum hydrocarbons sources, including any hydrocarbons intrinsic to the immediate urban environment includes n-paraffin (saturated and unsaturated carbon containing C_6-C_{32}) isoparafins, cycloparafins, aromatics, olefins, polynuclear aromatics and oxygenates. These are the major sources of biological oxygen demand (BOD) for oil refinery wastewaters.

Nevertheless, it is known that the majority of components in most oils are biodegradable. Since bioprocess can degrade the product, and microbial biomass can accumulate in critical locations. On the other hand, stimulating the natural process of biodegradation (bioremediation) offers an opportunity for reducing the environmental impact when oil refinery effluent escapes into the environment.

The present batch biodegradation investigations are undertaken to develop a biological treatment of oil refinery wastewaters using *Pseudomonas* sp. The various bioprocess parameters like time, pH, temperature and shaking speed are optimized for maximum biodegradation of oil refinery wastewater using *Pseudomonas* sp.

Materials and Methods

Procurement of Bacteria and Culture Preparation

The freeze-dried bacteria *Pseudomonas sp.* (MTCC-130) was procured from Microbial Type Culture Collection and Gene Bank, Institute of Microbial Technology, Chandigarh, India. Slant cultures were prepared with prescribed growth media: beef extract- 1. 0 g, yeast extract - 2.0g, peptone - 5.0 g, NaCl - 5.0 g, Agar 15.0g and distilled water 1.0 L. The growth medium is sterilized at 15 psi for 15 min in an autoclave. The slant culture is prepared at 37 C for 7 days incubation. Subculture is prepared within the period of 30 days.

The suspension culture is prepared with the following constituents per litre: NaCl- 5.0.g, KH_2PO_4- 1.0 g, $MgSO_4$-0.5 g, AlK $(SO_4)_2$. $12H_2O$- 0.01 g, $CoCl_2$-0.1g, Na_2 MoO_4. $2H_2O$- 0.01 g, and peptone- 5.0 g.

Collection of Petroleum Refinery Wastewater

The petroleum refinery wastewater is collected in a plastic container from indigenous source and stored in freeze. The effluents are analyzed for pollution parameters as shown in Table 53.1.

General method

500 ml of oil refinery wastewater with 200 ml of sterilized suspension culture media are taken in 2 L Erlenmeyer flasks (broths), 7 days age-old and 30 per cent (v/v) suspension inoculum culture is added to broths. Glucose 2.0 per cent (w/v) and l- alanine 0.20 per cent (w/v) are adjusted to the broths. The initial pH of the broths are adjusted to 3.5 by adding 1 (M) HCl acid or 1 (M) NaOH solution. Aerobic conditions in the system were maintained by putting non-absorbent cotton at the month of the broths. The temperature of the broths is maintained at 30 C by placing in a water bath with constant shaking of 20 rpm.

Table 53.1: Oil Refinery Wastewater Analysis

Parameters	*Magnitudes*
Oil and grease	9750
Phenol	1200
Sulfide	750
BOD_5 at 20° C	20575
Total suspended solids (TSS)	14600
pH	8.5

All are in mg/L except pH.

Effect of Time on Biodegradation

The general method is repeated accordingly. After 4, 6, 8, 10, 12 and 14 days of incubation, the broths are taken out and analyzed for various pollution parameters. The results were shown in Table 53.2.

Table 53.2: Effect of Incubation Time on Biodegradation of Oil Refinery Wastewater

Parameters	*Magnitudes*					
	DT 4	*DT 6*	*DT 8*	*DT 10 **	*DT12*	*DT 14*
Oil and grease	6250 (33.12)	5500 (43.59)	4200 (56.92)	3135 (67.85)	3050 (68.72)	3008 (69.23)
Phenol	950 (20.83)	760 (36.67)	525 (56.25)	329 (72.55)	315 (73.75)	300 (75.0)
Sulphide	560 (25.33)	450 (40.00)	370 (50.67)	257 (65/80)	240 (68.0)	210 (72.0)
BOD_5 at 20°C	16450 (20.10)	11050 (46.29)	8160 (60.34)	5025 (75.60)	5000 (75.70)	5000 (75.70)
Total suspended solids (TSS)	11820 (19.04)	9650 (33.90)	7200 (50.68)	4140 (71.65)	4000 (72.60)	4000 (72.60)

Optimum incubation time.

All values are in mg/L

Values in ( ) are per cent of biodegradation

DT 4, DT6, DT 8, DT 10, DT12, and DT 14: Incubation time after 4, 6, 8, 10, 12 and 14 days of inoculation respectively.

Effect of pH on Biodegradation

The general method is repeated for various initial pH of broths such as 3,5, 4.5, 5.5 and 6.5 respectively. The broths are taken out after 10 days of incubation (optimum time) and are analyzed for various pollution parameters. The results were shown in Table 53.3.

Effect of Temperature on Biodegradation

The general method is repeated for various temperatures of the water bath such as 30, 35, 40 and 45 C respectively. The initial pH of broths are maintained at 5.5 (optimum). The broths are taken out after 10 days of incubation (optimum biosorption time) and analyzed for various pollution parameters. The results are shown in Table 53.4.

Effect of Shaking Speed on Biodegradation

The general method is repeated for different shaking speed such as 20, 40, 60 and 80 rpm respectively. The initial pH of broths are maintained at 5.5 (optimum). The water bath temperature is maintained at 40 C (optimum). The broths are taken out after 12 days of incubation (optimum time) and are analyzed for various pollution parameters. The results are shown in Table 53.5.

Table 53.3: Effect of pH on Biodegradation of Oil Refinery Wastewater

Parameters	*Magnitudes pH*			
	3.5	*4.5*	*5.5**	*6.5*
Oil and grease	3250 (66.67)	3050 (68.72)	2910 (70.25)	3175 (67.43)
Phenol	345 (71.25)	320 (73.33)	302 (72.55)	350 (70.83)
Sulphide	270 (64.0)	250 (66.7)	220 (70.65)	290 (61.33)
BOD_5 at 20°C	4760 (76.86	4250 (79.34)	3755 (81.75)	4500 (78.12)
Total suspended solids (TSS)	3850 (73.63)	3450 (76.37)	3155 (78.40)	3950 (72.94)

Optimum pH value.

All values are in mg/L.

Values in ( ) are per cent of biodegradation.

Table 53.4: Effect of Temperature on Biodegradation of Oil Refinery Wastewater

Parameters	*Magnitudes Temperature, °C*			
	30	*35*	*40**	*45*
Oil and grease	2950 (69.74)	2850 (70.77)	2576 (73.60)	3285 (66.31)
Phenol	325 (72.92)	295 (75.42)	260 (78.45)	375 (68.75)
Sulphide	260 (65.33)	235 (85.61)	205 (86.70)	295 (81.77)
BOD_5 at 20°C	3550 (82.75)	2960 (85.61)	2735 (86.70)	3750 (81.77)
Total Sulphided Solid (TSS)	3,100 (78.76)	2,740 (81.23)	2,365 (83.65)	3,250 (77.74)

Optimum temperature.

All values are in mg/L.

( ) values are per cent of reduction.

Table 53.5: Effect of Shaking on Biodegradation of Oil Refinery Wastewater

Parameters	*Magnitudes Shaking Speed, rpm*			
	20	*40*	*60**	*80*
Oil and grease	2576 (73.60)	2350 (78.62)	2056 (82.67)	2785 (70.35)
Phenol	260 (78.45)	235 (80.45)	186 (89.54)	275 (76.75)
Sulphide	205 (86.70)	195 (87.61)	182 (89.83)	225 (81.77)
BOD_5 at 20°C	2735 (86.70)	2660 (88.61)	2535 (90.70)	3750 (81.77)
Total Sulphided Solid (TSS)	2365 (83.65)	2240 (85.23)	2165 (88.65)	3250 (77.74)

Optimum temperature.

All values are in mg/L.

Results and Discussion

Oil Refinery Wastewater Analysis

Table 53.1 shows the refinery wastewater analysis. All pollution parameters are analyzed by standard methods. Sulphide contains 750 mg/L. Oil and grease measures 9750 mg/L with BOD_5 at 20 C is 20575 mg/L respectively. All the parameters are in high levels. A total suspended solid is 14000 mg/L. The total suspended solid contains mainly organic and inorganic compounds like silica, mud etc. present in petroleum extracted from sedimentary basins.

Biological treatment process is primary meant for removal of pollutants like phenol, sulphide, oil and grease, BOD and the non-recoverable oil present in the effluents. Bacterial seeding and fertilization of the oily wastes with *Pseudomonas* sp. will accelerate biological degradation provided that dissolved oxygen (DO) are available in sufficient amounts.

Effect of Incubation Time on Biodegradation

The effects of incubation time on biodegradation of oil refinery wastewaters are shown in Table 53.2. In the present studies, biodegradation of refinery wastewaters is analyzed after 4 days of incubation. This incubation time is required by the bacteria *Pseudomonas* sp. for full growth and adaptation with tonic effluents of oil refinery. It is observed that the optimum time of biodegradation for maximum reduction of all the pollutant constituents is 10 days. The biodegradation of oil and grease is 67.85 per cent. The phenol biodegradation is 72.55 per cent with sulphide removal is 65.80 per cent (Table 53.1). Biological oxygen demand (BOD) is 5025 mg/L (75.60 per cent abatement) with biodegradation of total suspended solids (TSS) is 71.65 per cent (Table 53.2) after 10 days respectively. Other pollution parameters, which are present in wastewater, cause an increasement of BOD_5 and total suspended solids, are degraded by the bacteria *Pseudomans* sp. This is evidenced by the highest per cent reduction in BOD_5 and TSS (Table 53.2). For biological parameters optimization studies, 10 days is taken as optimum incubation time.

Effect of pH on Biodegradation

The effect of pH on bioremediation of oil refinery pollution load is shown in Table 53.3. The optimum pH is 5.5, slightly acidic medium of the bacterial environment. (Table 53.3) at which maximum biodegradation of oil refinery takes place. The maximum biodegradation in oil and grease is 70.25 per cent. Maximum phenol biodegradation is 72.55 per cent with maximum sulphide abatement is 70.65 per cent (Table 53.3) respectively. The BOD_5 at 20 C measured as 3755 (81.75 per cent biodegradation) with maximum suspended solids biodegradation is 78.40 per cent at pH of 5.5 (Table 53.3). Other pH values can biodegrade the pollution parameters to some little bit less (Table 53.3) extent comparing optimum pH of 5.5. Therefore, the optimum pH of 5.5 is taken for other biological parameter optimization.

Effect of Temperature on Biodegradation

The effect of temperature on bioremediation of oil refinery pollution load is shown in Table 53.4. The optimum temperature is 40 C at which maximum biodegradation of the refinery wastewater takes place. At optimum temperature, the abatement of all the pollution parameters is maximum (Table 53.4). The maximum bioremediation of oil and grease is 73.60 per cent. Maximum phenol biodegradation is 78.45 per cent with maximum sulphide reduction is 72.30 per cents. The BOD_5 at 20 C measured to 81.75 per cent with 83.80 per cent of total suspended solids (TSS) bioremediation at optimum temperature of 40 C respectively. Other temperatures reduce the pollution parameters of oil refinery wastewater to some little bit less extent than the optimum temperature. Every type of microbe has an optimum, minimum and maximum growth temperature. Temperatures below the optimum for growth depress the rate of metabolism of cells. Above the optimal temperature, the growth rate decreases and thermal death may occur.

Effect of Shaking Speed on Biodegradation

The effect of shaking speed on bioremediation of oil refinery pollution load is shown in Table 53.5. The optimum shaking speed is 60 rpm at which maximum biodegradation of the oil refinery wastewater takes place. At optimum shaking speed, the abatement of all the pollution parameters is maximum (Table 53.5). The maximum bioremediation of oil and grease is 82.67 per cent. Maximum phenol biodegradation is 89.54 per cent with maximum sulphide reduction is 89.83 per cents. The BOD_5 at 20 C measured to 90.70 per cent with 88.65 per cent of total suspended solids (TSS) bioremediation at optimum shaking speed respectively. Other shaking speeds reduce the pollution parameters of oil refinery wastewater to some little bit less extent than the optimum temperature.

Conclusion

Bioremediation of pollution load of oil refinery wastewater with bacteria *Pseudomonas sp.* is an effective treatment solution. All potential pollutants of oil refinery wastewater are biodegraded with the bacteria maximally. That is evidenced by major BOD_5 and total suspended solid bioremediation. Phenolic compounds and sulphide are also mostly degraded by bacterial treatment process. The oil and grease are biodegraded to a considerable extend with the optimum biological process parameters. As there is a great deal of strain-to-strain variation in the physiological and biochemical behavior of strains of *Pseudomonas* sp., optimized process parameters and biodegradation should vary from strain-to-strain. Highly active bacterial strains resistant to drastic and extreme conditions in oil refinery should be developed for maximum efficiency.

References

Brierley, C.L., Kelly, D.P., Seal, K.J. and Best, D.J., 1987, Desulphurisation of petroleum. In: *Biotechnology, Principles and Applications,* (Eds.) by Best, D.J. Higgins, I. I. and Jones, J., Blackwell Scientific Publications, London, U.K.

Ehrlich, H.L., 1982. *Geomicrobiology*. Marcel Dekker Inc, New York, USA.

Genetha, J.C., and Acevedo, F., 1985. *Trends in Biotechnology*. Blackwell Scientific Publications, London U.K.

Hossain, S.M. and Das, M., 2002. Optimization studies on pollution abatement studies on spend sulphite liquor with with *Phanerochacte chrysosporium. Indian J. of Environmental Protection,* 22(1): 26-32.

Hossain, S.M., and Das, M., 2002. Optimization studies on aerobic digestion of black liquor using *Phanerochacte chrysosporium* (MTCC-787). *Asian J. of Microbiol. Biotech. Env. Sc.,* 4(3): 401-407.

Hossain, S.M.., and Das, M., and Ibrahim, S.H., 2001. Aerobic studies on pollution abatement of sulfite pulp bleaching effluent using *Phanerochacte chrysosporium* (MTCC-787). *J. of Industrial Pollution Control,* 17(2): 191-200.

Hossain, S.M. and N. Anantharaman, 2004. Studies on biological pollution abatement of petrochemical industry wastewater. *Indian Journal of Environmental Protection,* 24(11): 845-848.

Hossain, S.M., N. Anantharaman, Begum, K.M.M.S. and Ibrahim, S.H., 2004. Studies on aerobic biodegradation of aromatic amine (aniline) using *Phanerochacte chrysosporium. Indian Journal of Environmental Protection,* 24(4): 422-427.

Hossain, S.M. and Anantharaman, N., 2005. Aerobic biodegradation of trichlorophenol using *Phanerochate chrysosporium*. *Chemical Engineering World*, 40(5): 88-91.

Khan, A.K., 1995. Treatment and disposal of wastewater from petroleum refineries. In: *Pollution Management in Industries*, pp. 104-144, (Ed.) Trivedy, R.K. Environmental Publications, Karad, India.

Mark, H.F., Othmer, D.F., Overberger, C.G. and Seaborg, G.T., 1982. *Kirk -Othmer Encylopedia of Chemical Technology*, 3rd ed, Vol.2, John Wiley & Sons, New York, USA.

Monticello, D.J. and Finnerty, W.R., 1985. Bacterial desulphurisation of petroleum. *Ann. Rev. Microbiol.*, 39: 371-380.

MOPNG, 1999. *Basic Statistics on Petroleum and Natural Gas*, Ministry of Petroleum & Natural Gas, Govt. of India, New Delhi, India.

Nero, V.P., 1999. Environmental analysis of petroleum hydrocarbons, In: *The Willey Encyclopedia of Environmental Pollution and Clean Up*, Vol.2, pp. 1274-1283, (Editor- in-Chief) Meyers, R.A. John Wiley & Sons, Inc, New York, USA.

Wichlacz, P.L., 1986. *Biotechnol Bioeng Symposium* No. 16, John Wiley and Sons Inc, New York, USA.

Index